TRANSPORTATION SAFETY
IN AN AGE OF DEREGULATION

Transportation Safety in an Age of Deregulation

Edited by

LEON N. MOSES

IAN SAVAGE

New York · Oxford
OXFORD UNIVERSITY PRESS
1989

Oxford University Press

Oxford   New York   Toronto
Delhi   Bombay   Calcutta   Madras   Karachi
Petaling Jaya   Singapore   Hong Kong   Tokyo
Nairobi   Dar es Salaam   Cape Town
Melbourne   Auckland

and associated companies in
Berlin   Ibadan

Published by Oxford University Press, Inc.,
200 Madison Avenue, New York, New York 10016

Oxford is a registered trademark of Oxford University Press

Library of Congress Cataloging-in-Publication Data
Transportation safety in an age of deregulation /
[edited by] Leon N. Moses and Ian Savage.
p.   cm.
Papers derived from a conference held at the Transportation Center
of Northwestern University, June 1987.
Bibliography: p.   Includes index.
ISBN 0-19-505797-X
1. Aeronautics, Commercial—United States—Deregulation—Congresses.
2. Aeronautics—United States—Safety measures—Congresses.
3. Trucking—United States—Deregulation—Congresses.
4. Trucking—United States—Safety measures—Congresses.
I. Moses, Leon N., 1924–   . II. Savage, Ian, 1960–   .
III. Northwestern University (Evanston, Ill.). Transportation Center.
HE9803.A35T73   1989   88-28728
363.1′24′0973—dc19   CIP

10 9 8 7 6 5 4 3 2 1

Printed in the United States of America
on acid-free paper

Contents

Contributors

Brock Adams is a Democratic United States Senator for the State of Washington, and a former U. S. Secretary of Transportation.

Martin Albu is the Licensing Authority for the North West Traffic Area, Manchester, Great Britain; and the former head of the Freight Policy and Road Haulage Division of the U. K. Department of Transport.

Richard J. Arnott is Professor of Economics at Boston College, Chestnut Hill, Massachusetts and Queen's University, Kingston, Ontario, Canada.

Elizabeth E. Bailey is Dean of the Graduate School of Industrial Administration at Carnegie Mellon University, Pittsburg.

Severin Borenstein is Assistant Professor of Economics and Public Policy at the University of Michigan, Ann Arbor.

Kenneth D. Boyer is Professor of Economics at Michigan State University, East Lansing.

Garland Chow is Assistant Professor of Transportation at the University of British Columbia, Vancouver, Canada.

Thomas M. Corsi is Chairperson of Transportation, Business and Public Policy, in the College of Business Administration at the University of Maryland-College Park.

Philip Fanara, Jr. is Chairperson of Finance and Insurance in the School of Business and Public Administration at Howard University, Washington, D.C.

Michael Glover is Principal Scientific Officer at the U. K. Transport and Road Research Laboratory, Crowthorne, Berkshire, Great Britain.

Ezra Hauer is Professor of Civil Engineering at the University of Toronto, Ontario, Canada.

Paul P. Jovanis is Associate Professor of Civil Engineering at the University of California-Davis.

Adib Kanafani is Professor of Transportation Engineering at the University of California-Berkeley.

Theodore E. Keeler is Professor of Economics at the University of California-Berkeley.

David M. Kirstein is a Partner of Beckman and Kirstein, a Washington, D.C., law firm.

Thomas Gale Moore is a member of the President's Council of Economic Advisors and a member of the faculty at the Hoover Institution, Stanford University, Palo Alto, California.

Leon N. Moses is Professor of Economics and Transportation at Northwestern University, Evanston, Illinois.

John J. Nance is an author, writer, attorney, and former airline pilot, based in Tacoma, Washington.

Clinton V. Oster, Jr. is Director of the Transportation Research Center at Indiana University, Bloomington.

John C. Panzar is Professor of Economics and Transportation at Northwestern University, Evanston, Illinois.

Nancy L. Rose is Assistant Professor of Applied Economics at the Massachusetts Institute of Technology, Cambridge.

Ian Savage is Assistant Professor of Economics and Transportation at Northwestern University, Evanston, Illinois.

Joseph E. Stiglitz is Professor of Economics at Stanford University, Palo Alto, California.

W. Kip Viscusi is Professor of Economics at Duke University, Durham, North Carolina.

Ramsey M. Withers is President and Chief Operating Officer of Government Consultants International, Ottawa, Ontario, Canada; and a former Deputy Minister of Transport Canada.

Martin B. Zimmerman is Associate Professor of Business Economics at the University of Michigan, Ann Arbor.

C. Kurt Zorn is Assistant Professor of Public and Environmental Affairs at Indiana University, Bloomington.

TRANSPORTATION SAFETY
IN AN AGE OF DEREGULATION

1

Introduction

LEON N. MOSES and IAN SAVAGE

This book is concerned with the implications for safety of two pieces of U. S. legislation—the Airline Deregulation Act of 1978 and the Motor Carrier Regulatory Reform and Modernization Act of 1980. These Acts had the effect of reducing the control of the federal government and of carrier rate associations on the conditions of competition in the airline and trucking industries. In particular, the regulatory reforms embodied in the two Acts allowed (1) greater freedom of entry into the two industries, (2) greater freedom of entry into, and of exit from, particular markets, and (3) greater freedom of individual rate making. The Acts significantly increased the influence of market forces on the prices charged for air and truck service and the profitability of individual firms. Increased rate competition among motor carriers had direct effects on the rates charged by railroads for the movement of high-value goods and had indirect effects on all other tariffs.

The regulatory reform bills were passed because lawmakers felt that increased competition would lead to more efficient operations and lower rates in the two industries while not compromising safety or quality of service. Some changes in quality of airline service were, in fact, hoped for. It was commonly believed that the suppression of price competition by the Civil Aeronautics Board (CAB) fostered competition in service quality variables that was highly uneconomical, such as too early replacement of aircraft and departures at major airports that were excessive in light of existing load factors. It was asserted that such quality competition drove up costs, which led to proposals to the CAB for relief in the form of fare increases. However, the positive effects of such increases (which were almost always granted) on the profitability of airline operations were soon dissipated by another round of quality competition and increases in costs of operation.

Lawmakers expected that the discipline of increased price competition would achieve cost economies by both airlines and motor carriers, because they would have to resist wage demands by unionized labor that exceeded increases in productivity. It was commonly believed that firms treated union wage demands as a pass-through; that is, they granted wage increases that exceeded productivity gains and clearly also exceeded what comparable labor received in unregulated, competitive industries. The pass-through philosophy grew out of two beliefs: (1) Regulatory bodies would grant increases in rates that would restore average profitability and (2) such rate increases would lead to actual increases in revenues and would restore profitability for the representative firm because demand was growing over time and was relatively price inelastic. In the main, these beliefs proved valid. Generally speaking, rate increases did temporarily improve profits, but at a significant increase in cost to users of airline and motor carrier services.

The framers of the motor carrier and airline bills hoped that a reduction in economic controls by government would increase price competition and bring benefits to users of the transport services provided by these industries. Clearly, that hope has been realized.

The rate benefits to users of airline service are very clear. Between 1977, when the CAB began to allow greater freedom of entry and increased price competition, and 1986, average revenue per passenger mile rose by only 30%, going from 8.3¢ to 10.8¢. In real terms, deflating by the Consumer Price Index, the cost of airline travel to passengers fell by 23%. Passenger miles increased from 226 to 366 billion, and enplanements increased by 52%. Of course, the increase in passenger usage was the result of an 8.5% increase in per capita real income, as well as of the reduction in the real cost of travel. It should be noted that in the nine years prior to the period of regulatory reform, average revenue per passenger mile increased by 17.7% in real terms (Air Transport Association, 1987). Part of the decrease in real air fares that occurred during the years of increased price competition was due to declining fuel prices. It is difficult to determine how much consumers would have benefited from such declines had they occurred in the regulatory era, but taking this into account, it has been estimated that deregulation brought approximately $16 billion annual benefits in current dollar terms (Morrison and Winston, 1986). This estimate of the magnitude of benefits has been questioned (Evans, 1987b); that there have been significant benefits is beyond dispute.

There is clear evidence that open entry and a reduction in the power of motor carrier rate bureaus to control rates led to an increase in competition and to reductions in the real cost of trucking service to shippers. The number of trucking firms increased by some 19,000 in the years since passage of the Motor Carrier Act. Between 1977 and 1980, years in which the effects of increased entry and competition on rates were already evident, revenue per hundredweight for truckload (TL) general freight traffic increased by 15.3% in real terms, whereas the increase was 6% between 1980 and 1984, the last year for which we were able to obtain these data. The figures for contract carriage are even more impressive. Between 1977 and 1980, real average revenue per ton-mile fell by 1.33% per year, whereas in the period 1980 to 1984 it fell by 3.99% per year (U.S. Department of Transportation, various years). Data obtained from the East/Central Motor Carrier Freight Bureau for less-than-truckload (LTL) freight show that the revenue per hundredweight increased by 4.7% from 1978 to 1980. In the years from 1980 to 1985 it was stable (Tye, 1987), again in part due to declining fuel prices. Nevertheless, it has been estimated that regulatory reform has brought significant logistical benefits to shippers (Delaney, 1987; Evans, 1987a).

Statistics in the above paragraph tend to understate the savings to shippers brought about by increased competition. In part they do so because they do not include data for the smallest class of common carriers, those that specialize in TL carriage. In this part of the motor carrier business the increase in competition has been the greatest. In addition, our use of the published tariffs of a rate bureau understates what has happened to actual rates. Almost all trucking firms offer discounts from such tariffs, which they publish as independent tariffs. The major trucking firms have established pricing strategy groups that evaluate the costs of providing service to different customers and the prices necessary to attract and hold business. Discounts from published tariffs are offered when they are needed to hold onto or acquire business. The practice of discounting has been so widespread that the General Accounting Office (GAO) was asked by Congress to investigate whether it arose from predatory incentives, in which case it might be viewed as anticompetitive. While supporting the finding of widespread discounting, the GAO concluded that it was procompetitive rather than anticompetitive (General Accounting Office, 1987).

There is also evidence that deregulation has reduced the profitability of the largest trucking firms. Data on operating ratios (the ratio of expenses to revenue), provided by *Commercial Car Journal* show that in the years between 1980 and 1985, the average operating ratio for the 20 largest trucking firms rose from 94.0 to 96.4. The squeeze on profits must have come from the revenue side because fuel prices fell over the period and labor cost increases were contained. These two items comprise the major part of total costs. The large carriers lost a great deal of their TL freight to small firms, presumably because the small carriers had lower costs and rates. The large carriers also had to cut LTL rates selectively.

Figures on airline usage and the real cost of airline travel, and comparable figures for the motor carrier industry, make it very clear that if the program of regulatory reform involved no disbenefits, it would have to be considered a tremendous success. However, regulatory reform has involved disbenefits, most clearly in the case of the airlines, because of reductions in many quality aspects of service.

Taken alone, such disbenefits as increases in travel time, in travel uncertainty, and in lost or damaged baggage would not have been enough to create the growing sense of public uneasiness and the increasingly popular opinion that deregulation may have been permitted to go too far; that the nation might be better served if government imposed some limits on the range within which rates could be changed, on entry of new firms, and on the freedom of carriers to change the markets they serve. The fear that deregulation may bring, or already has brought, significant increases in the hazards of travel to airline passengers and also to automobile users is the source of the uneasiness.

On first inspection there is something of a paradox in this public concern. Dr. Moore in Chapter 2 presents the accident records of both transportation modes in the years leading up to, and the years following, deregulation. It is clear that in both industries the accident rates are lower now than they were 10 years ago. The record stands, but many thoughtful people are concerned that the aggregate figures mask disturbing trends that might result in reduced safety in the future. Senator Adams in Chapter 3 discusses, in relation to the trucking industry, the kinds of concerns that have prompted public uneasiness. In general, these concerns can be summarized as four points.

1. One set of links is seen by many critics of deregulation as resulting from financial pressure on firms. That is, increased competition reduces profit margins and forces firms to reduce (1) wages and the quality of the personnel hired, (2) initial and careerlong training of personnel, (3) investment in maintenance, and (4) the rate of replacement of old, less safe equipment by modern, safer equipment. Financial pressure may also result in the adoption of unsafe procedures. A commonly held view has it that price competition leads to the framing of truck schedules that force drivers to violate speed laws, and also to violate driving and rest regulations in more serious ways than would be the case in a less competitive environment. Similar allegations are made about increased competition and the airline industry. It is held by some that pilots are pressured into making flights even when certain equipment is faulty, when they have not had sufficient and suitable rest, and when takeoffs violate weather limits.
2. A second alleged link involves new entrants. It is thought that new firms have inexperienced managers, tend to hire less well-paid and lower quality staff, and use old equipment that they do not properly maintain.
3. The third link is that deregulation has induced mode shifts that have safety implications. Lower rates by trucking companies for the transport of high-value goods has shifted some freight away from railroads onto the statistically less safe highways. On less traveled air routes, the adoption of hub-and-spoke operating patterns has caused the substitution of small, statistically less safe, aircraft for large jet aircraft. Low air fares may also have brought a safety improvement for those passengers attracted away from their statistically less safe automobiles.
4. The final link concerns congestion. Deregulation in both industries has been successful in increasing patronage. In the aviation industry this increase has been concentrated in the leading hub airports at peak times of day as banks of flights connect with each other. There is little doubt that the airports and the surrounding airspace have become very congested, leading not only to delays but also to air traffic controller errors and increased probability of collision.

This book is the result of a three-day conference, convened by the Transportation Center of Northwestern University in June 1987, that explored the relationships between the economic, including the regulatory, environment in which airlines and motor carriers operated and the degree of safety with which they operate today and may operate in the future. After the next two chapters, which set the stage for the debate, we adopt the following structure. First we present a group of chapters that explain the meaning of safety in transportation, why it is a public policy issue, and the difficulties in framing and evaluating measures to improve safety. We then have two modally specific sections—airlines and motor carriers—that empirically examine the four concerns just described. Finally, after an international comparison involving the deregulation experiences of Britain and Canada, we present conclusions and identify policy recommendations.

Following the final chapter is a short section designed to familiarize readers, especially those from outside the United States, with some background information on the regulation and subsequent deregulation of the airline and motor carrier industries, and also to describe the various government agencies involved in the safety regulation of the industries.

2

The Myth of Deregulation's Negative Effect on Safety

THOMAS GALE MOORE*

Virtually all studies of transportation deregulation show that the reduction or elimination of federal controls has produced large benefits for consumers, shippers, and the economy. Airfares and motor carrier freight rates have declined in real terms. In the main, service quality has either been maintained or improved, although there is more crowding in air travel. Improvements in freight transportation and reductions in freight rates have significantly reduced logistics costs for the economy. Contrary to predictions that small shippers, small towns, and remote communities would receive less service, the evidence suggests that service is as good as or better than it was before, although rates may have increased.

Deregulation has, however, brought charges that safety has been reduced in the transportation industries. Although theory suggests that safety might be lower in a competitive market than in a regulated one, empirical evidence shows that safety has not declined since the transportation industries were deregulated but has actually continued to improve.

Even though deregulation and partial deregulation have brought great benefits to the economy and to the consumer, some interests have been adversely affected. In the airline industry, organized labor has been the principal loser. To this day, the major airlines are attempting to bring down their inflated labor costs. A number of airlines have established dual pay schemes where new employees are paid less. In the motor carrier industry, both organized labor and the owners of some certificated firms have experienced losses. The value of the operating licenses granted by the Interstate Commerce Commission (ICC) to the owners of trucking firms has been driven toward zero. The Teamsters' Union has seen a fall in

---

* The views expressed are solely those of the author and do not represent the views of the Reagan Administration, the U. S. Government, or the Council of Economic Advisors.

membership as more trucking firms have become nonunion. Pressure has also mounted to reduce members' wages.

The losers in both the airline and the motor carrier industries have not been able to claim that deregulation has hurt the consumer or shipper; in fact, both have benefited significantly. The losers have asserted that safety has been compromised, although virtually all of their evidence has been anecdotal. Pilots, for example, have asserted that airlines are forcing them to work excessive hours under stressful conditions, thus endangering safety. While pilots are working longer hours, the Federal Aviation Administration (FAA) still restricts the number of flight hours per week. Moreover, anecdotal evidence suggests that prior to deregulation many pilots held second jobs and consequently did not use the time allotted for rest. Airline mechanics have asserted that maintenance is being neglected. Again, the evidence is anecdotal and may easily reflect unhappiness with increased attention to productivity and restrictions on pay.

In the trucking industry, both owners and teamsters have asserted that economic pressures are forcing truckers, especially owner-operators, to drive excessive hours and to neglect vehicle maintenance. Since most owner-operators never were subject to regulation, deregulation of certificated firms could not have reduced safety in their firms.

Analysis of the safety issue is complicated by the fact that the transportation industries continue to be subject to safety regulation. These controls have not been dismantled; in fact there have been efforts to strengthen such regulation. If safety regulation is effective, the question of whether economic deregulation reduces the incentives to provide safety is moot. However, it is reasonable to assume, as many observers do, that safety regulation is quite imperfect. Although such regulation can ensure that some of the most elementary safety precautions are taken, such as requiring that pilots have a minimum of training and experience, these precautions very likely would have been the norm in any competitive environment.

COMPETITIVE LEVELS OF SAFETY

Safety, like comfort, speed, and cleanliness, is a characteristic that most individuals prefer more of to less. However, as with other desirable attributes, consumers are unwilling to pay unlimited amounts to secure additional safety. Given a choice, they would purchase safety (comfort, speed, cleanliness) to the point where the marginal benefits, as they perceive them, are equal to the marginal costs in terms of what they must sacrifice to have more of these desirable attributes.

Economic theory tells us that a perfectly competitive market with perfect information would provide the optimum amount of safety for consumers. Firms would compete by offering more or less safety at differing prices. Consumers would then purchase the degree of safety that they found optimal. Multiple levels of safety would be offered, since consumers differ in the value they place on safety.

The optimum level of safety for consumers, of course, may not adequately reflect all social costs because third parties can be affected. Aircraft can fall on people or collide with other aircraft; trucks do run into passenger cars. However, a well-functioning market would force firms, through insurance rates and liability rules, to take third party costs into account.

In the real world we do not have perfect competition or perfect information. However, the importance of the deviation from perfection can be overstated. The airline and trucking industries are highly competitive. A reasonably large number of firms compete in most markets, and barriers to entry and exit in particular markets are low, especially for firms already in the industry. Competition has brought prices down and encouraged firms to compete on the basis of quality.

Information about accidents is widely available for the airline industry. Airplane accidents receive considerable publicity. For the motor carrier industry, regular shippers quickly become familiar with the safety records of motor carriers. A competitive market does not require that all shippers or passengers have knowledge of safety. A minority of passengers with reasonably good information can provide adequate market discipline. Consumers with little information often purchase goods and services from well-established firms that have reputations at stake. Because firms have strong incentives to maintain a good reputation, a traveler or shipper can better ensure safety by utilizing the best known, well-established firms. Consumers with more information on safety will purchase services from less well-established firms that they know to be safe carriers. Thus the market provides mechanisms to ensure a competitive level of safety without each participant having perfect information.

Information about new firms, however, may be costly to obtain. Neither shippers nor passengers can be confident that safety standards are as high as they are for more established carriers. Therefore, it may be cheaper for the government to rigorously enforce safety standards than to depend on consumers to obtain the information. However, market forces will also exert considerable discipline on new entrants. The cost of an accident to a new firm may easily be the failure of the enterprise.

Air Florida's demise soon after the tragic Potomac River accident in 1982 clearly demonstrates the high cost of an accident to the stockholders and managers of an airline. McDonnell Douglas's stock fell $200 million, or 22%, after a series of accidents involving its DC-10 aircraft make (Chalk, 1986). More recently, Arrow Air, whose aircraft crashed in Canada in 1985, filed for bankruptcy. The market exacts a high cost for an accident.

When information on safety is difficult to obtain, a firm can sometimes profit temporarily by underinvesting in safety. The firm can reduce its costs while consumers remain unaware that they are purchasing an inferior package, that is, less safety. At best this strategy can increase profits only in the short run. A firm that follows this policy may soon be out of business. Even in the short run, there are a variety of ways the market mitigates the tendency to "cheat" on a difficult-to-measure attribute.

Critics of deregulation claim that competition leads to a greater emphasis on price as a strategic variable. This results in pressure to reduce maintenance and to cut corners on safety. Maintenance and safety, however, are important in a competitive market. A firm that neglects maintenance is more likely to be involved in accidents, which will damage its reputation and drive shippers or passengers away. Moreover, the employees, especially pilots or drivers, have personal incentives to ensure their own safety. Even a firm in bankruptcy proceedings has a strong incentive to maintain safety, since a single accident may lead to the dissolution of the firm.

The claim that poor profits or losses will reduce safety is not plausible. Consumers value safety. They prefer a safer airline to a less safe one. Therefore, an airline can attract more passengers by providing more safety. Any profit-maximizing firm should invest in safety up to the point where the additional revenue generated from being a safer airline is equal to the additional cost of providing safety. The profit-maximizing level of safety is independent of the profitability of the firm. Thus, even if competition reduces profitability, it should not affect safety.

Unless a firm is an undercapitalized company seeking quick profits, it has little incentive to provide too little safety. In the airline industry, for example, if a firm has little equity, it will have to borrow its capital or rent its planes. The lender and the renter each have strong financial interests in ensuring that the firm follows safe practices, otherwise their investments are at risk. Renters, too, are likely to be subject to liability suits. In addition, each airline carries liability insurance. Insurance companies have a very strong profit incentive to monitor the behavior of their insurees. Moreover, pilots, other aircraft personnel, and truck drivers have their lives at stake.

In the motor carrier industry both the shipper and the insurance firm have a strong interest in safety standards. Fly-by-night firms have great difficulty in attracting cargo. Most shippers prefer to deal with carriers with which they have an ongoing relationship. Thus, quality, including safety, becomes well known to the shipper.

However, motor carriers and other users of owner-operators do face a problem. The safety practices of each owner-operator are costly to discover and monitor. Firms prefer, therefore, to use known owner-operators, in whom they have confidence. Nevertheless, an owner-operator who is in financial trouble or who simply wishes to increase the short-run return by taking risks can neglect maintenance, drive excessive hours, speed, or take drugs. Users of owner-operators have every incentive to weed out such people, but some of this behavior will go on. The potential profitability from taking risks is partially offset by the potential harm to the driver and to the equipment. These incentives to take risks exist with or without regulations. Actually, hired drivers are more likely than owner-operators to ignore maintenance and take chances with their equipment. In fact, this is why motor carrier firms continue to use owner-operators. In case of accident, the hired driver bears only part of the cost, while an owner-operator bears the full cost.

Even prior to deregulation, entry into the owner-operator business was virtually free. Owner-operators either contracted with certificated freight carriers or carried exempt agricultural goods. Since there were no barriers to entry, competition resulted in the elimination of any excess profits and provided carriers with earnings just sufficient to keep them in business. Since deregulation, it has become easier for them to secure their own certificates. This has broadened their market opportunities. It is alleged that increased competition since deregulation has put more financial pressure on owner-operators, but there is little reason to believe that this is the case.

While the market apparently provides incentives for the appropriate level of safety, regulation may have provided incentives for above-optimal levels of safety. Regulation of the transportation industries took the form of promoting a cartel-like fixing of rates and fares. However, the cartels were very imperfect. Capacity was not controlled, nor was quality. With agreement on rates and routes, competition took the form of nonprice competition.

Under regulation, carriers (air or truck) filed rates with their regulator (Civil Aeronautics Board [CAB] or, ICC). After a suitable period for comment, such rates became lawful unless suspended by the regulator. Since the regulator was concerned with the economic conditions of the industry, rates were fixed to provide adequate profits for even weak firms. Bankruptcies, especially in the air carrier industry, were considered undesirable.

With rates being set, ex ante, at very profitable levels (i.e., above marginal cost), carriers had strong incentives to capture more business. Better service, therefore, became the method of competition. Airlines were particularly noted for nonprice competition. Some airlines experimented with seductive costumes; other airlines competed in food quality; frequency of services was an important competitive dimension (U. S. Congress, 1975). As a consequence, profits were in the main competed away. The system created an unprofitable cartel, resulting in higher prices, more service, and lower load factors than in a competitive market.

Safety could have been one of the service dimensions in which airlines competed. At least for the airlines, however, explicit references to safety can be a sensitive issue. One airline cannot claim that it is safer than its competitors without suggesting that flying is dangerous and thus discouraging business. Indeed, safety is never mentioned directly in airline advertising.

Although regulation of the airlines might have increased safety beyond what a competitive market would have produced, individuals differ in their willingness to pay more for increased safety. The higher level of safety offered by the regulated airline industry might have pleased some passengers, whereas others would have preferred to pay less and take slightly greater risks. In this case, deregulation could be expected to result in a situation where some airlines provide higher levels of quality (safety) at premium prices and other airlines provide lower levels of quality (safety) at lower prices. Air travelers, consequently, would be able to choose the level of quality (safety) they desire at the fares offered. If deregulation resulted in an increased number of passengers who preferred lower levels of safety at lower fares, the socially optimum level of safety actually would be lower than it had been under regulation.

The motor carrier industry also had strong economic incentives to compete in the areas of safety and damage prevention. However, there are fewer inhibitions to advertising safety and damage prevention in this industry.

EMPIRICAL EVIDENCE

The quality of data on safety performance varies considerably among transportation modes. Unfortunately, information on the factors that contribute to safety, such as maintenance and training, is difficult to obtain. For motor carriers, the data on accidents are questionable. We present some of the available information in the following.

Aviation Safety

Air traffic safety apparently has improved in the postderegulation period, notwithstanding the air traffic controllers' strike in 1981, a sharp increase in air travel, and a burgeoning of airlines. Table 2.1 compares the prederegulation period 1971–1978 with the postderegulation period. Total accidents, fatal accidents, fatalities, accident rates, and fatal accident rates have all declined sharply. An exception may be rates for the charter carriers, which, while limited in what they could do by CAB regulation, were largely unregulated in routes and rates. It is impossible to conclude from these data that deregulation has impinged on the safety of the airline industry.

Figures 2.1 and 2.2 show that the improvement in safety is an extension of a trend that predates deregulation. Air travel has been getting safer since its beginnings. This good performance was achieved in spite of the fact that the number of air traffic controllers fell sharply in 1981 after the government fired the strikers. Even as late as July 1985, the number of controllers was 15% less than before the strike (Leyden, 1986). Over the same period the number of departures increased some 26%. Moreover, on average, today's controllers have less experience than the prestrike controllers. To compensate for the fewer number of controllers, the FAA has changed a number of practices to increase safety. There is now more time between takeoffs, planes are routed with more space between them, and airborne holding—that is, circling the destination airport until a landing slot becomes free—has been largely eliminated. These new safety practices

Table 2.1 Aviation Accidents

Category	Before Deregulation (1971–1978)	Transition After Deregulation (1979–1986)	Percent Change
Scheduled large jet air carriers			
Total accidents	257	152	−40.9
Fatal accidents	39	21	−46.2
Total fatalities	1,463	807	−44.8
Total flight hours	45,579,646	57,484,764	+26.1
Rates (per 100,000 flight hours)			
Total accident rate	0.56	0.26	−53.6
Fatal accident rate	0.09	0.04	−55.6
Fatality rate	3.21	1.40	−56.4
Nonscheduled large jet air carriers			
Total accidents	25	28	+12.0
Fatal accidents	5	6	+20.0
Total fatalities	589	334	−43.3
Total flight hours	1,824,337	2,210,624	+21.2
Rates (per 100,000 flight hours)			
Total accident rate	1.37	1.27	−7.3
Fatal accident rate	0.27	0.27	0.0
Fatality rate	32.29	15.11	−53.2
Air taxis and commuter airlines[a]			
Total accidents	1,484	1,399	−5.7
Fatal accidents	343	318	−7.3
Total fatalities	1,031	853	−17.3
Total flight hours	27,919,441	36,241,018	+29.8
Rates (per 100,000 flight hours)			
Total accident rate	5.32	3.86	−27.4
Fatal accident rate	1.27	1.07	−15.7
Fatality rate	3.80	2.44	−35.8

[a]Prior to 1975, commuter airline and air taxi statistics were not recorded separately. Therefore these categories have been grouped together for these 8-year comparisons.

SOURCE: National Transportation Safety Board (various years).

may have compensated for having fewer, less-experienced controllers, but they also have contributed to a dramatic increase in delays.

Delays unambiguously reduce the quality of service and have led to an increase in complaints. However, delays in themselves do not reduce safety. The FAA policy of increasing time between takeoffs is likely to increase safety. Moreover, the delays are, at least in part, a result of the great success of airline deregulation. Since deregulation, the number of passenger miles flown has jumped 62% (U. S. Department of Commerce, various years).

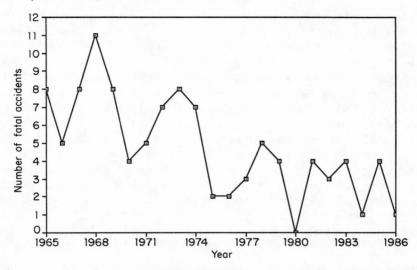

Figure 2.1 Fatal Commercial Aviation Accidents (Excludes Air Taxis and Commuter Planes). SOURCE: National Transportation Safety Board (various years).

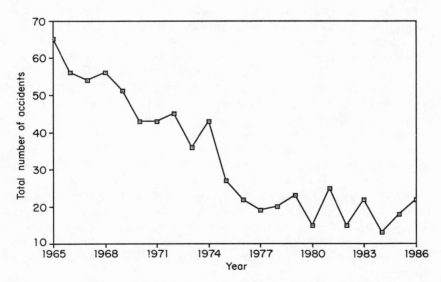

Figure 2.2 Total Scheduled Commercial Aviation Accidents (Excludes Air Taxis and Commuter Planes). SOURCE: National Transportation Safety Board (various years).

As already mentioned, critics of deregulation often argue that increased competition reduces profitability of some carriers to the point that they reduce maintenance expenditures. To examine that possibility we took a sample of 18 carriers, some very large and others very small, and did a regression analysis of their maintenance expenditure per plane on the average age of their fleet, whether they were primarily a cargo carrier, and their operating profits as a percentage of their airline investment. Age and the cargo carrier dummy variables explained half the variance, but the profitability variable, although positive, was not statistically significant.

In summary, there is no evidence in the data to support the hypothesis that deregulation of the airline industry has reduced safety. Safety as measured by number of accidents per departure or per passenger mile has been increasing steadily, probably since the first commercial flight. Whereas deregulation may not have actually increased safety, it certainly seems fair to assert that it has not decreased it either.

Motor Carrier Safety

Motor carrier safety is harder to measure than airline safety and may have been affected by other policy changes. Subsequent to the Motor Carrier Act of 1980, Congress passed the Surface Transportation Assistance Act of 1982, which opened up large portions of the nation's highway system to much larger trucks. Many observers have charged that such vehicles are inherently less safe than smaller trucks. On the other hand, trucking firms assert that since larger vehicles carry substantially more freight, the number of vehicle miles is reduced. Consequently, they claim that any increased tendency for larger trucks to be involved in accidents is more than offset by the fewer vehicle miles needed to move the nation's freight.

The evidence on large truck safety is inconclusive. A study done for the Insurance Institute for Highway Safety concluded that, "despite their greater load-carrying capacity, increasing use of doubles will produce more large truck crashes" (Stein and Jones, 1987). On the other hand, a report by the Transportation Research Board of the National Research Council concluded that "The increased use of twins will have little overall effect on highway safety because a reduction in miles of truck travel will approximately offset the small possible increase in accident involvements per mile traveled" (Transportation Research Board, 1986). A National Highway Traffic Safety Administration (NHTSA) study on large truck accident causation concluded that "Available evidence is conflicting on whether or not the accident rates per mile of travel differ between single-trailer and double-trailer combination trucks." (National Highway Traffic Safety Administration, 1982).

The data that do exist on trucking accidents are not very good. Accidents are typically defined and recorded in terms of the dollar severity of the accident or of bodily harm. As to the former, inflation tends to increase the number of reported accidents. Periodically the standard for reporting

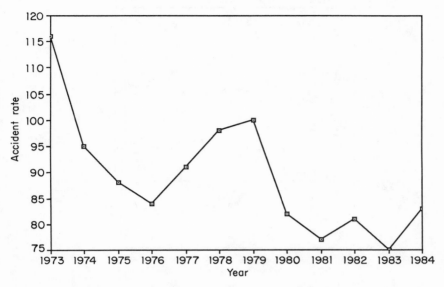

Figure 2.3   Truck Accidents per Billion Truck-Miles. SOURCE: Federal Highway Administration (various years).

accidents is revised upward to take into account the change in the value of the dollar. However, this makes time series data difficult to interpret. Moreover, the data published by the Office of Motor Carriers (OMC) of the Department of Transportation is based on voluntary self-reporting by truckers. Since there is no penalty for failure to report, the accuracy is hard to estimate. Studies indicate that reporting by private carriers is very low. In addition, the OMC data do not include intrastate truckers.

The poor coverage of the OMC accident data is indicated by the difference between the number of fatalities reported by the OMC and those reported by the Fatal Accident Reporting System (FARS), which also is produced by the Department of Transportation. For example, in 1985 OMC reported 2,676 fatalities whereas FARS reported 4,950 from heavy trucks alone. Most observers consider the FARS data to be superior.

The number of trucking accidents should be adjusted for any change in miles being driven. If there are more trucks driving more miles, there will be more accidents. Unfortunately, the data on truck-miles are simply rough estimates made by the Department of Transportation. For what they are worth, the data on truck accidents per 1 billion truck-miles (appearing in Figure 2.3) show a significant fall after 1979. Although 1984 and 1985 did show an increase, the rate was still below any preregulation year except 1976.

A better indicator of safety is truck fatality rates as reported by FARS. Figure 2.4 shows truck fatality rates per billion truck-miles for both OMC

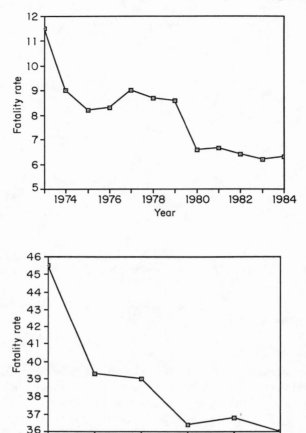

Figure 2.4 Truck Fatalities per Billion Truck-Miles. SOURCE: TOP: Federal High-
way Administration (various years); BOTTOM: number of fatalities: National
Highway Traffic Safety Administration Fatal Accident Reporting System Data
Base, and truck miles: Federal Highway Administration (various years).

and FARS data. As can be seen, since 1977 and especially since 1979,
fatality rates have declined. These data do not support a position that
deregulation has reduced safety. Figure 2.5 presents the same picture for
injury rates, based on data from OMC. Again, the data are inconsistent
with the hypothesis that deregulation reduces safety.

It is also argued that because of deregulation, trucks have become older
and less safe. The data do show that the average age of trucks in use was
8.1 years in 1986 compared with 7.1 years in 1980 (Motor Vehicle Manufacturers Association, 1986). However, the average age of trucks was
also 8.1 years in 1963 and 1964 and was 6.6 years in 1954. In other

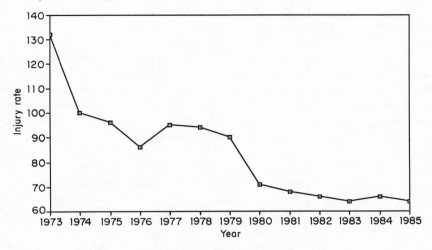

Figure 2.5 Truck-related Injuries per Billion Truck-Miles. SOURCE: Federal Highway Administration (various years).

words, the average age fluctuates considerably, and the 8.1 figure is consistent with the age range found in the regulated period.

As mentioned earlier, critics of deregulation often claim that the additional competition that results from reduced government controls lowers profits. If lower profits tempt firms to skimp on maintenance, the result will be an increase in accidents due to more mechanical failures. The OMC publishes data on accidents due to mechanical failure. These data, unfortunately, also depend on voluntary self-reporting. However, whereas the

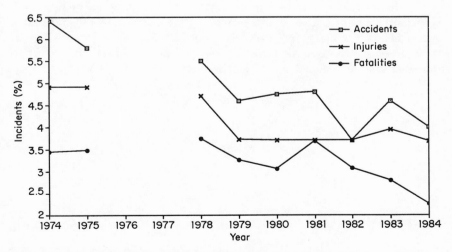

Figure 2.6 Percent of Reportable Incidents Due to Mechanical Defects. SOURCE: Federal Highway Administration (various years).

percent of accidents reported as being due to mechanical defects is un-doubtedly understated, the trend over time probably is a reasonable index of the true pattern of accidents due to mechanical failure.

Figure 2.6 presents the percentage of all accidents, accidents that result in injuries, and fatal accidents that are reportedly due to mechanical de-fects. These data are available for 1974 through 1984, with the exception of 1976 and 1977. As can be seen, while the data do fluctuate, since de-regulation in 1980 the general trend has been downward. There certainly is nothing in these data to indicate that mechanical defects have become a more important problem since deregulation.

In summary, although data on accidents and miles driven are poor for motor carriers, the figures that are available do not substantiate any claim that deregulation has reduced safety. The data, if anything, indicate that trucking is safer today than before deregulation.

CONCLUSION

Increased airline and motor carrier safety is simply part of a general trend toward increased safety in transportation and may have little to do with deregulation. For the purposes of this chapter, however, it appears that deregulation has not changed the trend toward greater safety.

Data fail to support the hypothesis that deregulation has reduced safety. Although it could be expected that regulation would increase safety above optimum levels, and thus a move to a competitive market might reduce measured safety, we find that not even this is true. Either safety regulation has been effective or competitive pressures have maintained safety at op-timum levels. We have no data to help us choose between these two com-peting hypotheses. I am skeptical that federal regulation can be and is sufficiently omnipresent and thorough to increase safety significantly. Nevertheless, for the purpose of this chapter, the question as to what has maintained safety does not have to be answered. This book asks whether deregulation has reduced safety. I believe it has not.

3
Deregulation's Negative Effect on Safety

BROCK ADAMS

Wendell Wilkie is not often quoted these days. He once said something, however, that I find very perceptive: "A good catchword can obscure analysis for fifty years." We haven't run the full 50 years yet, but one "catchword," deregulation, certainly has been able to obscure analysis. I hope here to dispel some of that obscurity and demonstrate that deregulation in the trucking industry has not been the panacea that its proponents claimed it would be. In fact, the economic benefits of deregulation have been uneven and the social costs have been underestimated. As I hope to show here, the major cost of deregulation, insufficiently accounted for in weighing the pros and cons of deregulation, has been a marked decrease in transportation safety.

Having been involved in transportation policy for many years from different perspectives, I have seen the deregulation debate from many angles. I first saw "deregulation" as a congressman in the early 1960s. Back then there was much talk about something we called "the iron triangle," which described an incestuous relationship among industry, congressional committees, and regulatory agencies. All three elements of this relationship recognized that the tighter their dealings with each other, the greater their power would be (Ladd, 1985).

Starting in 1935, an iron triangle developed in the trucking industry. For many years it provided safety and stability to a number of companies as the nation shifted to a highway transport system, but it also had a downside. It eliminated competition through rate bureaus, it restricted entry, and it stultified innovation by strictly regulating routes and commodities trucking firms could carry. The net result was that even as highways expanded, costs increased and services to new areas and new businesses were limited.

When President Kennedy assumed office, the economists who surrounded him suggested that he ought to break out of the triangle. He did try, but in the early 1960s iron proved to be stronger than economic theory. The triangle survived that first brief assault. By the 1970s, however, the iron triangle was facing more than economic theory: Regulatory stagnation was creating tremendous transportation and business pressures that threatened the collapse of the entire system. By the time of the Carter Administration, the nation had shifted much of its transport to the highways, yet every interstate highway seemed to be threatened by regulatory roadblocks. Examples of the nonsensical regulations of the time abound. Frozen dinners were exempt agricultural commodities unless they were chicken or seafood; crab shells were exempt, oyster shells were not. Certificates were approved with narrow restrictions. For example, one certificate permitted transport of paint hauled in 2-gallon cans but not in 5-gallon cans. The Interstate Commerce Commission (ICC) approved certificates with circuitous routing, forcing carriers to travel many extra miles and thus waste time, fuel, and money (Interstate Commerce Commission, 1978).

REFORM POLICY

President Carter decided that the old system of regulation needed to be relaxed. I agreed. Many of us at the Department of Transportation (DOT) believed that regulation should be reformed to meet new challenges. This reform was started with airlines and trucks. The vision was sound, but with the premature departure from the Cabinet of certain officials, the Administration shifted emphasis. Several of the president's economic advisors went much further in their zeal for change. They believed that only pure economic factors should shape transportation business decisions. Regulatory relaxation, reduction, and reform were replaced with "deregulation." The distinction is more than semantic; it is symbolic of a split between those who want to restrain government's role and those who wish to remove it. The victory of the latter group made the 1980s possible.

The Reagan Administration has taken the flexibility of the legislation crafted in the 1970s and created a policy in the 1980s that has turned regulatory agencies into ideologically driven debating societies whose mission is to eliminate themselves rather than eliminate unfair and unsafe practices.

Different people can see things differently—that's what democracy is all about. However, the way the Reagan Administration went about turning its philosophy into policy was wrong. A major motive of our founding fathers in creating the United States was to control interstate and foreign commerce. They knew all too well that a completely unregulated system of commerce would not encourage the development and free flow of goods and services. Despite its rhetorical praise for constitutional principles, the Reagan Administration abandoned a fundamental constitutional precept when it rejected the federal role in regulating interstate commerce for the good of the nation as a whole.

I am the first to admit that we needed to loosen the bonds of regulation. As I said in the 1970s, and as I say now, I am proud to have contributed to eliminating the sort of rules that once told truckers what routes they had to follow and what cargoes they could carry. There is a great difference between relaxing regulation and rejecting it, however, and the Reagan Administration has rejected regulation. Let me now turn to the effects of deregulation as it has been implemented by the Reagan Administration.

EFFECTS OF DEREGULATION AS CURRENTLY PRACTICED

Certainly the free market can do many things better than the government can do them. A free market, using money as the test, will allocate resources quickly and often efficiently. However, it allocates resources on the basis of the economics of the moment, without considering the broader needs of the nation. The bottom line, in the area of transportation, is a system that looks primarily at short-term gain, the "dollar at the dock," without considering the costs to society of unsafe practices, which add billions of dollars in the form of medical costs, lost personnel hours, and police and court costs. The net result of deregulation is a system that is true to a theory but that, in practice, is not as safe as it should be; that ends up charging more for a short haul than for a long one; and that often creates monopoly power at the management level while adversely affecting employees.

The primary mechanism through which deregulation is translated into a loss of safety is the incessant pressure of competition on truckers to reduce costs. Without the safety net of regulated rates and without stringent entry controls, truckers, particularly owner-operators, compete on the basis of price. There is no doubt that deregulation increased competition, but it did not create more business. Intercity truck tonnage has hovered at around 2 billion tons since 1970. All that grew was the number of carriers trying to carry that tonnage. In fact, the number of ICC-authorized carriers available to shippers jumped from 16,874 in 1978 to approximately 37,000 in 1986 (Interstate Commerce Commission, various years). As new entrants mushroomed, so did bankruptcies. In 1980, there were 382 carrier failures; in 1985 that figure was 1,533 (Glaskowsky, 1986). Using Dun and Bradstreet data, the American Trucking Associations estimate that 714 of the 1,533 trucking failures in 1985 were intercity carriers as opposed to local operations. This figure is approximately 13 times the number of such failures in 1978.

The pressure to reduce prices doesn't come exclusively from truck competitors. It also comes from purchasers, from shippers who use their overwhelming financial advantage to send a simple message: Carry it for half the price or I'll get one of six other firms to do so. The net result is that truckers without economic leverage have no choice but to lower their prices in an effort to become competitive and to cut their costs in any way they can. Truckers have cut costs in a number of ways. One way is by not buying new equipment. The age of the trucks on the road has increased since deregulation. Current data indicate that the *average age* of owner-operator equipment has increased from 3.5 years in 1979 to 5.7 years in 1984, and is now approaching seven years, the *maximum* age once accepted for highway tractors. Company-owned equipment has seen a similar trend (Baker, 1985; Page, 1987).

Aging equipment is bothersome. What is more disturbing is the fact that even with more and older trucks, and with the increased need for maintenance they create, spending on vehicle maintenance and parts has increased only 17% since 1978. It doesn't take a lot of experience in the industry to know that such a small increase is inadequate, or to see where it leads. When the trucks on the highways have aged and spending on new equipment has gone down, the small rate of growth in parts and maintenance spending means only one thing: We have shifted toward breakdown maintenance rather than preventive maintenance. The trucks on the road aren't as safe as they used to be because now most truckers don't have the money to put into either new equipment or repairs.

Recent data from roadside inspections corroborate the fact that many of the trucks now on our highways lack proper equipment and are inadequately maintained. The Office of Motor Carriers (OMC) indicates that in 1986, 213,700 trucks were placed out of service by federal and state inspectors because of serious safety defects. This figure represents 39.2% of the trucks inspected. In 1984, 48,279 trucks were placed out of service at inspection sites—approximately 30% of the trucks inspected (American Trucking Associations, 1986).

It isn't just that truckers haven't taken care of their physical equipment. It is also the kind of equipment they use. Ever since deregulation and the decline of the ICC, enormous pressure has been put on truckers to haul more and to haul more quickly. One manifestation of this trend is that there are now many more double-trailer trucks on the road. A study done by the Washington State Patrol and the Insurance Institute for Highway Safety has documented that double-trailer trucks are two to three times more likely to be in accidents than tractors pulling only one trailer (Stein and Jones, 1987). The use of longer combination vehicles (LCVs) has accelerated as a result of the Surface Transportation Assistance Act of 1982, which prevented states from barring LCVs from their highways in exchange for higher taxes on the trucking industry.

Equipment is only part of the story. Deregulation pressures have affected people too. The drive to cut costs and increase revenue has heightened the pressures on truckers to violate the hours of service rules, to ignore speed limits, and to haul loads that exceed statutory weight limits. The logbook system used to enforce the hours-of-service rules has become so permeated with abuse that these logs are now commonly referred to in the trade as "comic books." Nonstop stints of 16 hours at the wheel are not uncommon. The data on trucks hauling freight over the legal weight limit are equally disturbing. It is estimated that as many as half the trucks on the highway are over the weight limit, causing excessive wear on the highways and straining the ability of the equipment to operate safely. Driving was always tough; driving under deregulation is dangerous. And now that firms are having difficulty finding experienced drivers, dangerous is too mild a word (Abruzzese, 1987).

Given all these factors, it really isn't surprising that the number of truck accidents has increased. Unfortunately, it appears that the size of that increase is escalating. OMC figures indicate that in 1980 there were about 31,000 truck accidents reported; by 1985 the number had soared to more than 38,000. Even if one accounts for the increase in the number of miles driven by trucks, the accident rate has increased. The Office of Technology Assessment estimates that the rate of truck accidents increased from 2.34 accidents per million truck miles in 1981 to 2.69 accidents per million miles in 1985. I am concerned that with speed limits going up, both the number and severity of accidents will go up too, unless we increase the role of the federal government in regulation of safety.

While the safety impact of deregulation is acknowledged by many, we need to recognize that deregulation has had some adverse economic impacts. The entire justification for deregulation was free-market competition. To a certain extent, in terms of economic value at least, some shippers have realized very real advantages, but deregulation has been a mixed bag, even in pure economic terms. Competition in the less-than-truckload (LTL) segment of the industry, roughly 60% to 70% of the business for most successful truck lines, has become increasingly concentrated, dominated by large, well-financed, nationwide or regional carriers. These companies often use several subsidiaries to hire underfunded owner-operators. One has to wonder how the economic benefits of deregulation are going to be realized if the bulk of the industry becomes increasingly concentrated. While the LTL sector has remained fiercely competitive since deregulation, the potential for monopolistic rates exists, and we may see an increase in rates sooner rather than later if the economy remains strong. So to those people who claim that deregulation works, and who point to lower prices to prove it, I say that we haven't seen the final bill yet. However, we have seen enough to worry about what the tab ultimately will be.

DANFORTH-ADAMS BILL

There is "Adams on deregulation." There is also "Adams on reregulation." Let me sketch out a couple of areas of action. To begin with, I think we need to start now to address specific problems with targeted legislation. That is why Senator Danforth and I introduced the Truck and Bus Safety

Bill of 1987. This legislation focuses on some of the specific problems I have discussed here: the fatigue of drivers operating more hours than allowed under the federal regulations and the safety of some of the equipment being used. The Danforth-Adams Bill contains three principal provisions.

1. It requires the Secretary of Transportation to initiate rule-making proceedings on the need to improve brake performance standards. The Secretary is directed to investigate technologies such as antilock braking systems and brake compatibility. In a study released in May 1987, DOT concluded that efforts to improve truck brake systems should receive the highest priority of all equipment-related safety issues. The report estimated that brake performance could be involved as a contributing factor in up to one third of all truck accidents (U. S. Department of Transportation, 1987).
2. The bill directs DOT to consider ways to improve truck safety with respect to drivers' hours of service, including the possibility of using on-board monitors to record driving time, speed, and other information.
3. The bill would eliminate the so-called "Commercial Zone exemption." It makes little sense to regulate trucks on the highway but not when they operate in the dense traffic of our nation's metropolitan areas.

These sorts of specific changes will be helpful, but they will not be enough. We also need attitude changes that produce policy changes. We need more people at the ICC who are willing to do what needs to be done. Every president is entitled to nominate people to serve on the ICC or any other regulatory agency who share his basic values and beliefs. But we are entitled to have people in these agencies who abide by the law, understand operating problems, and are at least willing to listen to and try to understand other values and beliefs. That is one of the reasons that I introduced, with several of my colleagues, legislation designed to improve the way we appoint people to regulatory agencies by creating a Transportation Regulatory Commissions Nominating Commission to recommend to the president qualified applicants to fill any vacancies that develop.

We also need more people inspecting companies for proper maintenance, insurance, "fitness," and other safety-related variables. These sorts of inspections, and the punishments that can be imposed for noncompliance, can have a positive effect on safety. With respect to one requirement, insurance, the Department of Transportation estimates that as many as 25% of all carriers, regulated and unregulated, do not have adequate coverage.

I recognize that under the Motor Carrier Safety Assistance Program (MCSAP), established by the 1982 Surface Transportation Assistance Act and expanded by the Commercial Motor Vehicle Safety Act of 1986, roadside inspections and carrier safety audits have risen dramatically. OMC figures indicate that roadside inspections by state and federal officials have risen from 160,000 in 1984 to 524,000 in 1986. This is a noteworthy and positive development, but the MCSAP program alone will not solve our truck safety problem.

I do not advocate a return to a system that would regulate every route and rate, but I would like to see a system that would regularize standards of operation and impose safety and financial responsibility criteria for new entrants.

I began this chapter by quoting Wendell Wilkie. Let me close by paraphrasing another unlikely source, Heather Gradison, chairperson of the ICC. In testimony before the Commerce Committee in June 1987, Gradison indicated that she viewed the role of government in regulation as a "backstop" rather than as an umpire in on every play. I understand her argument, and find it incomplete. We all ought to understand that while an umpire doesn't tell a pitcher what to throw, an umpire *does* determine whether what was thrown was a ball or a strike. An umpire doesn't tell a football team how to line up on the field, but an umpire does determine if it has lined up offsides. And just like an umpire, in the trucking industry, government should not be telling companies how to run their businesses, but it must ensure that those businesses are run in a manner that protects the interests of the rest of the nation.

FRAMEWORK FOR ANALYSIS

The majority of this book is concerned with empirical investigations of the recent safety performance of the airline and trucking industries. However, these detailed analyses need to be set in perspective. Important theoretical issues should be addressed initially, such as why transportation safety might be linked to economic conditions, and the nature of the links. (The strength of these links will then be addressed in the later chapters.)

Panzar and Savage concern themselves with these issues in Chapter 4. They point out that safety is but one of many attributes of the transportation product; others are the frequency of service, comfort, and speed. The consumer weighs all these against the price of the service in the decision concerning mode of travel, or indeed whether to travel at all. Deregulation has had major impacts in allowing flexibility in pricing of transportation, and it is not inconceivable that some consumers may choose a lower price at the expense of reduced quality of travel in terms of comfort, travel time, and perhaps safety. Panzar and Savage go on to describe another feature of safety provision, the crucial role of government. This arises because the free market, left to itself, does not ensure the safety level that consumers desire. They offer three reasons for this, one of which is that consumers are often not in a position to evaluate the safety performance of carriers and cannot "vote with their feet" against unsafe carriers.

Moore has argued in Chapter 2 that despite such reasons, safety markets perform well. Large shippers are familiar with the safety records of the carriers they use, and airline passengers are informed of airline mishaps by intensive media reporting. The validity of Moore's assertions are an empirical question. Borenstein and Zimmerman in Chapter 5 attempt such an evaluation in the case of airline markets. They conclude that after an adverse effect on demand that lasts about two months following an accident, patronage returns to its former level. While open to several interpretations, this finding suggests that in the airline industry, a company's accident record does not send a signal to consumers. In these circumstances, as Panzar and Savage point out, there is the potential for safety regulation to improve social well being.

One should not be led into believing that government safety regulation is a panacea. Hauer, in Chapter 6, uses the record of safety regulations in the highway area to caution and raise doubts about their effectiveness. He concludes that part of the problem is that the evaluation of safety policies is often entrusted to the implementors of the policy, who have little incentive to identify policy failures. Issues also remain, as Panzar and Savage point out, about the proper role of government in the provision of infrastructure.

The section concludes with a chapter by Viscusi that takes a different approach to the safety issue. It focuses on injury rates to workers rather than consumers. Viscusi's analysis, which covers both the airline and trucking industries, fails to uncover a negative impact on safety.

4

Regulation, Deregulation, and Safety: An Economic Analysis

JOHN C. PANZAR and IAN SAVAGE

This chapter explores, at the theoretical level, the nature of the linkages between economic deregulation and safety in the airline and motor carrier industries. Perhaps our first duty is to explain why such an inquiry is necessary; after all, safety has not been deregulated in either industry. However, the level of safety provided is an important *economic* attribute of any transportation service. Therefore any change in public policy that affects the economic structure of an industry as dramatically as has deregulation can be expected to have important effects on both the industry safety level and the costs and benefits associated with achieving it.

Thus the question of interest is not whether economic deregulation has affected safety. Considering safety's economic importance, it would be remarkable if there were *no* effect. Rather, the question that must be addressed is: Given the changes in the marketplace resulting from economic deregulation, what is the appropriate policy response by government in areas related to safety? Answering this important question requires a thorough understanding of how safety levels are determined in the transportation marketplace. To achieve this understanding, it is necessary to recognize the important roles played by the decisions of firms, consumers, and government in determining the level of safety with which our transportation services are provided.

SAFETY IS AN ECONOMIC GOOD

We begin with the premise that transportation safety is an *economic* attribute of transportation service. That is, it is a characteristic of the service that is both desired by consumers and costly to provide. This may seem a rather unobjectionable statement, yet it will serve to take safety out of the realm of sacred cows and makes its determination an appropriate subject

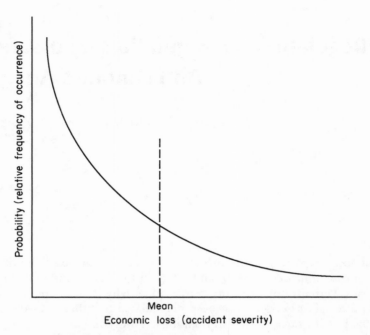

Figure 4.1   Accident Severity and Probability.

for economic analysis.[1] For as soon as one admits that safety is a desirable but costly product attribute, it no longer follows that the socially desired level of safety is the highest that is both technologically and humanly possible. Thus, without intending to trivialize the problem, we can quite accurately state that the problem of safety determination is, *in principle,* indistinguishable from that of determining the level of any other aspect of service quality, such as schedule frequency, on-time performance, or airline leg room.

However, before beginning our economic analysis there is an important semantic matter that must be addressed. Throughout our discussion we shall speak of "safety" as if it were an unambiguously defined unidimensional concept. In reality, of course, it is no such thing. Even if one were to define safety as "the probability that a trip will result in an accident," there would still be the difficulty that this probability would be different for trips made during different seasons of the year, on different carriers, over different routes, and so on. Add to this the fact that accidents differ widely in their severity, and the magnitude of the problem is apparent.

Nevertheless, we will refer to safety as if it were a well-defined scalar in order to keep the exposition intelligible. There is a meaningful scalar notion of safety that may be useful to keep in mind. On the horizontal axis in Figure 4.1 is plotted the amount of economic loss due to an accident.

On the vertical axis is plotted the frequency (e.g., per million ton- or passenger-miles) with which accidents of various severities occur. Obviously, accidents can vary in severity from small fender benders to disasters in which there is a major loss of life. It is not unreasonable to suppose that the latter are rare relative to the former, and this fact is reflected by the shape of the hypothetical frequency distribution in the diagram. Distributions such as these are often described by statisticians in terms of their mean. If one must have a scalar measure of safety, something inversely related to the mean of this distribution (i.e., the expected economic loss per million ton- or passenger-miles) is intuitively appealing.[2]

THE COMPETITIVE BENCHMARK: SAFETY DETERMINATION IN AN IDEAL WORLD

Initially we shall suppose that only one type of service, "very-safe," is offered by transportation carriers. Suppose also that very-safe service reduces the probability of an accident to the lowest level that is technologically and humanly possible. Whether particular consumers will choose to avail themselves of very-safe travel and transport will depend on a number of factors: the price of the trip, the value of the trip, the probability of an accident, the loss that would be incurred in the event of an accident, income level, the extent of the consumer's risk averseness, and so on. As with other economic goods and services, we would expect that, at any given price, some people would purchase the service and others would not. One would expect that more people would make more trips at a low price than at a high price if everything else (including safety level) were held constant.

Thus a *demand curve* for very-safe travel results from each consumer's self-interested decision of whether to purchase the service at the stated price. Mathematically, a demand curve describes the relationship between the price of service and the number of trips desired by consumers. The other factors determining consumers' purchase decisions also affect the demand curve. Factors, such as an increased accident probability (lowered level of safety), that reduce the net expected value of a trip to consumers will shift the demand curve inward to the left. That is, they will result in a smaller quantity being demanded at each and every price.

Similarly, it is possible to define a relationship between the market price and the number of very-safe trips that carriers are willing to offer. The higher the price, the more service carriers will be willing to provide. (And if existing carriers won't, new entrants that are attracted into the marketplace will.) This market *supply curve* for very-safe transport service will be affected by factors that determine the cost of providing the service: wage rates, equipment costs, fuel prices, and so on. Anything that makes it more expensive to provide the service tends to result in less service being provided at any given price: The industry supply curve shifts inward.

Given the usual textbook assumption that a large number of consumers interact with a large number of transportation carriers in an impersonal marketplace, an *equilibrium* will be attained. That is, a price will be reached at which the quantity of service that consumers desire will be equal to the quantity that carriers are willing to supply. Furthermore, given the assumptions we have made, this competitive equilibrium in the transport market will have all the desirable properties associated with free-market outcomes. Government intervention of any kind cannot improve matters. Thus, in spite of the fact that transport services are inherently risky, market interactions between fully informed consumers and firms result in an economically efficient allocation of resources in the market for transport services.

The scenario just described is, of course, much too simplistic. One might argue that the reason that government intervention was not required was simply because firms had no discretion concerning the level of safety provided. For some unspecified reason, every carrier provided the safest possible service. Therefore let us suppose that it is also possible to provide "not-so-safe" transport service to consumers. This level of safety might be the result of less frequent maintenance of equipment, a shorter training period for pilots, and so on. It is important to note that the actions taken by carriers that result in their service being not-so-safe also tend to make the service less costly to provide. Thus one can envision an industry supply curve for not-so-safe transport service that lies at every point to the right of that for very-safe service. That is, at any given price the industry would be willing to provide a greater quantity of not-so-safe service.

Obviously, not-so-safe service will be less valuable to consumers because of the greater probability of loss due to an accident. This decline in value may take the form of an increase in expected loss or an increase in premiums, should consumers choose to insure themselves against the risk of loss. Two points should be clear: (1) If not-so-safe were the only service offered, the quantity demanded at any price would be less than the quantity of very-safe demanded at that price. (2) If not-so-safe were simultaneously offered to informed consumers at the same price as very-safe, no one would purchase it. However, from our perspective, the interesting question is: What would happen if not-so-safe were offered at the same time as very-safe but at a significant discount?

While some readers may find it hard to believe that consumers would choose anything other than the safest possible service, we maintain that many consumers are prepared to accept lower levels of safety if the price is lower. A couple of examples should serve to demonstrate that this trade-off takes place every day. Consider the decision faced by a shipper. Employing a not-so-safe carrier will result in higher insurance premiums because of a higher probability of loss. Yet if the price differential is great enough to offset the higher premium (and the increased expected time and trouble involved in filing an insurance claim), the shipper will choose to patronize the not-so-safe carrier. When risk to life and limb are involved, the analysis is a bit more complicated. However, when it comes to intercity travel, large numbers of consumers choose to drive rather than fly. They do so primarily because it is cheaper, despite the generally recognized fact that air transport is by far the safer mode of travel.

Thus it is reasonable to expect that there will be consumer demand for both types of service, at least over some range of prices. The two demand curves will certainly be *interdependent,* since the demand for not-so-safe transport will be lower the lower the price of very-safe service, and vice versa. Yet this fact does not prevent competitive markets from attaining equilibrium in the usual way; at prices at which supply equals demand in both markets. Again, given the assumptions we have made, there is no government intervention that will improve on the outcome of the market. In this case, however, the market has done more than merely determine the equilibrium volume of traffic at a predetermined level of safety. Here, the informed choices of consumers and producers result in a market-determined overall level of safety that may be less than the amount theoretically possible. For whenever not-so-safe service is offered by carriers and purchased by consumers, "society" has, for economic reasons, chosen *not* to avoid some accidents. Nevertheless, in the ideal, perfectly competitive world we have described, this outcome is optimal. Government intervention, in the form, say, of prohibiting the offering of not-so-safe service, could only make society worse off.

There is nothing that limits the above analysis to the case of two possible safety levels. *As long as the market for each type of service is competitive,* the free market can be counted on to make society's determination of the overall level of safety in an economically efficient manner. Professor Corwin D. Edwards described the situation well in a report of the National Commission on Product Safety (1970):

The risks of bodily harm are not unreasonable when consumers understand that risks exist, can appraise their probability and severity, know how to cope with them and voluntarily accept them to get benefits that could not be obtained in less risky ways.

SOURCES OF "MARKET FAILURE"

The happy outcome just described depends on the assumption of *perfect* markets for transportation and related services. The list of requirements that must be satisfied by the perfect markets that inhabit the realm of economic theory is rather long and involved.[3] For our purposes it is important to mention two: that there be a large number of sellers for each type of service and that all market participants be fully informed.[4] Neither condition is satisfied in real-world transportation markets, and it is these *structural* and *informational* imperfections that provide the rationale for government action in the area of transportation safety.

Imperfect Information

In our discussion of consumers' choices of type of service, their ability to assess each carrier's probability of an accident correctly played an important role. However, in practice this information is seldom as readily available as that on, say, price or the extent of on-time performance. Thus a key ingredient in the formula that led to the optimality of the market outcome may be missing. If the consumer misperceives a carrier's probability of having an accident, we can no longer presume that his choice accurately reflects a careful weighing of the true economic costs and benefits.[5] One "solution" is to depart from the traditional caveat emptor liability rule implicitly assumed in the previous discussion and place responsibility for any damages due to an accident on the producer, who presumably has a much better idea of the true probabilities.[6] However, when information is imperfect, this approach will also fail to lead to the provision of the socially efficient level of safety.

Under a system of producer liability, the producer will seek to obtain insurance either because she is personally risk averse or, because of limited liability, insurance is required by society to ensure that the producer is truly responsible for her actions. However, when the carrier's accident probability is known only to the carrier (i.e., it is private information), the insurance market cannot be expected to solve the problem. To see why, suppose that a carrier's accident probability is a decreasing function of the level of maintenance effort it selects. Now let us examine the market determination of safety under a system of complete producer liability, compared with the socially desired level of safety. First it is important to note that, since maintenance is costly, there is an economic trade-off between the expected costs of accidents and maintenance costs. In general, society would not choose to reduce the probability of an accident to zero, even if it were technologically possible to do so. In a world without insurance, carriers that are liable for the full costs of their accidents have every incentive to select socially optimal levels of maintenance.

A problem arises, however, if carriers wish to insure themselves against the risk of loss. A carrier's accident probability is not a datum that the insurance industry can look up in its actuarial tables. It is determined, in our example, by the carrier's level of maintenance. A carrier's maintenance level is, in turn, selected by the carrier on the basis of the costs it faces for maintenance and the loss it would face in the event of an accident. Insurance plays an important role in efficiently allocating risk. However, in the current setting, it also has the unfortunate effect of diluting the incentives created by the producer liability rule. For example, if it were fully insured, a carrier would select that level of maintenance effort that minimized its operational costs, ignoring the effect of increased maintenance on the probability of an accident. The outcome would no longer be socially optimal. This problem is known in the economic literature as *moral hazard*. In effect, the holding of insurance reduces or removes the desirable incentive effects that a system of producer liability is designed to achieve, because the insured has an incentive to adversely affect risk through actions that are hidden from the insurer.[7]

Nor would the effects of imperfect information necessarily be eliminated even if consumers were fully informed and the liability rule were caveat emptor, with consumers assuming the responsibility for insuring themselves against possible losses. Consumers would then choose their carrier on the basis of the lowest *total* of transport and insurance costs. Under such a system carriers might still end up providing different levels of safety, due, for example to differing efficiency levels in their maintenance departments. If consumers were better informed about such differences than insurance companies,[8] patrons of the less safe carriers would purchase actuarially fair insurance policies, whereas those of the safer carriers would not. Such *adverse selection* would limit the efficiency of risk reduction possible via the insurance markets.

Thus imperfect information is a source of market failure in transportation industries. In fact, it can be argued that, if information were perfect, there would be no need for safety regulation at all! Carriers could be induced to provide efficient levels of safety via appropriate choices of liability rules and their enforcement through the legal system. However, as the foregoing arguments have suggested, liability rules alone will not lead to the provision of socially optimal levels of safety when information is imperfect. In such circumstances it may well be desirable to impose safety regulation. We now present an argument that explains the role of safety regulation in mitigating the moral hazard problem just discussed.[9]

Suppose that carriers have differing costs of performing maintenance on their equipment. Under a system of producer liability and private insurance, producers will choose maintenance levels that minimize their total costs. If information is imperfect so that insurance companies cannot observe a carrier's maintenance effort, premiums will be set on an industry-average basis. Therefore, since a carrier's premium is not affected by its maintenance level, each carrier will choose the level of maintenance that minimizes its operating costs, ignoring the costs of accidents. Less efficient carriers will undertake less maintenance than their more efficient rivals. The resulting levels of maintenance effort, and safety, will be too low relative to the socially efficient levels.

Now suppose that it is possible, at some cost, for the government to monitor the maintenance efforts of carriers periodically and to impose penalties if they do not meet established standards. Let us consider the outcome under such a regulatory regime. Through judicious choice of standards and penalties, the government may be able to increase the level of safety in the industry in a way that increases economic efficiency as well. Because the unregulated system cannot be relied on to provide the socially efficient level of maintenance and safety, it may be possible to induce increases in safety that raise overall efficiency in spite of the (socially nonproductive) expenditures on inspections and punishments that such a system requires.[10] Although such a system will not match the efficiency possible under perfect information, it may be the best that is possible in our imperfect world.

While it is, of course, impossible to argue that current systems of safety regulation are in any sense "optimal," it should be pointed out that many of the characteristics that would be associated with the stylized system just described are, in fact, reflected in practice. For example, under our hypothetical system some carriers might choose to adopt maintenance levels that meet or exceed the standards established by the government, while some would choose not to and occasionally be detected and hit with large fines. The important point to remember is that this type of regulatory system can be an optimal response to the market failures resulting from imperfect information.[11]

Imperfect Competition

In addition to the assumption of perfect information, the idealized competitive markets discussed above are also unrealistic in their requirement that all levels of safety be offered by a large number of competitive firms. This requirement cannot be satisfied in any real world market in which production is characterized by increasing returns to scale at small output levels. If firms must achieve some minimum efficient scale before they can operate at minimum unit costs, then it may be impossible for there to be a "large number" offering each of many different levels of safety, even if there are a large number of firms in toto. In the economics literature, this type of market structure is called "monopolistic competition."[12] In monopolistically competitive markets, there tends to be a trade-off between the number of different product varieties (levels of safety) provided by the market and the price of the service provided. Put simply, increasing the number of firms and product varieties means that each firm serves a smaller number of customers, realizes a smaller portion of available scale economies, and must charge a higher price in order to break even. How do the market equilibrium levels of product quality (safety) and variety in monopolistically competitive industries compare with the socially desirable levels? There is a considerable literature on this topic[13] which concludes that, in general, the market *cannot* be relied on to produce the socially optimal levels of product quality and variety.

The problems posed by increasing returns to scale or government-mandated entry barriers may be even more serious, however. In the extreme case, the service may end up being provided by a monopolist. How does the level of safety chosen by a monopolist (or small group of oligopolists) compare with the economically efficient level? There is also a substantial literature on the theory of quality choice under monopoly.[14] Again, the conclusion is that the quality (safety) level chosen will not, in general, correspond to that which would be socially desirable.

Analysis of another form of imperfect competition is particularly relevant for understanding safety level determination in the regulated transportation markets of the past. Since economic regulation took the form of price and entry regulation, carriers found themselves competing for market share in various quality dimensions, including safety. This situation has also been subjected to considerable economic analysis.[15] However, in this case it was possible to determine not only that the market equilibrium would not produce the socially optimal level of quality, but also the *direction* of the bias. Nonprice competition among price-regulated firms tends to result in quality levels *greater* than those that would be socially optimal. Since safety, like schedule frequency, is a costly quality attribute desired by consumers, there is every reason to expect that regulated transportation industries provided too much of it.

Finally, it should be pointed out that these inefficiencies in the provision of safety caused by imperfect competition would pose no significant problem in the absence of imperfect information and related problems in the insurance market. For then, any market equilibrium level of safety that did not meet the needs of passengers and shippers most efficiently could be mitigated or eliminated by the purchase of insurance. Also, the appropriate choice of liability rules could be used to shift the market equilibrium level of safety toward the economically efficient level without direct regulation of safety levels.

Congestion of Publicly Provided Facilities

Our discussion of the economics of safety determination has proceeded thus far under the assumption that the level of safety provided by a firm was something directly and completely determined by the firm's production decisions: maintenance, employee training, and so on. A moment's thought should remind the reader that there is an obvious missing ingredient in this description of safety determination in the motor carrier and airline industries. Trucks travel on publicly provided roads and highways and airlines use municipally provided airports and federally maintained air traffic lanes. Furthermore, these publicly provided inputs into the production of transportation services are also very important determinants of the safety of those services. No matter how well maintained the vehicle or well trained the driver, truck transportation will not be very safe if roads are not properly designed for the volume and type of traffic that they carry.[16] Similarly, the best designed airplane and the most experienced pilot would have little hope of surviving long in an environment in which airport runways are crumbling and the air traffic control system is woefully inadequate to handle the volume of traffic in the air.

These and related elements of the transportation infrastructure clearly play an important role in the determination of transportation safety. The fact that this infrastructure is publicly provided is, itself, a reaction to another potential source of market failure. Roads, airports, and so on tend to be natural monopolies, and would not provide their service to the public in an economically efficient manner if left in private hands. This, in turn, would mean that even ideally competitive transportation markets could not be expected to provide economically efficient levels of safety because of the distortion in the market for an essential input.

However, it must be recognized that public ownership of the transportation infrastructure does not automatically eliminate this potential source of economic inefficiency. It is still necessary to determine the correct capacity for these facilities and to allocate it in an economically efficient manner. Public ownership and provision of infrastructure facilities merely replaces inefficient private markets for these services with administrative processes that may themselves be characterized by varying degrees of inefficiency. While a large portion of the discipline of civil engineering is devoted to the complexities involved in the design and construction of such facilities, for our purposes it is sufficient to focus on a single dimension of the typical infrastructure facility: its capacity.

It has long been recognized that the capacity of a transportation facility such as a highway must be tailored to the volume of traffic it carries if the transportation system is to operate efficiently. Too little capacity and the roadway becomes overly congested, increasing the time-cost component of transportation costs. On the other hand, installation of too much capacity involves a social loss. Economists have also recognized the desirability of using *congestion prices* to allocate available capacity most efficiently.[17] The desirability of congestion pricing arises because adding an additional vehicle to the roadway (or airway) typically imposes delays (and associated time costs) on other users of the facility. This externality results in a divergence between the private cost borne by the marginal user and the social cost of his use, that is, the total increase in travel time borne by all the other users.[18]

However, in addition to this time-cost externality, congestion is likely to give rise to a safety externality. This follows from the simple observation that, other things being equal, accidents are more likely to occur when highways and airways are crowded than when they are not. Thus the capacity of the infrastructure and the way that access is priced can be expected to have an effect on safety levels because they determine the level of congestion of the transportation system. Improper decisions in this area can be an important source of governmental failure.

SAFETY DETERMINATION IN THE REAL WORLD

The discussion of the previous section explains why the determination of transportation safety should not be left entirely to the marketplace, despite the fact that safety is an economic good. We identified three primary causes for the government to assume a role in determining transportation safety:

imperfect information, imperfect competition, and infrastructure congestion. In this section we shall examine the ways in which economic forces and government actions jointly determine the level of transportation safety.

First, it is important to recognize that the level of safety we observe in transportation markets is an aspect of "economic equilibrium" in those markets. That is, it is determined by marketplace interactions between consumers' desire for safe transportation and the costs at which profit-seeking firms can provide it, given the transportation infrastructure provided by the government and the safety rules and regulations it imposes. Therefore it is incorrect to think of safety as something that is unilaterally selected by the government. All that government can do is to establish the framework in which the economic decisions of market participants determine the equilibrium level of transportation safety. This framework has three parts: economic rules and regulations, the publicly provided transportation infrastructure, and the government's safety surveillance activities, that is, the system of safety rules and regulations combined with the inspection and penalty policies used to effect compliance with those rules.

Thus governmental influence on safety is ultimately indirect. Our earlier discussion should have made this point clear with respect to economic regulations and publicly provided infrastructure. Clearly, neither directly determines safety levels, though they may have important roles to play. But even governmental safety rules and regulations do not *directly* determine what the equilibrium levels of safety will be in the marketplace.

At one level this point, too, is obvious. After all, there is no gauge on a truck or an airplane that allows an inspector to determine the vehicle's "safety level" and correct it if it is too low. As we have already stated, the probability of an accident is determined by the whole range of the firm's investments in maintenance and training, as well as the quality of the infrastructure that its vehicles use. Safety rules and regulations cannot begin to specify precise levels for all of these variables; they can only establish standards for those that are both important and readily observable. Yet one could argue that determining the levels of key safety-related expenditures would also determine accident probabilities (and safety), at least to a tolerable approximation.

Unfortunately, this view ignores the problems posed by imperfect information, the fact that the levels of a firm's safety-related investments cannot be *costlessly* observed by the enforcement agency. The fact that each vehicle inspection is a costly use of social resources means that the monitoring is likely to be less than all pervasive. Not every vehicle will be inspected, and as a result, not all firms will be in compliance all of the time. Thus the government never has complete control over the level of safety-related expenditures, even when standards are set, periodic inspections occur, and fines and penalties are levied for noncompliance.

Thus, as mentioned earlier in this chapter, the government must solve a complicated economic problem to perform its surveillance role effectively. A complete discussion of the issues involved in constructing an efficient surveillance system would fill a book.[19] For our purposes it is sufficient to recognize that the more money effectively spent on surveillance activities, the greater the realized level of safety in the industry. *Safety is an economic good.*

The second way in which the government influences the market level of safety is through its infrastructure investments. While the efficient design and construction of transportation systems is a highly sophisticated discipline, the important insight is that the greater the amount efficiently spent on the infrastructure, the safer the transportation system, other things being equal. Similarly, the pricing of airport landing rights will have the effect of decreasing or increasing safety to the extent that it encourages or discourages peak period congestion.[20]

Finally, it is necessary to recognize that the economic environment also plays an important role in safety determination. This follows from the fact that changes in the economic climate will tend to affect firms' decisions, including their safety-related expenditures. Of course economic deregulation has been a major change in the economic environment during the modern history of the transportation industries. As discussed earlier, the system of price and entry regulation that prevailed prior to deregulation encouraged nonprice competition, and it is reasonable to conclude that safety and other quality attributes were probably overprovided. Similarly, *exogenous* changes in the economic environment that are not a direct result of government action, such as the growth of aggregate demand, may also have significant effects on market safety.

AN ANALYTICAL FORMULATION OF SAFETY DETERMINATION

We are now in a position to discuss safety levels in transportation markets as being influenced by three "variables": the economic environment and government expenditures on surveillance and infrastructure. If we let E, B, and I represent these influences, respectively, we can describe safety *(S)* determination in a stylized algebraic equation:

$$S = h(E,B,I) \qquad\qquad (4.1)$$

Our purpose here is not to endow the process of safety determination with spurious precision, but rather to clearly indicate that the safety levels[21] actually realized in the marketplace are the result of decisions made by profit-seeking firms and optimizing consumers, *given* the economic environment and governmental expenditures on surveillance and the infrastructure. The function $h(\cdot)$ summarizes a host of complex interactions, some of which have been discussed above. Note that we have written E in boldface to indicate that it represents a long list of factors that, in toto, describe an "economic environment."

We now summarize our stylization of safety determination for a given economic environment. The government budget process determines the amount of money made available to the agencies responsible for surveillance and infrastructure activities. Ideally this process takes place with some knowledge of what those agencies can hope to achieve, given varying amounts of money. That is, the budget makers are, it is hoped, capable of making intelligent choices between dollar expenditures and realized levels of safety, because that *is* what they are doing, like it or not. If they are informed and conscientious, their decisions will at least approximate the socially optimal trade-off between dollars and transportation safety.

The surveillance and infrastructure agencies determine the most effective ways to spend their budgets. The solutions to problems they face may be quite complex, and the strategies employed to solve them may change as the underlying economic environment changes. For example, if the economic environment is such that the market is served by a few large firms, the optimal surveillance strategy may involve placing much of the inspection and data collection obligation on the firms themselves, with the agency monitoring their day-to-day activities relatively infrequently, relying on hefty violation fines to enforce compliance. This strategy may not be nearly as effective if a change in the economic environment (e.g., economic deregulation) results in an industry that consists of a large number of small firms, each with far less experience and financial resources. Furthermore, this change is likely to make it more costly to achieve the original level of safety, or equivalently, only a lower level of safety may be achievable with the same expenditure.

It is now possible to present a simple analytical description of the activities of those governmental bodies whose job it is to regulate safety and provide the transportation infrastructure. We have assumed, realistically, that these agencies take the economic environment as given when performing their functions. To make our analysis as simple as possible, we also assumed that, for example, the Federal Aviation or Highway Administration allocates any given budget B in an optimal way, given the economic environment of the industry it is responsible for regulating. Agencies responsible for the infrastructure are also assumed to spend their available budget, I, in the most effective way. This "governmental efficiency" assumption allows us to use the surveillance and infrastructure budgets, B and I, as determinants of safety (arguments of $h(\cdot)$), rather than a long list of variables describing the levels of each of the hundreds of distinct activities in which these agencies engage. Having established this framework, it is natural to suppose that $h(\cdot)$ is increasing in both B and I, other things being equal.

We have further argued that, at least in principle, it is possible to determine budget levels for these agencies that are optimal for a given economic environment. We will denote these optimal budgets for surveillance and infrastructure investments as $B^*(E)$ and $I^*(E)$, respectively. It is important to remember that these are merely shorthand representations of the outcomes of complicated and economic calculations that attempt to measure the economic value of safety.

A FRAMEWORK FOR EVALUATING THE SAFETY EFFECTS OF ECONOMIC DEREGULATION

We will now attempt to discuss the effects of economic deregulation on transportation safety using this framework. Let E_1 represent the economic environment under economic regulation. The level of safety in the marketplace is also dependent on expenditures on surveillance and infrastructure investments. We will suppose that these were optimally chosen for the regulated environment E_1. That is, $B_1 = B^*(E_1)$ and $I_1 = I^*(E_1)$. Then the level of safety under economic regulation would be given by

$$S_1 = h(E_1, B_1, I_1) \tag{4.2}$$

Now suppose economic deregulation occurs and a new level of safety S_2 is realized in the marketplace. While the comparison between S_1 and S_2 is the subject of much of the empirical research in the remainder of this volume, for the sake of argument we will suppose $S_1 > S_2$. (If careful scrutiny of the available evidence revealed that the inequality should be reversed, the public debate would soon be over.)

What should one make of such a decline in safety? More important, what should be the policy response? The first point to recognize is that a decline in safety, even a significant one, does *not* necessarily mean that economic deregulation was a mistake. Remember, *safety is an economic good,* and Morrison and Winston have estimated the economic benefits to airline consumers and firms resulting from deregulation as being on the order of $16 billion per year.[22] Thus, at a minimum, it would be necessary to attempt to balance the decline in safety against these benefits. However, it is even possible that an observed decline in safety might represent an efficiency gain in and of itself!

To see how this could be, consider again the ideal world discussed previously, in which there are only two possible levels of safety. Suppose, initially, that the government regulated the market price of transportation and set that price at a level high enough to ensure comfortable profits to firms offering very-safe service. This would leave consumers who were unable or unwilling to pay the regulated price out of the market. (While firms would clearly be willing to offer less costly not-so-safe service at the regulated price, they would find no takers.) If price regulation were removed, not-so-safe service would begin to appear in the market, drawing its customers from both those who were shut out by the high regulated price and others who simply prefer not-so-safe service at a low price to very-safe service at a higher price. If one were to compare the average level of safety in this market before and after deregulation, it would clearly have declined. The moral? Economic deregulation gives the market an opportunity to better serve consumers' needs. Sometimes this may mean giving them an opportunity to purchase *less* of a desirable attribute (safety) at a lower price.

Unfortunately, we do not live in this ideal world, and it may turn out that the sources of market failure discussed earlier are sufficiently important that the drop from S_1 to S_2 is legitimate cause for concern. How should one go about evaluating the seriousness of the drop and formulate appropriate policy responses? First, it is important to remember that the economic environment is only one determinant of the level of safety. For example, suppose one observed the change in safety level after deregulation but before any adjustment in surveillance and infrastructure expenditures. Then, letting E_2 denote the economic environment under deregulation, this level of safety would be given by

$$S_2^0 = h(E_2, B_1, I_1) \tag{4.3}$$

Equation 4.3 reveals the fallacy involved in attributing the problems associated with this hypothetical drop in safety entirely to economic deregulation. Recall that the expenditures on surveillance and the infrastructure were assumed to be set at the levels optimal for regulating safety in an industry subject to economic regulation, that is, $B_1 = B^*(E_1)$ and $I_1 = I^*(E_1)$. After deregulation these expenditure levels may no longer be optimal. Changes in the economic structure will, in general, require adjustments in both. For example, it is reasonable to suppose that it is more costly to enforce safety regulations in an industry served by a large number of small firms than the reverse. Also, the increase in traffic resulting from lower postderegulation prices may tax the capacity of the infrastructure. Both effects suggest that, after economic deregulation, any given level of safety can be achieved only with greater levels of expenditures. The same arguments will apply even if our initial postderegulation observation of safety, $S_2 > S_2^0$, occurred in the context of some small increase in surveillance and infrastructure expenditures, for example, at levels $B_2 > B_1$ and $I_2 > I_1$.

Clearly, then, any concern over the safety effects of economic deregulation should not be predicated upon the initially observed safety level S_2. Rather, the first order of business for policymakers is to move as rapidly as possible to implement those policy changes needed to increase the efficiency of safety determination in a postderegulation world. We have discussed the importance of public investments in surveilliance activities and the infrastructure for the process of safety determination. However, another policy decision must logically come first. That is, before embarking on the difficult and costly process of adjusting surveillance and infrastructure expenditures to levels that are optimal for the deregulated environment, it makes sense to consider making changes in the economic environment that may enhance the efficiency of the existing infrastructure. The policy change we are referring to is congestion pricing of infrastructure facilities.

We have already discussed the potential impact of congestion on transportation safety. Particularly in the aviation industry, given the changes in industry structure following deregulation (hub-and-spoke networks, etc.), safety problems arising from congestion are likely to have increased. Of course one way to reduce congestion is to expand the capacity of the infrastructure. Another way, however, is to allocate the existing capacity more efficiently by using congestion pricing. By charging carriers more for access during peak periods, congestion pricing tends to spread the demand over the (daily or weekly) cycle in an economically efficient way. Use of congestion pricing does not mean that one can forgo needed capacity expansion. But under congestion pricing such expansions may not need to be as large, and the policy of pricing capacity can help provide a market test of when capacity expansion is economically efficient.[23]

In terms of our analytical framework, implementation of congestion pricing would change the economic environment from E_2 to E_2'. As a result, safety would increase to $S_2' > S_2$, even without any change in the level of surveillance and infrastructure expenditures, that is, $h(E_2', B_2, I_2) > h(E_2, B_2, I_2)$—an economic environment that includes congestion pricing will increase the level of safety achievable from a given level of expenditure on surveillance and infrastructure.

Only after evaluating the efficacy of a move to congestion pricing can the policy question of appropriate infrastructure and surveillance expenditure levels be addressed. Therefore, for purposes of our discussion, we will assume that congestion pricing has been introduced.[24] The next order of business for policymakers would be to adjust expenditures on surveillance and the infrastructure to levels appropriate for the new economic environment E_2', that is, deregulation with congestion pricing of infrastructure facilities. For purposes of illustration, we will suppose that the increases in traffic and in entry and exit that accompany economic deregulation will require increased expenditures on both the infrastructure and surveillance activities. Of course these are actually empirical questions that are addressed in many of the other chapters in this volume. In terms of our analytical model, this assumption means that $B_2^* = B^*(E_2') > B_2$ and $I_2^* = I^*(E_2') > I_2$. Then the benchmark level of safety in the postderegulation world would be $S_2^* = h(E_2' B_2^*, I_2^*) > S_2' > S_2$.

A simple diagram may aid understanding of these points. In Figure 4.2, safety increases along the vertical axis and surveillance and infrastructure expenditures along the horizontal axis. The points S_1, S_2, S_2', and S_2^* are drawn in line with the assumptions made in the above discussion. After deregulation, the initial observation is of decreased safety and a small increase in expenditures on surveillance and the infrastructure. A move to congestion pricing results in an increase in safety (to S_2') without any increase in safety-related public expenditures. Full adjustment to the post-deregulation situation might then be represented by a northeast movement

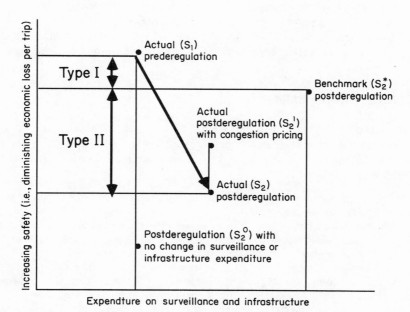

Figure 4.2   Possible Analytical Interpretation of the Effects of Deregulation on Transportation Safety.

to S_2^*, involving both increased safety and an increase in infrastructure and surveillance expenditures.

Figure 4.2 also provides a convenient framework for decomposing any initially observed postderegulation decline in safety into two parts. We will call the change from S_1 to S_2^* the "type I" effect. It represents the inherent change in safety that results from going from a transportation sector subject to economic regulation to one characterized by price and entry deregulation *combined with optimal policy responses with respect to congestion pricing and the levels of infrastructure and surveillance expenditures.* The second, "type II," component represents the safety improvement, $S_2^* - S_2$, that will result from appropriate policy responses to the problems caused by economic deregulation. For policy purposes, it is clearly only the type I change that should be of concern. That is, it is the magnitude of that decline (if any) that should be weighed against the demonstrable benefits of economic deregulation. The type II decline is not caused by deregulation, but rather by the (hopefully temporary) inadequate policy response *to* deregulation.

Our argument makes clear what the priorities should be in the policy debate. First, consideration should be given to the possibility of making some adjustments (for example, congestion pricing) in the postderegulation economic environment that might improve the terms of the safety expenditure trade-off. Then the remaining necessary adjustments in surveillance and infrastructure expenditures must be determined and the resulting effects on safety observed. Only then will policymakers be in a position to carefully consider the question of whether any remaining decline in safety outweighs the manifest benefits of economic deregulation.

CONCLUDING REMARKS

Our objective in this chapter has been to provide a framework that the reader can use to interpret the policy arguments and empirical evidence presented in succeeding chapters. Our main arguments can be summarized as follows:

1. Safety is an economic good. In an ideal competitive world, its determination could be left to the marketplace.
2. In the real world, imperfect information, imperfect competition, and the importance of the publicly provided infrastructure mean that government has an important role to play in safety determination.
3. Any observed change in safety following economic deregulation must be evaluated in the context of changes that deregulation may have caused in the effectiveness of surveillance agencies and the adequacy of the infrastructure.
4. Before reconsidering economic deregulation, policymakers must address the adequacy of surveillance and infrastructure expenditures and the efficacy of economic policies that would enhance their effectiveness.

NOTES

1. Of course this fact has been recognized by economists for some time. See, for example, Lave (1968) and Oi (1973).

2. We must caution the reader that other properties of this distribution (e.g., its variance) may also be important in evaluating transportation safety.

3. See, for example, Hirshleifer (1988) for a complete statement.

4. A third condition, that there be no *economic externalities,* that is, effects on third parties, is also violated in many transportation contexts. Innocent bystanders who are not parties to the transportation transaction are often affected by highway and air accidents. However, we shall argue that these important effects would not, in and of themselves, require safety regulation by the government. In the absence of imperfect information, these effects could be internalized through appropriate liability rules.

5. An important issue in the theoretical literature on the economics of imperfect information is the extent to which consumer ignorance can persist in a long-run equilibrium situation. Rational consumers will attempt to infer a firm's quality level from the price it charges or other market data. See, for example, Nelson (1970) and Shapiro (1982, 1983a, 1983b). In the present context, one piece of market data the consumer might, in principle, draw reliable inferences from are the insurance rates associated with patronizing any particular carrier. However, as subsequent discussion will show, if the insurance industry is always perfectly informed about any particular carrier's accident probability, liability rules can be used to effectively eliminate market failure.

6. See Spence (1977) for an analysis of the optimal policy response in this situation.

7. See, for example, Shavell (1979).

8. As might well be the case, for example, when the "consumer" in question is the shipping manager of a large manufacturer.

9. Our discussion follows that of Hansson and Skogh (1987). See also Shavell (1984a, 1984b).

10. The design of such a regulatory system is itself a nontrivial economic problem. Standards, inspection strategies, and penalties all have an impact on the level of economic efficiency achieved by the system. For an analysis of the economics of "crime and punishment," see Becker (1968).

11. In our discussion we have assumed that it is the government that does the inspecting and imposes the penalties on companies found to be in violation. Of course it is possible to envision private insurers performing these functions, and indeed, for some types of risks they do. However, government action in this area is likely to be more efficient to the extent that there are substantial scale economies in the inspection process. Also, the coercive power of the state may make it possible to impose penalties more efficiently.

12. Chamberlin (1965) is the classic reference.

13. See, for example, Dixit and Stiglitz (1977) and Koenker and Perry (1981).

14. See, for example, Spence (1975).

15. See Douglas and Miller (1974), Schmalensee (1977), and Panzar (1979).

16. See Chapter 6 in this volume for a discussion of safety issues in highway engineering.

17. The classic references are Vickrey (1971) and Strotz (1965).

18. One could take the view that this externality, and the resulting wedge between private and social costs constitutes a market failure and is the true source of the adverse effect on safety discussed below. However, we take the view that this externality would be *internalized* if private firms were providing infrastructure facilities, perhaps via some form of congestion pricing. Furthermore, the natural monopoly characteristics of infrastructure facilities would remain a source of market failure, even if congestion were not an issue.

19. See, for example, the "Liability, insurance and safety regulation" issue of *The Geneva Papers on Risk and Insurance*, Vol. 12(43), April 1987.

20. See Chapter 12 in this volume for a detailed discussion of this issue.

21. As we stated at the outset, safety is not a simple unidimensional concept. The probability of an accident and the severities of the accidents that do occur may both vary widely, even within a single well-defined transportation market. As we indicated in this chapter, this dispersion in safety levels may result from firms' efficient responses to the varying risk preferences of their consumers. For concreteness and clarity of exposition, we will continue to talk of *the* market level of safety as if it were a scalar. The reader is invited to view S as representing a measure inversely related to something like "the expected loss due to accidents per million ton (or passenger) miles," if that interpretation is more satisfying.

22. Morrison and Winston (1986).

23. See Chapter 12 for a complete discussion of the advantages of congestion pricing.

24. The policy debate surrounding congestion pricing may, in fact, turn out to be quite heated, because some carriers and interest groups may perceive that they will be disadvantaged by the change. However, it is difficult to think of any *economic* argument against the change, and we will not discuss the politics of the situation any further.

5

Losses in Airline Demand and Value Following Accidents

SEVERIN BORENSTEIN and MARTIN B. ZIMMERMAN

Central to the debate over the need for government regulation of product safety is the incentive that the private market provides for producing safe goods and services. This incentive comes from the costs imposed on firms responsible for unsafe products. With the economic deregulation of airlines, the industry has become a focus for this debate. Using a sample of nearly all fatal U. S. airline accidents between 1960 and 1985, we attempt in this chapter to quantify the costs that airlines incur due to crashes.

Though we begin by examining the losses incurred by shareholders when a serious accident occurs, a major focus of this work is the reduction in demand that an airline might encounter as consumers respond to an accident. Several previous studies attributed much of the lost equity value following an airline accident to declines in future demand for the firms' products (see Chalk, 1986, 1987, and Mitchell and Maloney, 1988), but none measured this effect directly. We have found that little, if any, loss of airline equity value after accidents can be reliably attributed to consumer response. Furthermore, the losses in equity values themselves are quite small.

DESCRIPTION OF THE SAMPLE

The base sample for the analysis is every fatal accident aboard a U. S.-certified air carrier between 1960 and 1985 that involved a fatality and some damage to the aircraft. The sample excludes air taxis and commuter airlines. Demand or stock market data were not available for every firm at all times during the 27-year period, however, so the demand and financial studies omit some of these accidents. Stock market data were available only beginning in the middle of 1962, so accidents during 1960, 1961, and part of 1962 are excluded from the estimation of financial losses. Our study of equity value loss is also limited to firms listed on either the New

York or American Stock Exchanges. Seventy-four accidents met these criteria.

Estimation of airline demand and the effects of accidents was broken into four periods, 1960–1965, 1966–1971, 1972–1977, and 1978–1985. The first period covers the time of transition from piston engine to jet aircraft. The next period was relatively stable in both prices and technology. The third period covers the time of the Domestic Passenger Fare Investigation, which imposed new and stricter policies on discount fares and route entry, as well as increased use of wide-bodied aircraft. The final period is taken to be the deregulation era, though the exact starting date can, of course, be questioned. In this period, prices are taken to be a (endogenous) decision variable for the airlines, rather than exogenously set by the Civil Aeronautics Board, as was the case in earlier years. Although 26 airlines are included in the overall demand study, some were excluded in some periods due to an insufficient number of months in which the carrier was in full operation. In the four periods there are, respectively, 15, 14, 12, and 12 airlines that experienced accidents and for which sufficient data are available.

RESULTS OF THE STATISTICAL ANALYSIS

Average Effect of Accidents on Shareholder Wealth

We used the standard event study approach to analyze equity value changes (see Fama, Fisher, Jensen, and Roll, 1969; Borenstein and Zimmerman, 1988). Table 5.1 presents the average abnormal returns (i.e., returns after adjustment for marketwide movements in stock prices) for the 74 accidents included in the financial analysis from one day before news of the accident reached the market to seven days after. (Chance and Ferris, 1987, and Mitchell and Maloney, 1988, also examined equity value changes following airlines accidents.) The third column shows the cumulative abnor-

Table 5.1 Financial Returns Associated with Airline Accidents

Day	Abnormal Return (AR)	Cumulative AR	% Firms with −AR	t Value	z Value
−1	.00104		52.703	0.314	−0.022
0	−.00940[a]	−.00940[a]	64.865	−2.818	−3.087
1	−.00031	−.00971	52.703	−0.093	−0.108
2	.00098	−.00873	51.351	0.295	0.199
3	−.00102	−.00975	56.757	−0.309	−0.024
4	−.00302	−.01277	62.162	−0.910	−1.086
5	.00090	−.01187	48.649	0.271	0.069
6	.00089	−.01098	54.054	0.267	0.434
7	.00348	−.00750	50.000	1.050	0.756

[a]Significantly different from zero at the 1% level.

Table 5.2   Estimated Impact of Accidents on Airline's Demand

Variable	1st Month	2nd Month	3rd Month	4th Month	Total
1960–1977: 78 Observations					
Weighted mean	−0.4%	−1.0%	−0.1%	−0.2%	−1.8%
Mean	−0.8%	−0.9%	1.1%	2.8%	2.2%
t value	−1.00	−1.00	0.39	1.15	−0.14
z value	−0.65	−1.20	0.14	0.47	−0.86
1978–1985: 13 Observations (Ordinary Least Squares)					
Weighted mean	−1.4%	−5.2%	−4.7%	−3.9%	−15.3%
Mean	−1.6%	−3.1%	−1.6%	−1.0%	−7.4%
t value	−0.56	−0.86	−0.45	−0.35	−0.86
z value	−0.80	−1.63	−1.26	−1.05	−1.77
1978–1985: 13 Observations (Two-Stage Least Squares)					
Weighted mean	−1.0%	−4.8%	−4.7%	−0.3%	−10.7%
Mean	−1.3%	−4.9%	−1.0%	5.3%	−1.8%
t value	−0.09	−0.18	−0.06	0.02	−0.06
z value	−0.36	−0.93	−0.88	−0.21	−0.64

mal returns for the accident day and the following trading days. The t and z statistics, tests for whether the abnormal returns are significantly different from zero, are presented for the daily returns. (z statistics give less weight to very "noisy" observations and thus are probably better indicators of statistical significance.)

Crashes are associated with, on average, a 0.94% loss in the equity value of the firm on the first trading day. This loss is statistically significant at the 1% level and translates to an average loss in firm value of $4.5 million. On average, the information appears to be totally absorbed in the stock price on the first trading day after the accident. Abnormal returns are small and statistically insignificant after that date.

Effect of Accidents on Demand

Analysis of the demand effect of accidents is analogous to analysis of the equity value effect. Demand is estimated as a function of price, consumer income, seasonal variables (monthly dummy variables), and a trend term. "Abnormal demand" due to a crash is then the residuals from the regression estimation during the month of the accident and the three succeeding months, that is, the percentage deviation from expected demand during these months.

Table 5.2 presents estimates of the average abnormal demand following accidents. Due to the variation between firms in the precision of the estimates, particularly in the two-stage least squares estimates of the deregulation period (price treated as endogenous), Table 5.2 shows both the unweighted average effects and the weighted average where the weights are the inverse of the variance of each estimated effect. The "Total" column indicates the average of the sum of the abnormal demand effects in the crash month and in three succeeding months as a percentage of the firm's average monthly demand.

Throughout the regulation period, accidents, even the most catastrophic ones, appear to have very small average effects on the demand that the airline faces. Even the most negative estimates are quite small relative to the fluctuation caused by other factors. As a point of comparison, the peak-to-trough seasonal variation in demand (with February being the trough and July or August the peak for most carriers) is estimated to average more than 30% during these periods. Furthermore, the standard errors of the demand regressions, which represent the random demand component, range from 3% to 31% and average 9%.

Since deregulation, consumers' responses to crashes may have increased, though the sample for this period includes only 13 accidents and these may not be representative of the "typical" crash. The estimated effects are generally not significant at conventional levels, but they do indicate a pattern of negative response to crashes that tapers off after approximately two months. The total loss of demand due to an accident is estimated to average 10.7% of one month's traffic when estimated by two-stage least squares. When the deregulation period demand regressions are estimated by ordinary least squares, assuming that price is exogenous, the estimated average demand effect of a crash is 15.3% of one month's demand.

Still, the danger of inferring a systematic demand response from these 13 observations is highlighted by a closer look at the accident of Air Florida in January, 1982. For many reasons, this crash of a Boeing 737 in Washington D.C. that killed 74 of the 79 people on board seems to be among the most likely candidates for significant adverse consumer reaction. Air Florida was a new airline without an established safety record. The accident was quickly and conclusively deemed to be the result of pilot error. Furthermore, the error was probably related to inadequate pilot training by the airline regarding flight procedures in subfreezing temperatures. Due in part to its location and the heroic rescues that took place, the crash received extensive media coverage, including mention during the President's State of the Union address two weeks later. Finally, most of Air Florida's traffic was carried on the highly competitive routes between northeastern cities and Florida, so travelers could fairly easily switch airlines. Still, the cumulative abnormal demand in January through April of 1982 is estimated to be *positive*, though not significantly so. Despite a slumping economy and virtually the same prices as a year earlier, Air Florida carried 30% more passengers in the first quarter of 1982 than in the first quarter of 1981. Though Air Florida had lower load factors in the first quarter of 1982 than a year earlier, the load factors for February and

March 1982 were above that for December 1981, before the crash and historically a month of high demand for north-south travel. The January load factor was 1% below December.

If there were a consumer response, variations in the response might be explained by various characteristics of the accident or the airline involved. We examined a number of possible explanatory variables including the number of on-board fatalities in the accident; the size of the airline as measured by revenue passenger miles flown in the month before the crash; whether the airline was primarily responsible for the accident; the recent accident records (fatalities in the one-year period ending the day of the crash) of the firm and of the industry; and two measures of the extent of newspaper coverage of the crash. These measures are the number of days of front page coverage and the total number of articles appearing in the *New York Times* in the two weeks after the crash (newspaper coverage variables for New York crashes are taken from the coverage of the *Chicago Tribune*). None of these variables appears to be strongly associated with the demand reaction when analyzed through simple correlations or multivariate regression.

Explaining the Change in Shareholder Wealth

Table 5.3 presents an attempt to relate the loss in firm value to variables expected to influence the magnitude of that loss. Direct costs (uninsured tort liability, loss equipment value, and expected changes in future insurance rates) are represented by the total number of people killed or seriously injured in the accident. The number of passengers seriously injured is included based on the observation that a tort settlement or award for a

Table 5.3 Analysis of the Causes of Firm Value Change,[a] Least Squares Regressions

Observations: 67 accidents	
Degrees of freedom: 64	
Dependent variable: Change in total value of outstanding stock on day of event	
Intercept	0.200 (0.419)
Direct costs (total number of people killed or seriously injured)	−0.038[b] (0.011)
Revenue change	−0.001 (0.016)
R squared	0.150
F (3,64)	5.647[b]

[a] Standard errors in parenthesis.

[b] Significantly different from zero at the 5% level.

seriously injured person is often as high or higher than for a wrongful death suit. An estimate of indirect costs, that is, lost revenue, was derived from the estimated abnormal demand that the airline experienced in the months following the crash. (Because the revenue change is estimated from demand regressions, this explanatory variable is measured with error. The results presented ignore this bias. Borenstein and Zimmerman [1988] presented similar regressions using an instrumental variables correction.)

The change in firm value is regressed on the estimated lost future revenue and the number of fatalities plus serious injuries. (Because the volatility of airline stocks differs quite a bit among airlines, these regressions are corrected for heteroskedasticity.) Table 5.3 shows that estimated lost revenue is not significantly related to the change in firm value and is very close to zero. The number of fatalities plus serious injuries does appear to be a significant cause of the change in equity value. The parameter estimate for the change in value of the outstanding stock on the day of the event, significant at the 5% level, indicates an equity loss of $38,000 per life.

CONCLUSION

We have examined the losses suffered by airlines in connection with a crash. On average, crashes are associated with a statistically significant 1% loss in equity value, or about $4.5 million. Since most crashes in the sample involved total destruction of the aircraft and an average of more than 40 fatalities, the average firm loss appears to be well below the total social costs of the accident. This is consistent with the airlines' practice of carrying insurance against the direct costs of major crashes, and indicates that the insurance is only partially experience rated. The investigation of consumer response found that before deregulation, travelers did not respond to crashes to an extent that is statistically discernible. In the period since deregulation, consumer responses may have increased. The small number of accidents and greater demand variability for a firm since deregulation make these effects difficult to estimate with precision.

It must be stressed that all of these results are conditional on the level of federal safety regulation, and the outstanding safety record that the industry currently maintains. The relatively small overall effects indicate that consumers view accidents as providing little new information about the safety of an airline and the industry. Though that perception may be changing, it is generally consistent with recent reports from airlines that have experienced safety problems in the deregulation period. Thus, Delta Air Lines had few cancellations following a spate of near-accidents in the summer of 1987, almost all of which were due to errors by flight crews or maintenance personnel (*New York Times,* July 25, 1987). Vocal dissatisfaction with an airline due to poor performance in other areas also does not seem to indicate much avoidance of the carrier (*Wall Street Journal,* November 19, 1987).

6

The Reign of Ignorance in Road Safety: A Case for Separating Evaluation from Implementation

EZRA HAUER

GIVING THE THUNDER AWAY

One focus of this book is the need for safety regulation and surveillance. The voice heard tends to be the voice of the regulators and their advisors. Between writer and reader there is tacit understanding that "regulators" means some public agency and "regulatees" in this case are the trucking industry and the airlines. It is thus clear from the beginning who are the cops and who are the robbers and also who is supposed to do what to whom. Into this harmony of agreement I intend to add a dissonant note. I will claim that the regulators need to become the regulatees.

Consider, for example, the recently rejuvenated Motor Carrier Safety Assistance Program (MCSAP). The aim is to increase roadside inspections in order to detect and correct truck safety defects, driver deficiencies, and other unsafe practices. The federal government covers 80% of the cost. In 1986, with its first grant, a certain state supported 22 Department of Transportation (DOT) positions and 12 troopers; 7,185 vehicles were inspected, 4,651 were placed out of service. For 1987 that state program received three times as much money. This allowed the hiring of 10 more inspectors, 3 more troopers, and the purchase of a mobile inspection facility.

Since there is genuine public concern about truck safety, the regulator (DOT) is seen to respond, and big trucking associations endorse the MCSAP. All seems to be in order. That so many trucks need to be put out of service points to a real problem with their state of repair. One element, however, is not mentioned. The costs to the public and the costs to the truckers are not negligible—$50 to $60 million every year in federal funds alone. Is MCSAP causing a measurable improvement in the state of repair of trucks on the road? Are the accident reductions achieved commensurate with the costs? Nobody knows.

Since the present program is relatively new, this lack of knowledge is perhaps understandable. What is disturbing is the apparent lack of a serious plan to ascertain the effect of MCSAP. There are numerous programs of this kind (such as periodic motor vehicle inspection, speed limit enforcement, road salting, and demerit point systems) that have been in effect for many years, and that are also unevaluated in terms of costs and benefits. If MCSAP follows in the footsteps of its predecessors, money will continue to be spent without any knowledge about corresponding savings in accidents. This might be good public relations (politics) but it is definitely not safety management.

I reserve the term *safety management* for that class of activities for which safety consequences matter, for actions that rest on substantive knowledge. Such knowledge allows the manager to anticipate what course of action will have what safety results. Thus, the main difference between public relations and management, so defined, is knowledge. To have a successful public relations program, one does not need to anticipate safety consequences or know the accident reduction rate that has been achieved. The reward for public relations is popularity, goodwill, and the tangible benefits of these. In contrast, for introduction of a program of safety management, there has to be anticipation of safety benefits; the continued existence of a safety management program can be justified only if such benefits materialize.

If knowledge is the kingpin of safety management, why is so little known about the safety effects of the many costly programs? First, in many cases the effects are objectively difficult to know. Second, there is little incentive to know. By now, many political and professional careers depend on programs such as MCSAP. The discovery that the safety results are miniscule, that with the same money many more lives could be saved by doing something different, would be an unpleasant one. Even if an evaluative study is carried out by those who have implemented MCSAP, one can only expect encouraging results to be publicized. There are few advantages in reporting failures.

I mention MCSAP only because it is recent, topical, and typical. It may indeed by a useful program. My concern is not with MCSAP in particular but with the evaluation of programs such as this and dissemination of the acquired knowledge. Since no one is attracted by the prospect of political, institutional, professional, or personal embarrassment, those who initiate and implement programs of road safety have no real incentive to find out what works and what does not. Thus, what is done for road safety decade after decade, generation after generation, seems to be more in the nature of public relations than actual safety management.

In the remainder of this chapter I attempt to support this belief. I begin by demonstrating that the amount of road safety delivered to society is largely determined by what public bodies do. It is thus important to examine their role in delivery of road safety—one cannot focus exclusively on regulation of the road user, on the argument that most accidents are caused by human error. I then raise questions about how little is known about the technology of road safety production.

Finally, I attempt to state the diagnosis with clarity, describe some of the symptoms that are visible from my vantage point, and provide a generic remedy. The remedy has two elements. First, there must be an obligation to determine the road safety repercussions of costly programs. Second, those bodies that are in charge of initiating, implementing, and operating safety programs should not be the bodies responsible for the scientific evaluation of the effect of the programs.

WHO DETERMINES THE LEVEL OF ROAD SAFETY?

Those who debate the need for more or less regulation of safety have the following question in mind: What should public bodies do in the interest of safety to influence the behavior of truckers, airline personnel, drivers, or other private parties? This point of view reflects the belief that safety is largely determined by what the truckers, airline presidents, maintenance mechanics, or car drivers elect to do; that accidents are caused by truckers working too long hours, by shoddy maintenance, by drivers who drink or take risks. This is a decidedly one-sided point of view. It tells us more about the viewer than about the system being observed. Alternatively, the safety of transport systems—I will speak mainly about the road system—can be viewed as determined largely by those who build the systems and are responsible for their operation. One can argue with persuasion that the choices the road user can reasonably be expected to make are perhaps secondary in importance from the point of view of managing road safety. Both points of view have merit. My predisposition is toward the second argument, and I will therefore attempt to buttress it.

Road accidents are a by-product of transport activity on a road system. I use the term *road system* as a collective one for roads built to certain standards, intersections spaced at certain distances, vehicles designed with given features and crashworthiness, the prevailing rules of the road, the procedures for driver licensing, the level of enforcement by the police, the luminaires and radar guns, the traffic signals and breathalyzers, defensive driving courses and child-seat loan programs, snowplows, ambulances, and railroad crossing gates. The safety-related actions, decisions, programs, and features that go with the road system are what most people understand by the term *road safety delivery*. The public bodies that orchestrate all this are states, municipalities, bureaus of motor vehicles, police forces, engineering professions, motor vehicle manufacturers, legislators, the federal administration, and the like.

Thus, when one asks *how* road safety is delivered, or what the accident rate is, the following kinds of questions are implied: Why are highway lanes 12 feet wide and how many more accidents would occur with 11-foot lanes? Why do we have four-legged intersections every 300 feet, and would there be fewer accidents with T intersections every 400 feet? Why do vehicles have lap belts in the rear seat instead of a harness? Why do we now allow drivers to turn right on red? Why do we license drivers at age 16 rather than 17, or allow people to drink at age 21 rather than 18? Why do we set speed limits that are exceeded by a majority of drivers and then hire police officers to chase them down?

One can thus think of road safety as largely determined by the road system, which in turn is shaped mainly by public bodies. It is therefore natural, when focusing on delivery of road safety, to point the searchlight at the public bodies and their safety-shaping activities. The common assertion that most accidents are caused by the human element, and that therefore primary attention has to be devoted to the behavior of the road user, not the road or the vehicle, is inadequate. The following illustrates this point.

Suppose that I decide to stop suddenly because a traffic signal has turned amber, and someone hits my car from behind. If I had proceeded through the intersection, or been less abrupt in braking, the accident perhaps would not have happened. Similarly, the driver behind me could have braked faster, or followed me less closely. Between the two of us we must admit to having caused the accident.

The operative principle I use in admitting to having been partly the cause of this accident is the speculation that had I behaved slightly differently, the accident might not have occurred. The police officer uses the same principle, blaming the other driver for following too closely. This principle invites us to extend the causal chain to more remote events that might have averted the accident. Thus, for example, the signal could have been of the actuated kind, which usually changes to amber when no vehicles are near the stop line. To regress further, had there been a freeway I could take to go to work, I would not have been driving on the surface streets and been at that intersection in the first place. Thus, calling "cause" what is easy to see and proximate to the accident is too narrow a view. I cannot single out the two drivers as having caused the accident while the decision of the traffic engineer, the guidelines by which the engineer works, and the politics of freeway building are not considered a part of the causal web.

It seems clear that the causes of an accident are many, and the causal web has a time dimension to it. The tendency to single out the event most proximate to the accident as the cause is perhaps understandable but is indefensible on logical grounds. Above all, it is useless as a guide to action.

The preceding illustration shows the users-versus-public bodies dichotomy in a new light. Public bodies prefer to tell the users how to behave rather than have users question the safety of the systems which they, the public agencies, have built. It seems to me that popular support for the activities of public bodies derives in some measure from the fact that most road users are convinced that the road safety problem is of their own making—that they are the cause of accidents. This belief casts public bodies as the good guys and puts them into the role of regulators, and casts road users as the bad guys and thereby into willing regulatees.

We are now groping near the roots of the problem at hand. It is human nature to seek the cause of a crash in the events immediately preceding it. Therefore, we are predisposed to blame the driver. In addition, we are exposed daily to a multitude of little insults to our egos and small encroachments on our freedom of movement, which are inflicted on us by those with whom we have to share the road. It is therefore natural to react with irritation against the rude, poorly trained, risk-taking, maniac, idiot (you choose the adjective to fit your mood) driver, trucker, pedestrian, bicyclist (you choose the target of your ire). For these reasons road users generally see as justified and are inclined to support the various crackdowns, the tightening of licensing, the stiffening of penalties, the regulation of this, and the surveillance of that. The general public cannot be expected to seek cause in the remoter events and in the invisible or long-forgotten decisions by anonymous professionals and retired politicians that created the conditions for the occurrence of the accident in the first place.

This aspect of road-user psychology gives public bodies the popular support they require. The same aspect of road-user psychology tends to distort the actions public bodies undertake. First, public bodies find it easy to do what is supported: crackdowns, tightening, stiffening, and so on. Second, since the elements of the road system that crucially affect safety are further down the causal web and escape the road-user's attention, public bodies do not feel compelled to consider the safety repercussions of their actions. This is why it is convenient for public bodies to promote and cultivate the notion that most accidents are caused by the human element.

I do not claim that what users do does not affect safety. They can drink and drive or refrain from doing so; they can use bald tires on trucks or have a regular program of replacement. My intent in focusing on what public bodies do derives from my desire to restore balance, to give prominence to what is seldom mentioned. Most of those who speak and write about road safety are accultured into the cause of public bodies. Road users do not attend conferences, do not issue public statements, and do not write reports or scientific papers. Therefore I point the searchlight in the direction of the public bodies.

An additional reason for focusing on what public bodies do and know has to do with what we expect to be a sensible sequence of purposeful action. It makes little sense to discuss the need for more or less safety regulation and surveillance without first understanding what level of safety to build into the system. I will claim that the level of safety the public bodies build in gets in by default. This claim will be given substance in the next section.

In summary, one should look at what public bodies do with fresh eyes. It is important to ask whether they do their safety job right.

THE REIGN OF IGNORANCE

In the previous section I attempted to establish that the actions and inactions of public bodies exert a large influence on the level of road safety in society, and that the delivery of road safety is their job. I also claimed that the predisposition to focus on regulating the behavior of the road user derives from the natural but shortsighted concept of cause, which places attention on the easy-to-notice actions of those involved in an accident and diverts attention from the earlier, less readily apparent but widely influential actions of those who were instrumental in creating the circumstances that contributed to the accident.

In this section I want to make two points. First, that the cost of road safety delivery by public bodies is large. Second, that too often the safety results achieved by these large expenditures remain largely unknown. The first point is easy enough to make. We need only recall our daily encounters with familiar road safety programs. The pervasive nature of actions taken to deliver safety and the enormity of their social cost in money, time, and freedom should be self-evident.

That the safety effects of many programs, standards, or actions are largely unknown is a more difficult point to make. One would have to consider programs one by one and, for each, assess the extent to which its safety effect is known. The ensuing opportunities for disagreement and obfuscation are endless. Moreover, it is an irritating argument to make. The sources of the irritation are manifold. First, we are predisposed to trust public bodies and to rely on the knowledge of professionals. A line of reasoning seen to diminish this trust is likely to be associated with subversive intentions. Second, most of those reading these comments are themselves professionals working directly or indirectly with public bodies. The thesis that professionals and public bodies often do not know what they are doing is bound to offend. Third, those who are professionally associated with the delivery of road safety cannot help but develop an emotional commitment to their work. Their work becomes their cause, and they themselves identify with the safety movement. Any suggestion that the safety effects of traffic signals, speed limit enforcement, or stiffer penalties for drinking drivers is not known is perceived as a direct threat to these programs and the institutions or professions that manage them—a stab in the back to the cause of road safety and to the safety movement.

Still, however irritating, one must engage in this line of reasoning. Such reasoning should not only be endured or graciously tolerated but contributed to, supported, and cultivated. In the long run, the delivery of road safety rests more on the ability of professionals to foresee the effect of their actions than it does on the strength of the convictions of today's custodians of safety programs.

My own critical views about the amount of factual knowledge that is available in the field of road safety delivery rest on years of study. As I moved from one inquiry to another and began to notice how shallow are the foundations of what passes for knowledge, I gradually realized that ignorance about the safety repercussions of the many common measures is not the exception. How can I effectively share with the reader personal opinions acquired laboriously and gradually? How can I impress on the reader that my purpose is not mischief, that there is substance to my views and therefore they merit sympathetic consideration? I know of no other way than to give an anecdotal account of some experiences that helped to shape my belief.

In 1982, Congress asked the National Academy of Sciences to study the safety cost effectiveness of geometric design standards to be used in the rehabilitation of highways. The request was motivated by a real problem. A lot of money was to be spent on roads that were badly in need of repair. Were the Federal Highway Administration to insist that highways be upgraded to exacting modern geometric design standards, fewer miles of road could be rehabilitated than if existing roads were simply resurfaced.

A committee was set up. To assist it, several experts in the field were hired to review what is known about the relationship between safety and some key highway features. The answers sought by the committee were not for minutiae of design but such basic questions as: How many accidents are saved by widening narrow lanes or shoulders? How many by reconstructing a sharp curve, or by making the sight distances on a crest longer? When the reports of the experts came in, a sorry picture emerged. The committee was led to conclude that, "Despite the widely acknowledged importance of safety in highway design, the scientific and engineering research necessary to answer these questions (i.e., about the relationship between roadway geometry and safety) is quite limited, sometimes contradictory, and often insufficient to establish firm and scientifically defensible relationships" (Transportation Research Board, 1987, p. 76). Coming after half a century of modern road building, this is an alarming finding.

Important questions beg to be answered. How is it that even now we do not know the relationship between safety and road design? It might be understandable that an initial decision had to be made on a hunch and, while motivated by concern for safety, had to be based on other considerations. However, one might rightly expect that immediately following a meeting at which decisions were made in the name of safety but without any factual knowledge of it, appropriate steps would have been taken to ensure that the next revision of standards would be based on more definitive knowledge. Why was there no such follow-up? How is it that the various standards committees, which over the generations have changed in personal composition, continue to feel free to set standards of practice on matters they believe to affect safety, but without the benefit of knowing the expected safety results of what they are prescribing? How is it that in the course of more than half a century of modern road building, and in the light of the large number of accidents that have occurred on these roads and could serve as data, the professions that build and operate the road system have failed to learn much about the safety consequences of their professional activities.

The reader may be surprised to learn that in North America highway design practice, engineers never have to examine explicitly and estimate the annual number of accidents expected to occur with design alternative A and compare it with what can be expected to occur with design alternative B. In fact, the engineers could not do so even if they had to or wanted to. Neither the training they receive nor the extant published and accessible information enables them to do justice to this task. Regrettably, the existence of written standards free highway designers from this bother. They can and do design highway without explicitly analyzing the safety consequences of design decisions.

The establishment of a highway design process, which seems to have evolved around the manuals (American Association of State Highway and Transportation Officials, 1984; Roads and Transportation Association of Canada, 1986), has fatally wounded highway engineering. It has relieved the engineer of the need to determine the consequences of design and thereby severed the artery that feeds the need to know. It seems to me that the ability to foresee the consequences of one's design are at the very foundation of all engineering, indeed of all human endeavor.

Diverse strands of the argument now begin to form an intriguing pattern. Some standards such as lane width, shoulder width, sight distance on crests, are often written without any knowledge about the relationship between these features and road safety. On other matters such as curvature, grades, and intersection sight distances, some knowledge exists but it affects design in an indeterminate manner. I think most designers strive for mild grades, large radii, and long sight distances. However, inasmuch as these cost money to attain and there is only vague guidance on their safety benefits, the cost-benefit comparison cannot be determined. Thus, whereas the concern with safety of those involved is genuine, the safety-related activities of highway geometric designers are akin to a ritual. Engineers design roads either without knowledge of what affects safety and how, or without using such knowledge explicitly. The designed roads are considered safe. One might ask how a road or intersection can be nominally safe and yet have accidents occur on or at it with statistical regularity. Here the user comes in handy. If accidents occur on a road that has been declared safe, human error must be the cause. This closes the circle. The myopic and unsound concept of accident causation discussed earlier serves to explain how something proven unsafe by repeated accident occurrence can be declared safe by dogma.

In the preceding section I claimed that once a road system is built with certain lane widths, grades, intersection density, and so on, the number of accidents that will occur on it has been largely determined. In this section I have said that the standards and warrants that give the road system its shape are usually written with safety in mind but without quantitative knowledge of the link between decisions and their safety consequences. There are other considerations, such as construction cost and delay, for which the link between cause and effect is better known. Naturally, the level of safety that is eventually built into the road system is determined by these other considerations. The cumulative effect of this process implies that the overall level of safety (in the United States some 40,000 killed and 3 million injured every year) is determined, so to speak, by default.

I have singled out highway and traffic engineers not because they are less concerned about road safety or less knowledgeable than others but because this is the field I know. There is gross injustice in this focus. Other public bodies with responsibilities for road safety know even less about the safety effect of their actions than engineers do. For example, the police have an unshakable faith in the safety effect of speed enforcement. I followed the empirical evidence on this until about 1981 and found no grounds for or against this belief. The enforcement of speed limits created a dip in the speed distribution in the vicinity of the cruiser; if the cruiser stayed put for a few days, it created a short-lived halo effect. However, the effect of speed enforcement on accidents was not known. Nor was there substantive knowledge on the general deterrence aspects of speed enforcement. How much is required to maintain the status quo in speed limit observance? What would be the result of increasing or decreasing the level at which enforcement currently exists? Nobody seems to know. In spite of the enormous social burden, say, 600,000 annual speeding convictions in a Canadian province with 6 million drivers, a level of enforcement maintained consistently for over a decade or so, still 60% to 70% of all drivers exceeded the speed limit. Would halving the enforcement be noticed? How many lives would twice the current enforcement save? Nobody knows.

Recently, a very substantial accident reduction has been publicly attributed to a speed enforcement crackdown. On inquiring about the factual basis of this claim, it turned out that it was based on a most superficial examination of gross accident statistics carried out in-house. No changes in speed were measured on the roads, no formal study was carried out. To claim that what has been done worked well is good public relations. Repeated exposure to public relations statements of this kind gradually gives them a ring of objective fact. When such statements come from different sources, even the spokesperson will begin to believe in the folklore so created.

THE REMEDY

The following summarizes the points made so far:

1. Once a road system is established, the level of road safety is largely determined.
2. The road system is produced largely by what public bodies do, and what they do is costly.
3. In spite of the large cost, the safety effect of the actions of public bodies often remains unknown.

4. Even when usable knowledge about safety effects exists, it has only a weak link to what designs are implemented and what decisions are made.
5. The delivery of road safety to society is more akin to public relations than to knowledge-based management.

Why is it that public bodies are more concerned with looking good than with doing good? A specific answer would be quite involved. However, in general, because so many of us work for public bodies directly or indirectly, the answers are discouragingly familiar.

We wish public bodies to respond to people's concerns. However, unlike the private sector, which usually responds to what people wish to buy, public bodies have to be sensitive to less tangible matters. Society's desire for road safety is one such matter. Thus, for example, the MCSAP program, mentioned earlier, is a current response to people's concern about the behemoths with which they have to share the road.

Because we wish public bodies to be responsive to those needs for which the marketplace does not provide, such responses often being their main raison d'être, we must expect that the need to be seen as responsive is for public bodies of paramount importance. It is not an aberration, it is their essence. If so, should we be surprised by their passion to look good? This is what we expect of them, and this is what they are rewarded for.

For a public body, a response means that a program is initiated, funded, and implemented. Once a program is born, people's careers and jobs are tied to it. Many then develop a personal stake in the program, its success, and its continued existence. If it looks good, their personal interest will be served; anything that shows the program in a bad light is a personal threat of a political, professional, and economic nature. Thus, the second motive for the need to look good is the self-interest of those who are tied to the public body and its programs. It would be as unwise to disregard the importance of self-interest for public bodies as it would be unfair to condemn it. We expect people to be motivated by self-interest in all walks of life; we even think reliance on self-interest is efficient and possibly noble.

Our problem resides in the distinction between the *need to do* (for road safety and any other program) and the *need to be seen* as doing it. We would like public bodies to do good, but their natural tendency is to act so as to seem to be doing so. The former requires knowledge-based management, the latter needs good public relations. It is unfortunate that there is a profound discord between the *need to do* and the *need to be seen* as doing, for example:

Need to Do	*Need to Be Seen Doing*
Judge success by how many accidents are saved, by how much misery is eliminated.	Judge success by the popularity of what is being done.
Measure the extent of accident reductions attributed to programs so that, as experience builds, so does substantive knowledge.	Claim that accidents are being saved (research into the actual safety effect may turn up adverse statistics).

Publish only valid results, whether success or failure, so that others can learn what is useful and what does not work.	Have evaluations done in-house by persons over which the public body has control, so that the decision of what to make public is made with institutional responsibility in mind.
Work with professionalism based on science.	Work with imagery and form; design by ritual and make decisions by cargo-cult science.

No real-life public body comes in the pure form of either all public relations or all knowledge-based management. The brief checklist just provided is helpful in deciding in what proportions the two ingredients are mixed in a specific case. If, for example, a public body attempts to find out whether its programs work but does so in-house, the mixture is perhaps half and half. If no attempt is even made to measure the effect of the programs, it is all public relations.

It is this discord between the need to do and the need to be seen as doing and succeeding that more than anything else, supports the reign of ignorance. Since ignorance of fact is what gives freedom to managers of images, whereas knowledge of fact restricts freedom, it is naive to hope for progress toward knowledge-based road safety management by doing more research, devising better methods of analysis, or throwing more money at the problem. The roots of the problem reach much deeper—they have to do with the nature of public institutions. Any attempt to make progress toward road safety management must be aimed primarily at institutional arrangements.

I believe my diagnosis is correct in its broad outlines but perhaps too sharp in contrast and lacking in important detail. Thus, for example, it would be an exaggeration to claim that all safety-related programs are initiated as a cynical response to a public relations opportunity. Some programs are introduced because professionals in the employ of a public body see an opportunity to increase safety in a cost-effective manner. Even though the well-intentioned professional usually has to grasp at straws when the cost effectiveness of the suggested program needs to be supported by facts, he or she will do the best that can be done to provide those higher up with valuable information. Thus, at least in the beginning, there is genuine concern with safety management. The public relations concern begins to dominate at subsequent stages, when implementation and evaluation decisions need to be made.

The results of the natural tendencies of public bodies to look good and to claim success, when accumulated over many public bodies and long periods of time, are the current paucity of factual knowledge and proliferation of misinformation. It may be difficult for professionals to accept the broad validity of this claim on first reading. However, some contemplation of their professional life experiences, of the contracts their employers insisted they sign, of the channels and procedures through which their findings and recommendations must go, of the criteria by which these are examined and their eventual fate, should offer sufficient proof.

What, then, is the remedy? Of course we wish to preserve the responsiveness of public bodies to people's concerns. Thus, for example, public bodies should continue to take initiatives such as described early in this chapter. It is folly, however, not to recognize that public bodies, just like the private sector, are composed of persons motivated by legitimate self-interest that, in this case, acts against the conduct of evaluative research and the accumulation of substantive knowledge on how to manage road safety. This recognition points the way to the solution, which is to create requisite checks and balances. Since no public body can be fairly and reasonably expected to act against its self-interest, it should not be asked to do so. In this case the checks and balances will be created by two actions. First, the functions of initiating, implementing, and operating a program should be separated from the function of measuring the program's impact on safety. Second, finding out whether costly programs are effective should be made mandatory. This work should be carried out by professionals employed by institutions within which the rewards are completely independent of the success or failure of the programs being evaluated.

Thus, the State Police should not be put in charge of finding out whether the enforcement of speed limits reduces accidents; the Office of Motor Carriers or the Federal Highway Administration should not be saddled with the decision whether to evaluate MCSAP. However, the safety effect of these programs must be evaluated lest one is willing to accept shamanism as a proper way to deliver road safety to society.

The most common objections to the separation of the functions of program initiation and implementation from program evaluation are cost, practicality, and duplication. The added cost is that of creating a new layer of supervisory bureaucracy, the independent evaluators. The impracticality objection comes from the fact that such supervisory bodies would lack the required expertise, since most experts are now in the employ of the initiators and implementers. The concern about duplication stems from the fact that similar duties are already performed by existing institutions such as the General Accounting Office and the Auditor General.

I think the objection on the grounds of added costs is spurious. Without evaluation, money can be spent in perpetuity on programs of little merit, as is now the case. Which is more costly? The question is not *whether* to evaluate but *who* should be doing so.

The claim that most experts are now in the pay of public bodies that initiate and implement programs, and consequently that no independent body has the expertise to perform solid evaluations, is to be taken seriously. Not only does it correctly describe the prevailing state of affairs, it serves to underscore the grip public bodies now have on information about the safety effect of their programs. They control the experts.

Of course, when speaking of change, it is precisely the present state of affairs one wishes to alter. Were the conduct of independent evaluations for costly programs to become mandatory, were the budget for this activity to become commensurate with the need, the staffing difficulties would diminish in time. Some experts would migrate from the public bodies for which they now work; consultants and university researchers would be less dependent on the public bodies that now control research budgets; others would respond to the signals of demand.

The last concern is that of duplication. If the Auditor General or the General Accounting Office whose independence is beyond doubt, already examines the effectiveness of some programs, the first part of what I propose, separation of evaluation from implementation, already exists. To set up new institutions could be wasteful. Indeed, new arrangements are not required if existing institutions could expand their scope of activity as well as their jurisdiction and responsibility to undertake the job proposed in the second part—mandatory evaluation of costly programs. If this is not feasible or desirable, other institutional arrangements need to be made.

Freeing evaluation from the malforming institutional and personal self-interest will unlock the gates to progress toward knowledge-based management. It will provide for delivery of road safety to be founded on a constantly improving understanding rather than on stagnating and unaided common sense. It will ensure that programs that have a good chance of failing will not get started; that programs that involve large costs will have the process of evaluation built into them; that what does not work will not be endlessly and needlessly tried.

Inasmuch as most highway and traffic engineers work for public bodies, the same solution applies here. However, there is an added element in this equation: the profession. The profession of highway and traffic engineering is also a public body, but it is not of the same kind as a state DOT, the engineering department of a metropolis, or the Federal Highway Administration. If the profession want to claim professional responsibility in road safety, it should take steps that such responsibility be carried out on the basis of empirical fact and scientific knowledge, not on the basis of collective engineering judgment. After all, what distinguishes a professional from a layperson, what gives the professional's advice weight and authority, is the possession of relevant factual knowledge over which the layperson has no mastery. In the absence of relevant factual knowledge, professionals have to resort to the waving of manuals, and their influence is correspondingly diminished.

I have no simple prescription for the reform of the profession. A useful stimulus for change would be the explicit requirement that all engineering designs and recommendations that affect road safety be accompanied by an engineering analysis of their expected road safety repercussions. This would at once confront the engineer with the need to know. It would also jolt the profession into recognizing that practicing professionals do not have the time or the ability to do research into road safety—that neither personal experience nor engineering judgment is the source of substantive knowledge in road safety; that one cannot build professional knowledge without establishing a premeditated organizational set-up for that purpose. The creation of substantive knowledge is a long-term process; its structuring requires careful thought.

There is another reason for requiring a new habit of thought—the air of disillusionment in the road safety community. Much of what common sense suggests has been tried; much of what has been tried, except for the packaging and protection of car occupants, cannot be shown to be effective. I think this pessimism is unjustified and premature. Did we condemn the efficacy of medicine because spells, leaches, and exorcism had no demonstrable healing effect? Of course not, the response was to develop a knowledge-based profession. The engine of progress was the experiment, the measurement, the planned clinical trial. Talent, money, and resolve were found to deliver science-based health care. The use of spells and exorcism in medicine has much diminished; we even seem to know when the medical use of leaches can be helpful.

Not so in road safety. Comparison of the delivery of road safety with the delivery of health care is not farfetched. In road safety we are somewhere at the leaches, spells, and exorcism stage. Instead of disillusionment, our response should be the resolve to embark on an era of science-based safety management. It alone promises to deliver results.

7

The Effect of Transportation
Deregulation on Worker Safety

W. KIP VISCUSI

The focus of this chapter is the effect of the recent deregulation of the transportation industry on worker health and safety. This choice of focus was made essentially because of our greater knowledge of forces affecting worker safety than consumer safety and the empirical magnitudes of these forces. In large part because of extensive data gathered with respect to job injuries, a number of studies have assessed the determinants of workplace safety, whereas far fewer studies have looked at product safety (Viscusi, 1984). In addition, work accident statistics and Occupational Safety and Health Administration (OSHA) inspection statistics are available for use in assessing job accident performance to date.

Job-related accidents are an important component of all transportation accidents. Even viewed as one component of the safety performance of a firm, worker safety has a large role to play. Indeed, this aspect of safety may be the most important safety component in some instances.

The examination of worker safety has general implications to the extent that the same kinds of factors that affect worker accidents affect other accident categories similarly. In the case of airlines, the operating procedures of the crew, the nature of the capital stock, and other factors that may contribute to injury will also tend to cause "product" injuries to passengers and external risks to others. In the case of trucking, there are no risks of personal injury to consumers of the product, but pedestrians and occupants of other vehicles are at risk. Driving practices, vehicle maintenance, and other factors likely to cause injuries to drivers also are likely to generate situations that impose risks on others. Even if the main source of injuries were vehicle maintenance rather than vehicle operation, one would expect firms with high injury rates in carrying out maintenance functions to have high accident rates in operations.

It should be emphasized that we are concerned here with changes due to deregulation rather than with determinants of absolute levels of risk. Thus, we are not greatly concerned with whether trucking accident rates are higher than airline accident rates, nor is job safety of primary interest. What is important for the discussion in this chapter to be of general relevance is whether changes in job safety resulting from deregulation are reflected in similar changes in other aspects of a firm's safety performance, such as third party accidents.

THE THEORETICAL REFERENCE POINT

The Basic Model

As a basis for discussing likely paths of influence, it is helpful to set up a simple theoretical model that summarizes the essential economic structure driving a firm's safety decisions. This model is intended primarily to illustrate the main economic mechanisms of influence within the context of a specific economic framework. Readers wishing to focus on the substantive implications of the model can proceed directly to the next section. The empirical aspects are addressed later in the chapter.

Consider a firm that is in compliance with OSHA standards so that the expected costs of regulatory violations can be ignored (Viscusi, 1979b). Let π be the firm's expected profits, where the firm is assumed to be risk neutral. If we set the price of the firm's output equal to 1, then the value of its production is the output level $F[L,s(q)]$, where F is the production function, L is the labor input, $s(q)$ is the safety level, and q is the safety-related input. One would expect that firms with higher levels of safety will be more productive, since there will be fewer interruptions in the work and less need to replace injured workers. Greater safety will, of course, be costly, but we assume that these increased costs will be subsumed into the unit cost of safety inputs, which we will denote by c. Whereas a full general model of long-run decisions would let the firm also pick its entire technology and capital stock, here we are focusing only on shorter run safety investments such as ventilation equipment and changes in operating procedures.

The firm's choice of the safety input q will determine the overall safety level $s(q)$ that prevails, but this relationship is not direct. Given the nature of the work environment, workers' precautionary actions are intervening factors that will also be consequential. In general, greater levels of q imply a higher level of safety outcome. However, there are bizarre conditions that can lead to the opposite result, for example, if workers have a greater incentive to exercise care because their environment is unsafe (Viscusi, 1979b).

Workers will demand a compensating differential for unsafe jobs, so that the wage will be a function of the safety level. Thus, the firm's wage rate will be set by $w[s(q)]$, where dw/ds is negative. There are additional accident costs to the firm. These will be denoted by the variable v. If the value of $s(q)$ is scaled so as to represent the probability that a worker does not have an accident, then the expected accident costs will be given by $v[1 - s(q)]L$. These accident costs include several components: merit-rated workers' compensation costs, the costs of training new workers, and reputational costs associated with accidents.

It should be noted that the structure of the model would not be greatly different if our focus were on product safety rather than job safety. In that case, q would be a product quality input, which would lead to a higher price being paid for the product rather than a lower wage. In addition, the scale of the accident costs would be determined by the level of the output rather than by the number of workers. For both product risks and job risks, the firm can make some safety outlays that generate financial benefits both through the market's valuation of safety (compensating wage or price differentials) and the expected accident costs (workers' compensation, product liability awards, and losses in reputation).

Firms in compliance with OSHA standards pick their levels of safety input q and labor input L to

$$\text{Max } \pi = F[L, s(q)] - Lw[s(q)] - cq - v[1 - s(q)]L \qquad (7.1)$$

After some simplification the pertinent first-order conditions are

$$w = F_L - v[1 - s(q)] \qquad (7.2)$$

and

$$c = s_q(F_s - Lw_s + vL) \qquad (7.3)$$

Here and elsewhere below, subscripts denote derivatives. Firms out of compliance with OSHA standards face an additional cost—expected regulatory penalty $G(q, q^*, L)$—based on the discrepancy between the level of safety input of q that they have selected and the safety input mandated by regulation. If $q < q^*$, the firm is out of compliance and $G(q, q^*, L)$ is a positive expected cost. The penalty level may increase with the size of the work force L exposed to the risk. For $q \geq q^*$, the value of $G(q, q^*, L)$ equals zero and the problem reduces to that given in Equation 7.1

The optimization problem for firms out of compliance is consequently to select L and q to

$$\text{Max } \pi = F[L, s(q)] - Lw[s(q)] - cq - v[1 - s(q)] - G(q, q^*, L) \quad (7.4)$$

The first-order conditions are similar to Equations 7.2 and 7.3, except that a regulatory penalty term has been added. Thus, we have

$$w = F_L - v[1 - s(q)] - G_L \tag{7.5}$$

and

$$c = s_q(F_s - Lw_s + vL) - G_q \tag{7.6}$$

Nature of the Theoretical Influences

The firm's choice of its labor and safety inputs is governed by factors that accord with one's expectations. In conventional labor choice models, the firm hires workers until the marginal product equals the going wage for the job. This result continues to hold except that the marginal product must be reduced by the expected accident costs per worker in the case of firms in compliance with OSHA standards (Equation 7.2). In the case of firms in violation of OSHA standards, the marginal product must also be net of the marginal effect of an additional worker on the expected penalty assessed for being in violation of the regulation.

The firm's selection of the optimal safety input level, for fixed values of L, is governed by similar types of factors. In this case the price of safety capital is set equal to the net marginal increase in profits that result. As indicated in Equation 7.3, for firms in compliance these benefits consist of the increase in safety resulting from the greater safety input, which in turn is multiplied by the sum of the effect of greater safety on output, the value of greater safety in terms of reducing the wage bill, and the effect of safety on the firm's accident costs. For firms out of compliance with OSHA standards, greater safety inputs also reduce the expected penalty that they will be assessed (Equation 7.6).

It is also useful to assess the substantive theoretical implications that will play an important role in our subsequent discussion of transportation deregulation. One of these implications is firms' abilities to produce safety. In particular, there may be important differences in the manner in which safety inputs of q are translated into safety outcomes $s(q)$. As will be discussed in our review of empirical patterns, newer firms and smaller firms may be less efficient in producing safety. If that is the case and a firm has a smaller value of s_q for any given level of q, then the optimal level of the safety input chosen will be reduced and the firm will adopt a riskier technology.

Matters are a bit more complex if there are safety differences across firms due to the nature of the work force. Suppose that a firm has a disproportionate share of inexperienced employees who tend to be more prone to job accidents than experienced workers. If the greater riskiness is only at that particular firm, the firm will take this accident cost into account and hire fewer workers. If the firm is no riskier than other enterprises but the particular workers are hazard prone and can be identified as such, they will be paid less.

There are also other factors relating to firms' characteristics that will affect the accident cost. Large firms will face greater accident costs to the extent that their workers' compensation insurance premiums are merit related or governed by a self-insurance arrangement. In addition, large and established firms may incur potentially a larger reputational loss from an adverse safety record to the extent that they have more of a good reputation at stake. An offsetting influence is that large firms have more stable reputations that are less likely to be destroyed by a single adverse event. For example, American Airlines was able to survive a catastrophic accident at Chicago in 1979, but newcomer Air Florida went out of business following a much publicized accident in Washington D.C.

This possibility of shutting down alters the structure of the economic problem, since firms may not have to bear the brunt of the accident costs in the event of a major catastrophe. Consider a new airline that experiences a major crash that kills 100 people and imposes a product liability burden in excess of $50 million, which is by no means out of line with possible product liability awards. If the airline is on the brink of unprofitability, as many are, and if the loss is not covered by an adequate insurance plan, shutting down or reorganizing the bankrupt firm may be the preferable option. This influence tends to raise the possibility that deregulation will increase the number of financially pressed firms, firms that may have less of an incentive to be concerned with catastrophic accident losses associated with their actions.

PRINCIPAL EMPIRICAL PATTERNS

Worker Age and Experience

A principal effect of transportation deregulation is that it gives rise to the entry of new firms, which in turn may hire workers not previously employed in the industry. This entry of firms is not an incidental consequence of deregulation. In the case of trucking deregulation, for example, a major purpose of deregulation was to promote new entrants. Indeed the number of new trucking firms has increased. However, the increase is in truckload operations. The number of large less-than-truckload firms has declined.

The shifts in the composition of employment resulting from deregulation could raise accident rates for a variety of interrelated reasons. First, the workers hired are likely to be young men, who tend to be more accident prone in general. The highest accident rate group for all individuals under age 75 is the age range 15 to 24 (National Safety Council, various years). Increasing the number of such workers can increase a firm's riskiness if there are important age-related factors that boost workers' riskiness within a particular set of job activities. However, it should be borne in mind that National Safety Council statistics pertain to all accidents rather than just to job-related ones. Since younger people spend more time in risky activities such as hunting, swimming, and driving cars, they can be expected to have a higher overall accident rate. This does not necessarily imply a high job-related accident rate.

A second factor pertaining to worker characteristics is job experience. Workers with low levels of experience at the firm or in similar jobs elsewhere tend to be more prone to accidents. Although cross-sectional accident statistics suggest that inexperienced workers tend to be more accident prone, the causality is not clear-cut. High-accident environments lead to more injuries and higher turnover, so that even if less experienced workers were not prone to accidents one would observe more of these workers in high-risk jobs. Indeed, in many firms the new hires are traditionally assigned to the riskiest and most unpleasant tasks.

The most compelling evidence pertaining to the role of the age and experience mix of the work force is time series evidence. As industries experience a growth in employment, there is a substantial increase in the frequency of job accidents of all types (Viscusi, 1979b, 1986). This effect is observed even after taking into account differences in weekly work hours and overtime hours that are cyclically related. It should be noted, however, that some of this increase may be due to changes in operating procedures or in the capital stock due to actions taken by the employer rather than the worker. What interests us here is the combined impact of these influences rather than their precise decomposition.

The powerful procyclical nature of job accidents is a phenomenon that can be observed in any firm experiencing growth in employment. The new entries that arise from deregulation and the expansion in operations of existing firms both tend to raise the job accident rate. The greater part of this effect will be transitory, however. As new firms age, their work force will age and become more experienced as well. Even in the long run there may be some moderate increase in accidents, however, to the extent that there is more entry into and exit from the industry than there might be in the absence of deregulation.

Unionization

One effect of deregulation is that new nonunionized firms may enter the industry. If these firms become unionized in the long run, there may be only a short-run effect on safety in the absence of unionization. What this effect will be is difficult to ascertain.

In terms of unions' effect on safety outcomes, there has been no definitive study in the literature. Overall, workers in jobs covered by collective bargaining agreements do not differ greatly from other workers with respect to the risks that they face (Viscusi, 1979a). This result does not distinguish the incremental effect of unions, since the targets of unionization may be the higher risk firms.

The ability to hire nonunion workers offers one potential advantage to employers, because nonunion firms are less encumbered by detailed work rules. In the railroad industry in particular, the role of rules in affecting the allocation of employees is legendary. Freedom from these constraints enables firms to allocate workers more efficiently in all dimensions, including those related to safety.

A much-studied effect of unions is their impact on workers' relative wages. Unions increase union workers' relative wages by 10% to 15%, and a possible response by the firm is to diminish work quality, including job safety, since the existence of a wage premium enables the firm to attract workers.

One factor that tends to offset this influence is that unions alter the composition of the wage package as well. In particular, the premium that union members receive for hazardous jobs is considerably greater than that received by nonunion members (Viscusi, 1979a). By altering the compensation mix in this fashion, unions provide a financial incentive to firms to prevent a deterioration in the level of safety.

Unions also serve a number of other safety-enhancing functions. They often provide job information to workers; they represent workers in grievance procedures relating to work quality; and they bargain on behalf of workers' interests in job safety. Indeed, because of unions' reflection of the preferences of inframarginal workers, unions give greater emphasis to safety concerns than do markets, which are driven by the preferences of marginal employees (Freeman, 1976; Viscusi, 1979a, 1983).

On balance, unions create two sets of conflicting influences on safety. The increased prevalence of nonunion firms resulting from deregulation should enhance safety by reducing the constraints on the firm and eliminating the incentive to diminish job quality in view of the higher wage level that unionized workers receive. An offsetting influence arises because of unions' direct effect on safety conditions and their efforts to alter the structure of wages. On the basis of the literature, there is no clear-cut prediction regarding the union-related effect of deregulation on safety.

Age of the Capital Stock

Much can be said about the likely effect of the firm's capital stock on risk levels. The most fundamental starting point for this discussion is how deregulation will affect the capital mix. One's initial intuition might be that new entrants will purchase new capital. In most industries, the capital stock of existing firms is not mobile, and installing new equipment is the only viable option. New entrants in the transportation industry need not always purchase new equipment to enter, and exceptions to our prior beliefs appear to be especially likely. Particularly in the trucking and airline industries in which the major capital investments, trucks and planes, are highly mobile, one might observe the phenomenon of new entrants purchasing old trucks and planes from existing firms, repainting them with their corporate logos, and beginning operation. Indeed, this reliance on existing equipment as the capital base may be the norm rather than the exception. For the sake of concreteness, in the discussion here I will assume that the new entrants have older capital stock. In the absence of firm empirical support, this assumption should be regarded only as a working assumption for purposes of exposition. Clearly, there is a need to augment this discussion with more refined evidence on the nature of firms' capital investments.

Firms with more recent vintage capital stock generally have safer technologies because it is less expensive to increase the safety of new technologies than to retrofit old technologies with safety features. This safety cost differential is recognized by government agencies as well. OSHA has occasionally adopted grandfather clauses to make its regulations pertinent only to equipment introduced after the promulgation of a standard. The U. S. Environmental Protection Agency (EPA) has gone even farther in such differentiations, as some analysts have bemoaned the "new source bias" in the stringency of EPA standards.

Overall, for the transportation industry one would expect that newer firms would tend to be higher risk enterprises to the extent that they have older capital stock. Although this effect may be muted in the long run as new firms become established, there is likely to be continuous new entry once an industry has been deregulated, so that the age mix of the firms in it will continue to be influenced by deregulation. The long-run equilibrium may embody a riskier capital stock because of the greater flux of firms into and out of the industry.

Firm Size

The new entrants that come into the market after deregulation tend to be smaller than the established firms. Several economic factors directly related to size should tend to make small firms riskier ventures than their larger counterparts.

First, workers' compensation benefits for small firms are based more on the rating for their industry group than on their own performance. Larger firms are more likely to self-insure or to be merit rated, which creates an additional safety incentive. Second, product liability burdens should follow a similar pattern, since liability insurance can be merit rated more easily for large firms. Third, small firms are better able to avail themselves of bankruptcy options in the event of a catastrophic accident.

Finally, economies of scale are present in the production of job safety (Viscusi, 1979a). Larger firms may also be more efficient in producing safety because such production can involve substantial indivisibilities. Smaller firms may not have sufficiently great safety needs to justify, say, the establishment of an industrial hygiene department or of a product liability group in its legal department. For this, as well as the other reasons just cited, greater firm size per se should enhance safety. A decline in risk level with firm size is borne out empirically once we control for aspects of the job other than the simple risk–firm-size correlation (Viscusi, 1979a).

WORKER INJURY RATE TRENDS

Although experience under deregulation has not been extensive, there has been an opportunity to observe at least the initial injury performance in the deregulation years. The relatively short time frame is not ideal, but it is not likely to be a major impediment to analysis. Many of the starkest effects arising from deregulation are those associated with the transition to the deregulation equilibrium: the entry of new firms; the hiring of young and inexperienced workers; and the more general influence of changes in the nature of operations, such as shifts in flight schedules and routing patterns. Given the important influences of these factors on safety, any dramatic effects of deregulation are likely to be evident in the five years of postderegulation injury reports that are available.

An alternative possibility to this reasoning is what I term the "1987 Boston Celtics basketball team phenomenon." After relying on a starting lineup that logged an excessive amount of playing time with no adverse effect on the victory percentage for the season, the Celtics became riddled with injuries in the playoffs apparently because of the physical strain. Similarly, airlines and trucking firms may be eroding their margin of safety and may be near the beginning of an upswing in accident rates (see Chapter 13 in this volume). The statistical analysis that follows provides an assessment of accident performance to date only and should not be regarded as evidence that no new problems will emerge.

The data base that I will rely on is the U. S. Bureau of Labor Statistics (BLS) injury and illness data (U. S. Department of Labor, various years). This source has been widely used to assess the performance of OSHA, the value of compensating differentials for job risks, and the effect of workers' compensation. The reasons for the attractiveness of this data base for empirical work are clear. Since 1971 all firms above a minimal size have been required by law to make annual reports of their injury records to the U. S. Department of Labor. The first year of publicly released data was 1972.

Three injury measures will be considered. The first is the overall injury and illness rate per 100 full-time employees. This measure includes all injuries experienced by workers. Its comprehensiveness raised classification problems, particularly in the early 1970s. How severe must an injury be before it must be recorded? It should be emphasized that the illnesses included are only those that are job related. Job-related illnesses should be driven by the same kinds of economic factors as are injuries. The bulk of all adverse outcomes that are recorded are accidents. In 1982, for example, the occupational injury incidence rate per 100 workers for airlines was 13.3 and the injury and illness rate was 13.6, or only 0.3 greater.

The second measure includes only those injuries and illnesses that result in at least one lost day of work. The incidence rate of lost workday cases provides a much more consistent measure of injury trends. The final measure is the rate of total lost workdays. This index weights each lost workday case by a number of days lost and as a result constitutes a severity-weighted injury measure.

Trucking Risk Levels

Trucking injury rates of all types are double the national average for the private sector. The risk trends for the trucking industry are summarized in Table 7.1 for the 1972–1985 period. By almost any standard, trucking is a very hazardous occupational pursuit.

Selection of the critical regulatory event date is somewhat arbitrary, since the Interstate Commerce Commission took a variety of deregulatory actions in 1980 and 1981. In addition, the advent of the Motor Carrier Act on July 1, 1980, was a major legislative effort to relax the controls on the trucking industry. The ICC subsequently greatly increased its approval of operating licenses (Moore, 1986). I will use 1981 through 1985 as the principal deregulation period for purposes of analysis, recognizing that there was a continuum of deregulation actions around that time.

Results in Table 7.1 show no sharp break in any of the injury rate series in 1981. All trends continued much as before. There was, however, a sharp increase in all measures of the injury level in 1984 and 1985 compared with 1983. This shift coincides with the dramatic increase in intercity trucking in 1984 (U. S. Department of Commerce, 1986, p. 589).

The growth trends in the annual injury growth rates reported in Table 7.2 do not yield a much different picture. Overall injuries and illnesses

Table 7.1 Worker Injury Rate Trends for the Trucking Industry (Standard Industrial Code 421)

| | Injury Rate Frequency Measure/100 Workers | | |
Year	Overall Injuries and Illnesses	Lost Workday Injuries and Illnesses	Total Lost Workdays
1972	16.6	7.3	115.2
1973	17.1	8.0	143.6
1974	17.9	8.6	156.0
1975	14.7	7.6	146.3
1976	15.2	8.1	161.9
1977	14.9	8.4	162.8
1978	16.3	9.5	182.1
1979	15.7	9.5	192.2
1980	14.8	9.1	191.9
1981	14.7	9.0	187.8
1982	14.2	8.8	196.8
1983	13.3	8.1	189.3
1984	14.6	9.3	213.3
1985	14.0	8.6	216.5

Table 7.2   Trucking Injury Growth Rates Over Key Time Periods

	Percent Annual Growth in Injury Rate Frequency Measure (per 100 workers) for Trucking Industry (SIC 421)		
Period	Overall Injuries and Illnesses	Lost Workday Injuries and Illnesses	Total Lost Workdays
1972–1980	−1.42	2.79	6.59
1972–1985	−1.30	1.27	4.98
1981–1985	−1.21	−1.13	3.62

SIC, Standard Industrial Code.

declined at a somewhat slower rate in 1981 through 1985, but this pattern may be largely due to initial overreporting of total injuries during the first years of the new injury-reporting system.

Both of the lost workday series show a downward shift in the injury growth rate in these years. As already stated, this pattern is quite general and is by no means compelling evidence of a beneficial effect of deregulation. The most that can be concluded is that there is no clear-cut adverse effect of deregulation exhibited in the risk series.

Air Transportation Risks

The second industry to be examined is the air transportation industry, whose risk patterns are summarized in Table 7.3. For each measure, the levels of the risks involved are well in excess of the private sector average. Air transportation is a relatively high-risk industry, but not an extreme outlier.

In the case of airlines I will explore the influence of two possible critical dates of deregulation. In October, 1978, the Airline Deregulation Act was passed; thus one possible deregulation period is 1979 through 1985. A more relevant period is likely to be 1981 through 1985, the period in which the Civil Aeronautics Board (CAB) had no authority over routes. Other deregulation dates could also have been selected because of the continuing nature of deregulation in the late 1970s. For example, there was a relaxation of operating requirements in 1976; the ending of CAB authority over fares in January, 1983; and the introduction of discounts and super-saver fares in 1977 and 1978 (Kaplan, 1986). Our more general concern is whether there has been an upsurge in the risk level in recent years, where it may be most accurate to view the deregulation period as a broad band of dates rather than any single critical year.

The data in Table 7.3 reveal a pattern of decline in overall injuries and illnesses, particularly after 1975, and an inverted U-shaped pattern for the two lost workday series until the 1984–1985 period, patterns similar to those exhibited by trucking. The only prominent candidates for an upsurge in risk levels due to deregulation are the rises in all three injury categories in 1979 and the increase in total lost workdays in the 1984–1985 period. The 1979 increase is fairly modest in all three cases and not clearly out of line with what might be expected on the basis of random fluctuations in injuries due to changes in industrial activity. The year 1979 did see an increase in almost 50% in the number of certificated route air carriers. This increase in number of carriers continued through 1981. No parallel increase in accidents through 1981 is evident, however. The most plausible explanation for the 1979 risk increase is that it reflects the familiar procyclical nature of accidents. The number of revenue passengers enplaned increased in 1979 by 15%, reaching a level that was not surpassed until 1983. The substantial increase in accidents in 1984 may be due, at least in part, to cyclical factors as well. In 1984 there was a 7.5% increase in the number of passengers enplaned and an 11% increase in revenue planemiles (U. S. Department of Commerce, 1986).

The growth rate statistics reported in Table 7.4 reinforce this general impression that there was no apparent acceleration in risk levels in the deregulation period. For the overall injury rate series, the decline in accident levels was greater in the 1979–1985 and 1982–1985 periods than in the 1972–1978 and 1972–1981 periods. For the lost workday incidence series there was a decline in the risk level in 1979–1985 and 1982–1985, whereas in the two earlier periods just noted the risk level was rising. For the rate of total lost workday series, the growth rate in accidents was less in 1979–1985 than in 1972–1978, and the growth rate from 1982–1985 was only slightly above that in 1972–1981. No catastrophic effect of deregulation is apparent. In the airline industry, as in the case of trucking,

Table 7.3 Worker Injury Rates for Air Transportation (SIC 45)

| | Injury Rate Frequency Measure/100 Workers | | |
Year	Overall Injuries and Illnesses	Lost Workday Injuries and Illnesses	Total Lost Workdays
1972	13.7	6.8	69.3
1973	11.7	6.0	67.3
1974	14.4	6.9	77.6
1975	15.0	7.4	85.1
1976	14.2	7.4	89.6
1977	14.0	8.0	97.2
1978	13.4	8.4	95.8
1979	13.7	8.6	102.4
1980	13.3	8.2	105.0
1981	13.5	8.3	103.1
1982	13.6	7.6	101.4
1983	12.7	7.3	97.7
1984	13.1	7.5	107.8
1985	13.1	7.4	116.3

SIC, Standard Industrial Code.

Table 7.4 Air Transportation Injury Growth Rates over Key Periods

	Percent Annual Growth in Injury Rate Frequency Measure (per 100 Workers) for Air Transportation Industry (SIC 45)		
Period	Overall Injuries and Illnesses	Lost Workday Injuries and Illnesses	Total Lost Workdays
1972–1978	−0.37	3.58	5.55
1972–1981	−0.16	2.24	4.51
1972–1985	−0.34	0.65	4.06
1979–1985	−0.74	−2.47	2.14
1982–1985	−1.23	−2.63	4.63

SIC, Standard Industrial Code.

the injury rate trends are not clearly out of line with what one would expect on the basis of secular trends and cyclical forces within the industry.

REGRESSION ANALYSIS OF RISK TRENDS

Although examination of the risk trends is suggestive, the trends do not take into account explicitly the factors that influence accident rates other than general long-term trends and deregulation.

To explore the effects of deregulation further, it is instructive to formulate and estimate a multivariate model that controls for other influences and assesses whether there are any statistically significant effects of deregulation.

Individual Industry Regressions

The first analysis will be of the industry-specific risk trends. The number of observations is quite limited, since industry risk data are available on an annual basis only from 1972 through 1985. As a result, specification of the individual industry regressions must necessarily be parsimonious. The model used is described in detail in Viscusi (1979b, 1986). The formulation I will adopt is

$$\text{RISK}_{it} = \alpha + \beta_1 \text{RISK}_{it-1} + \beta_2 \text{ \% change in employment}_{it} + \beta_3 \text{DEREG}_{it} + u_{it} \tag{7.7}$$

where RISK_{it} = risk variable for industry i in year t

$RISK_{it-1}$ = lagged dependent variable

$DEREG_{it}$ = dummy variable that takes on a value of 1 in years after deregulation and is 0 otherwise

The $RISK_{it-1}$ variable captures the effect of the riskiness of the industry's technology and the effect of past investments in health and safety capital on safety. The expected sign of β_1 is positive, and its magnitude will equal 1 if risk levels are time invariant.

To capture cyclical influences, I have included percent change in the industry's employment, which typically performs better than other cyclical proxies, such as work hours or overtime hours. The expected sign of β_2 is positive, since work accidents are procyclical.

The main variable of interest is the deregulation dummy variable DEREG, which has been constructed using the same deregulation time periods as used previously in this chapter. If deregulation has led to an increase in accidents, β_3 will be positive, whereas if deregulation has improved safety, β_3 will be negative.

The risk variables used in the analysis took several forms. The three BLS risk series were utilized: the overall injury and illness rate (IR), the rate of injuries and illnesses involving at least one lost workday (LW), and the rate of total lost workdays (LWDAYS) (all per 100 workers in the industry). Because of possible changes in the productivity of the industry due to deregulation, additional denominators were also used. In the case of airlines the number of airplane departures (National Transportation Safety Board, various years) served as the output measure, and for trucking the numbers of ton-miles (in billions) served as the capacity measure (Transportation Policy Associates, various years).

The risk variable must necessarily be nonnegative. An ordinary regression with the risk variable as the dependent variable would not take this into account, possibly leading the equation to predict negative injury rates. To recognize this nonnegativity constraint, the dependent variable is transformed using either a log-odds or a logarithmic transformation. For the risk probability per worker variables IR and LW, the variable used is the log-odds of the risk, so that for IR the risk variable would be of the form $\ln[IR/(100-IR)]$. The logistic transformation has well-known statistical properties if one assumes that accidents are generated from a Weibull distribution. In addition, each of the risk variables was transformed by taking its natural logarithm.

Table 7.5 reports results for trucking. There are no statistically significant deregulation effects in either direction for this industry. Indeed, the coefficients closest to significance at standard confidence levels are negative rather than positive.

The airline results in Table 7.6 reinforce this general pattern, as none of the deregulation variables is statistically significant in any of the equations estimated. On the basis of individual industry regressions, then, deregulation shows no adverse effects.

Pooled Time Series and Cross-Section Analysis

In addition to the individual industry runs, it is possible to estimate the effects of deregulation using a pooled time series and cross-section regres-

Table 7.5 Summary of Regression Results for Trucking (SIC 421)

Dependent Variable	Risk Measure/Worker Results, Deregulation Dummy Variable Coefficient (SE)[a]	Risk Measure/Ton-Mile Results, Deregulation Dummy Variable Coefficient (SE)[b]
Injury rate (log-odds)	−0.009 (0.055)	—
Lost workday rate (log-odds)	−0.027 (0.028)	—
Injury rate (log)	−0.008 (0.046)	0.001 (0.049)
Lost workday rate (log)	−0.025 (0.026)	−0.027 (0.031)
Lost workdays (log)	0.021 (0.035)	0.019 (0.052)

SIC, Standard Industrial Code; SE, standard error.

[a] Each equation also includes an intercept, a lagged dependent variable, and the percent change in the industry's employment.

[b] Each equation includes the foregoing variables as well as percent change in ton-miles.

Table 7.6 Summary of Regression Results for Airlines (SIC 45)

Dependent Variable	Risk Measure/Worker Results, Deregulation Dummy Variable Coefficient (SE)[a]		Risk Measure/Departure Results, Deregulation Dummy Variable Coefficient (SE)[b]	
	1979–1985	1982–1985	1979–1985	1982–1985
Injury rate (log-odds)	−0.033 (0.041)	−0.047 (0.042)	—	—
Lost workday rate (log-odds)	−0.031 (0.057)	−0.048 (0.046)	—	—
Injury rate (log)	−0.028 (0.035)	−0.041 (0.037)	0.040 (0.045)	0.027 (0.051)
Lost workday rate (log)	−0.028 (0.053)	−0.044 (0.043)	−0.035 (0.071)	−0.039 (0.053)
Lost workdays (log)	0.010 (0.055)	0.012 (0.043)	−0.035 (0.071)	−0.039 (0.053)

SIC, Standard Industrial Code; SE, standard error.

[a] Each equation also includes an intercept, a lagged dependent variable, and percent change in the industry's employment.

[b] Each equation includes foregoing variables as well as percent change in departures.

Table 7.7 Pooled Time Series and Cross-Section Regressions with Risk/Worker Variables

Independent Variables	*Dependent Variable Coefficient (SE)*		
	Ln [IR/(100-IR)]	*Ln [LW/(100-LW)]*	*Ln (LWDAYS)*
Intercept	−0.838	−0.695	1.428
	(0.286)	(0.182)	(0.446)
Lagged dependent variable	0.598	0.711	0.687
	(0.130)	(0.063)	(0.098)
Percent change in employment	0.635	0.768	0.976
	(0.301)	(0.238)	(0.250)
Truck Industry dummy variable	0.145	0.013	0.192
	(0.083)	(0.044)	(0.060)
Rail Industry dummy variable	0.080	−0.033	−0.027
	(0.066)	(0.040)	(0.040)
RAIL-DEREG	−0.082	−0.075	0.042
	(0.054)	(0.042)	(0.043)
TRUCKS-DEREG	−0.051	−0.020	0.035
	(0.048)	(0.036)	(0.043)
AIR-DEREG	−0.020	−0.025	0.041
	(0.044)	(0.036)	(0.043)
R^2	.90	.91	.96

SE, standard error; IR, overall injury and illness rate; LW, lost workdays; LWDAYS, total lost workdays.

sion. To increase the sample size, I have included the railroad component of the transportation industry. In the first specification, I simply estimate the pooled version of Equation 7.7. The new varibles are dummy variables for two of the three industry groups (DTRUCK, DRAIL) to capture industry-specific risk differences. Thus, the equation is of the form

$$\text{RISK}_{it} = \alpha + \beta_1 \text{RISK}_{it-1} + \beta_2 \text{ percent change in employment}_{it}$$
$$+ \beta_3 \text{DTRUCK}_{it} + \beta_4 \text{DRAIL}_{it} + \beta_5 \text{RAIL-DEREG}_{it} \qquad (7.8)$$
$$+ \beta_6 \text{TRUCKS} - \text{DEREG}_{it} + \beta_7 \text{AIR} - \text{DEREG}_{it} + u_{it}$$

Because the output measures are not comparable across industries (e.g., ton-miles and departures are not in comparable units), pooled results for the transformed output-based risk variables are not appropriate.

Representative results for Equation 7.8 are reported in Table 7.7. Other specifications with different dependent variables yielded results that were identical in terms of their general spirit. In addition, although the AIR − DEREG variable in Table 7.7 was for 1979 through 1985, the 1982– 1985 AIR − DEREG variable performed similarly. There are no positive and statistically significant deregulation coefficients. There is no evidence of an adverse deregulation effect. Indeed, the only statistically significant coefficient is for railroads, and it is negative, that is, shows a safety improvement.

The final specification is a model with fixed time- and year-specific effects. Thus, a separate dummy variable is included for each particular year. Although this specification precludes the inclusion of DEREG variables, one can analyze the pattern of coefficients over time to assess the deregulation effect.

This RISK equation is of the form

$$\text{RISK}_{it} = \alpha + \beta_1 \text{ RISK}_{it-1} + \beta_2 \text{ percent change in employment} \\ + \text{DMODE}_i + \text{DYR}_t + u_{it}$$

(7.9)

where DMODE_i takes on a value of 1 for transportation mode i and 0 otherwise, and DYR_t (dummy time variable) takes on a value of 1 for year t and 0 otherwise.

The results in Table 7.8 fail to indicate disturbing temporal patterns. The time-specific coefficients that are significant in recent years tend to be negative and no greater in magnitude than those in earlier years. This formulation yields the same general conclusion as the others, which is that there has been no apparent adverse effect of deregulation on worker safety.

In terms of assessing the factors that influence transportation safety, it is also instructive to assess the effect of OSHA inspections on risk levels. The principal OSHA variable is the number of OSHA inspections per 1,000 workers in the industry, where this variable enters with a one-year lag. This formulation is identical to that in Viscusi (1986), as we simply add OSHA_{it-1} to Equation 7.9.

Due to the lag in the availability of OSHA inspection data, an analysis including OSHA cannot be more recent than 1984, so that the sample is 1973–1984. The inclusion of an OSHA variable does not alter the general implications we have found concerning the effect of deregulation on safety. Therefore the complete set of results of the current regression is not reported.

The results in Table 7.9 indicate that OSHA inspections do have a statistically significant negative effect on all but one risk measure in the transportation industry. The only exception is the rate of total lost workdays, for which the coefficient is negative and exceeds the standard error but is not statistically significant. Thus, worker safety is responsive to these inspections and perhaps other contemporaneous governmental actions correlated with OSHA activities.

OSHA COMPLIANCE STATISTICS

As a final measure of riskiness let us consider the rates of compliance with OSHA standards. These statistics provide a good index of the safety inputs supplied by the firm, both through its capital investments in safety and work practices. The particular measure that I will examine is the fraction of firms found to be in compliance with OSHA inspections.

Although this is an attractive measure of riskiness, its usefulness is hampered by changes over time in the content of the measure. In fiscal year

Table 7.8 Pooled Time Series and Cross-Section Regressions with Risk/Worker Variables

Independent Variable	Ln $[IR/(100-IR)]$	Coefficients (SE) Ln[LW/ $(100-IW)$]	Ln (LWDAYS)
Intercept	−0.763 (0.342)	−0.983 (0.297)	2.645 (0.453)
Lagged dependent variable	0.682 (0.155)	0.661 (0.100)	0.375 (0.108)
Percent change in employment	0.716 (0.503)	0.781 (0.356)	1.071 (0.346)
DYR_{74}	0.151 (0.080)	0.129 (0.056)	0.155 (0.057)
DYR_{75}	0.080 (0.093)	0.142 (0.064)	0.148 (0.074)
DYR_{76}	0.120 (0.078)	0.117 (0.052)	0.137 (0.055)
DYR_{77}	0.080 (0.076)	0.143 (0.052)	0.160 (0.054)
DYR_{78}	0.126 (0.076)	0.197 (0.055)	0.207 (0.059)
DYR_{79}	0.096 (0.076)	0.153 (0.060)	0.239 (0.063)
DYR_{80}	0.056 (0.081)	0.097 (0.062)	0.293 (0.078)
DYR_{81}	0.086 (0.085)	0.117 (0.060)	0.255 (0.082)
DYR_{82}	0.082 (0.095)	0.106 (0.064)	0.292 (0.086)
DYR_{83}	0.003 (0.089)	0.056 (0.059)	0.201 (0.078)
DYR_{84}	0.079 (0.082)	0.119 (0.052)	0.265 (0.058)
DYR_{85}	0.029 (0.081)	0.067 (0.054)	0.285 (0.074)
Truck Industry Dummy Variable	0.117 (0.093)	0.065 (0.057)	0.381 (0.063)
Rail Industry Dummy Variable	0.072 (0.077)	0.007 (0.050)	−0.019 (0.031)
R^2	.89	.93	.98

SE, standard error; IR, overall injury and illness rate; LW, lost workdays; LWDAYS, total lost workdays; DYR, dummy time variable.

1977 there was a major change in the OSHA inspection strategy, with increased emphasis on serious violations. In 1981 the Reagan Administration left its imprint on the inspection process by scaling back the number of inspections and increasing their targeting. These and other changes in inspection policy make the compliance statistics a less stable measure of risk than the BLS injury statistics.

Table 7.10 summarizes information by industry regarding the fraction of inspected firms in that industry that was found to be in compliance with

Table 7.9 Summary of $OSHA_{it-1}$ Coefficients in Pooled Time Series and Cross-Section Regression[a]

Dependent Variable	Coefficient (SE)
$Ln[IR/(100-IR)]$	−0.114 (0.058)
$Ln[LW/(100-LW)]$	−0.089 (0.038)
$Ln(IR)$	−0.102 (0.051)
$Ln(LW)$	−0.083 (0.035)
$Ln(LWDAYS)$	−0.046 (0.036)

SE, standard error; IR, overall injury and illness rate; LW, lost workdays; LWDAYS, total lost workdays.

[a] Other variables included are identical to those in Table 7.8.

Table 7.10 OSHA Compliance Rates by Industry

	Fraction of Inspected Firms in Compliance	
Year	Trucking (SIC 421)	Air Transportation (SIC 45)
1972	.47	.65
1973	.36	.38
1974	.35	.36
1975	.27	.40
1976	.32	.39
1977	.46	.51
1978	.51	.53
1979	.53	.55
1980	.57	.59
1981	.53	.60
1982	.45	.53
1983	.38	.51
1984	.57	.70
1985	.54	.69

SIC, Standard Industrial Code.

OSHA standards. In the trucking industry there was a temporary dip in compliance rates in 1982 and 1983, but this drop coincided with more intensive trucking inspections. Finally, in the case of air transportation there was a drop in compliance rates in the 1982–1983 period, but this increase may be misleading since the number of serious violations declined substantially in 1982. Examination of OSHA compliance rates indicates no apparent cause for alarm in the wake of deregulation.

CONCLUSION

Job safety is an important component of transportation safety and will tend to capture the types of economic influences likely to affect all, including consumer and third party, aspects of transportation safety. In terms of the economic forces set in motion by transportation deregulation, almost all factors suggest that there should be a drop in safety as a consequence of deregulation. Moreover, if there is such a decline it is likely to be larger in the initial years of deregulation than after the industry settles down into the postderegulation equilibrium.

Examination of several BLS accident rate series fails to indicate any cause for alarm in terms of major departures from expected accident trends. There has been no apparent upsurge in accident levels in the postderegulation period. Analysis of risk trends is complicated both by the secular decline in accidents and the procyclical nature of accidents, which appear to be the driving influences. A variety of regression specifications fail to reveal any adverse effects of deregulation.

The absence of a dramatic upswing in accidents should not, however, be a cause for either complacency or a conclusion that transportation safety does not remain an important issue. The level of risk is quite high, for both trucking and airlines. Even a modest increase in accidents due to deregulation that cannot be readily disentangled from the other forces at work may involve a large number of deaths. Additional large-scale catastrophes, such as more major airline crashes, could also generate the need for additional analysis of safety concerns.

One should not dismiss out of hand the claim by pilots and trucking operators that safety conditions are worsening. The adverse effects on the accident rate may simply not be apparent, particularly if it is a deterioration in workers' physical capabilities that is involved.

The costs of giving transportation deregulation an entirely clean bill of health in the safety area may be substantial. What can be concluded at this point is that the dire predictions regarding safety have not been borne out. There is no evidence of any significant decline in job safety, and thus far the safety deterioration predicted by the critics of deregulation has not materialized.

THE AIRLINE INDUSTRY

Chapter 1 identified four areas of potential safety concerns following deregulation: (1) the impact of carrier profitability on safety performance, (2) the safety performance of new entrants compared with that of existing carriers, (3) safety implications of modal shifts, and (4) whether the provision of infrastructure has kept pace with expansion in demand.

In the present section each of these issues is examined for the airline industry. Rose, in Chapter 8, analyzes the safety-profitability relationship; Kanfani and Keeler, in Chapter 9, look at the relative safety performance of new entrant carriers; and Oster and Zorn, in Chapter 10, focus on whether passengers on routes where smaller commuter aircraft have replaced jet aircraft face increased risks.

In general, all these authors find little evidence that would be a cause for concern. Nevertheless, there are many thoughtful observers of the industry who believe that past accident statistics should not be used to forecast safety trends in the industry. Nance, in Chapter 13, comes to the conclusion that deregulation has increased the potential for unsafe operations and accidents in the future. He bases this conclusion on a study of negative trends in training and aircraft maintenance. He argues that these trends are reducing the stock of safety investment that was built up in the days of economic regulation, and that such a reduction will lead to increased accident rates. Data do not exist that would allow econometric testing and evaluation of the Nance hypothesis.

A serious problem area is the increasing congestion at major airports, where deregulation's success in attracting patronage has pushed runway and air traffic control systems to their capacity at peak times. Congestion has led to the well-publicized increases in delays and, allegedly, in the number of near collisions on the ground and in the air. While one solution to the problem would be to provide more capacity, some argue that the existing capacity can be used more effectively by spreading the peak in traffic. Bailey and Kirstein, in Chapter 11, believe that the best short-run policy for reducing congestion is to inform passengers of the delays expected at peak times and at certain airports. Currently airlines' published schedules are unduly "optimistic." Arnott and Stiglitz in Chapter 12 favor a system of variable landing fees that would result in much higher ticket prices for passengers who travel at peak times at congested airports.

These chapters represent a subset of the nearly 20 aviation papers presented at the Northwestern Conference, by leading experts from academia, industry, labor, government, and consumers. All of these papers contained substantive information that can add to our understanding of the issues being considered. Chapter 14 summarizes these additional insights.

8

Financial Influences
on Airline Safety

NANCY L. ROSE

Whether airline safety has been compromised by deregulation is a matter of great concern. Despite substantial media attention, relatively little systematic research has been done on possible links between deregulation and safety conditions in the industry. Graham and Bowes (1979) investigated the link between firms' financial condition and their accident rates, maintenance expenditures, and service complaints; Golbe (1986) analyzed the relationship between accidents and profitability. Both studies used a short time series of prederegulation data. Neither found much support for the argument that reduced profitability lowers airlines' safety performance. Advanced Technology (1986) analyzed bivariate correlations between financial measures and carriers' inspection ratings in the Federal Aviation Administration's (FAA) 1984 National Air Transportation Inspection Program. While the study found some relationship between financial indicators and inspection failures, its methodology had substantial shortcomings, particularly in that it failed to control for other factors that may affect inspection ratings.

In this study I investigate the relationship between accident rates and financial performance, controlling for operating characteristics of carriers that may affect their accident probabilities. This is a test of one channel through which deregulation may affect safety, *not* a test of the effect of deregulation per se. Evidence suggests that deregulation may have *improved* the industry's financial condition, which means a financial performance—safety link could lead to increased safety since deregulation. This chapter discusses economic models of firms' quality or safety choices and assesses their implications for the airline accident—financial performance relationship; analyzes the aggregate accident data for U. S. certificated scheduled carriers over the period 1954 through 1986 (this longer time period than in earlier studies permits more precise estimates of both the accident trend rate under regulation and the effect of deregulation on the level or direction of that trend); explores the determinants of accident rates for individual carriers using data on 26 carriers; and presents a statistical model of the determinants of individual accident rates. The results provide evidence that less profitable airlines may have higher accident rates, other things being equal.

MODELS OF THE DETERMINANTS OF AIRLINE SAFETY DECISIONS

The effect of financial performance on safety levels is ambiguous on theoretical grounds. A number of private incentives encourage safety, independent of direct government intervention.

1. Insurance companies base the premium rates for liability insurance on an assessment of risk. These companies have a strong incentive to monitor safety performance of carriers and to increase premiums for firms that take on additional risk. This effect suggests that reducing safety expenditures may *increase*, not decrease, total costs once insurance premiums are taken into account.
2. Firms have an important stake in maintaining a reputation for providing safe service in order to attract and retain business. Airlines that develop a reputation for being less safe will, other things being equal, be likely to lose passengers to safer competitors. Some empirical research has analyzed the strength of these reputation incentives—see Barnett and Lofaso (1983) on passenger responses to a series of crashes involving McDonnell Douglas DC10 aircraft; Chalk (1985, 1986) on the effects of fatal accidents on aircraft manufacturers' profitability; and Borenstein and Zimmerman (Chapter 5 in this volume) on share price of airline stocks and traffic responses to airline fatalities.
3. Employees have strong incentives to monitor safety, particularly when linked to maintenance of equipment or operating procedures. Pilots and flight attendants, because of their extensive exposure to an airline's flights, are at highest risk from any decline in air safety. These groups are likely to resist reductions in safety, or at minimum, to require compensating wage differentials for higher risk exposure. (The strength of this effect depends critically on the alternative employment opportunities for these employees, for if their wages are above their opportunity wage, reduced safety may reduce their economic rents rather than leading to higher wages.)

These private incentives do not guarantee the optimal level of safety. In particular, asymmetric information about safety may lead to less safety than is socially desirable. Asymmetric information is present when a firm has knowledge about its own level of safety that insurance firms, employees, and customers cannot or do not obtain. Under these circumstances, the firm may have an incentive to provide less safety than customers or employees desire. Models by Akerlof (1970), Klein and Leffler (1981), Shapiro (1982, 1983b), Allen (1984), and numerous others examine the effects of asymmetric information on product quality or safety choices by firms. However, in theses models, firms do not unambiguously provide *lower* safety. Firms may choose to *overprovide* safety, depending on the values of various parameters. In Shapiro's (1982) model, for example, a firm may choose higher quality (safety) to improve its reputation.

The reputation models do not focus directly on possible links between the *level* of profitability and firms' safety choices. A second class of models, based on the effects of limited liability and bankruptcy risk, may better reflect the focus of many critics of deregulation. Given limited liability laws, the "downside" risk to shareholders of a bankrupt firm is limited to the amount of their equity holdings. This creates an asymmetry in shareholders' expected returns, with unconstrained positive returns but limited negative returns. Firms that are near insolvency might choose, for example, to reduce maintenance expenses and gamble on no increase in accidents in an effort to avoid bankruptcy. Bulow and Shoven (1978) and Golbe (1981) described a model of the bankruptcy decision and its effect on stockholders' preferences for the firm's risk behavior. Even in this type of model, however, the attractiveness of increased risk through reductions in safety expenditures is ambiguous. If bankruptcy costs are high, firms will try to avoid actions that raise the risk of bankruptcy.

The tendency for failing air carriers to exit the industry by merger or acquisition rather than by liquidation of assets may provide another incentive to maintain safety. If an airline increases its accident rate by reducing its safety expenditures, its value to potential acquirers is likely to be lower because the accidents will have eroded its goodwill or reputation. This could counteract the incentives provided by limited liability. While more complex models of carriers' responses to financial distress may provide some predictions for the direction of financial influences on safety decisions, the issue is one that seems likely to be resolved only through empirical investigation. It is to this type of analysis that I now turn.

AGGREGATE DATA ON AIRLINE SAFETY

An analysis of the potential effect of deregulation on safety logically begins with a comparison of the performance of the industry before and after deregulation. In this section I explore the behavior of industrywide accident measures over time and investigate the relationship between profitability and safety at the industry level.

Accident Rates as a Measure of Safety

Although we cannot observe safety directly, a number of proxies for safety levels are available, including number of accidents or "incidents" (such as near-midair collisions, runway incursions); FAA inspection results and number of citations; and levels of safety inputs such as maintenance, training, and operating procedures. Each measure has particular advantages and disadvantages; depending on the questions of interest and the data availability, more than one of these measures might be incorporated in a statistical analysis. For a variety of reasons, accidents seem the most appropriate measure for the present analysis (see Appendix 8-1 for a description of what constitutes an accident). First, air carrier accidents, fatal and nonfatal, reflect safety *outcomes,* which seem of most concern to passengers and policy analysts: What is the probability a flight will be involved in an accident? FAA inspections and citations are more indicative of safety *inputs,* as are maintenance expenditures, training programs, and the like. If we understood how various safety inputs translate into reduced accident probabilities, the use of input measures might be meaningful. We could, for example, construct a safety index with appropriate weights on different types of inputs. Because little is known about this transformation of inputs into safety performance, however, actual performance may provide the most reliable measure of the level of safety.

Second, accident reporting and detection are quite accurate, particularly for more serious accidents involving larger airlines relative to reporting of incidents and detection of safety violations through FAA inspections. Incident reporting is less consistent over time and across carriers than is accident reporting: Judgment of what constitutes an incident is subjective; detection of nonreporting is difficult; and levels of incidents may be sensitive to "campaigns" that attempt to improve reporting rates. Safety-conscious carriers may appear to have higher incident rates than less safe carriers, simply because the former encourage more complete reporting. Similarly, FAA inspections are unlikely to detect all violations, and the number of citations and fines may depend critically on the intensity of the FAA's enforcement activity, which is unlikely to be constant over time. For example, the FAA's recent record-setting fines of major trunk carriers may reflect lax safety procedures by the airlines, a decision by the FAA to signal its "seriousness" about enforcing safety violations, or even weaker enforcement though not necessarily higher safety in earlier periods.

Third, accidents may be more appropriate to this study's focus on air carrier safety rather than on air system safety. The substantial majority of accidents are attributable to causes that may be under the control or influence of air carriers, such as pilot or crew error, maintenance deficiencies, and inadequate training. Incidents are likely to include a higher proportion of events that may be partially or wholly attributable to air traffic controller errors. Incident trends therefore may be more sensitive to the aftermath of the 1981 air traffic controllers' strike than are accident indices. Although these incident trends may reflect system safety, they are not properly attributed to economic deregulation or air carrier safety decisions.

Finally, aggregate accident rates will yield fairly precise estimates of accident probabilities, particularly given the large number of flights in each year. U. S. major certificated air carriers (FAA Part 121 carriers, a category that excludes commuter and air taxi operators) currently operate well over 5 million flights logging 3 billion aircraft-miles per year. This substantial exposure of U. S. air carriers, combined with annual total accidents in the range of 15 to 25 during recent years, allows us to identify the probability of an accident with a great deal of precision. This feature of the aggregate statistics is frequently overlooked. When an individual carrier is small relative to the frequency of accidents, one more or one less accident may substantially change its accident rate, although this will average out across carriers. Although this sensitivity often is cited as an argument for disregarding accident statistics, the infrequency of accidents for individual carriers does not invalidate studies based on carrier accident rates. The statistical noise associated with these small numbers may make it more difficult to estimate precise relationships between accidents and other factors, which reduces the power of statistical tests but does not bias their results.

Given these considerations, the analysis here uses accidents and accident rates per thousand departures to measure airline safety. I measure accidents by the number of aircraft accidents rather than by passenger-based measures of accident incidence, such as passenger fatalities. This corresponds to asking the question: What is the probability that a flight selected at random from the pool of available flights will be involved in an accident? One accident that kills all 200 people on board may have different implications for the safety of the system than 20 accidents with 10 fatalities each. The present analysis implicitly assumes that the second scenario reflects lower safety levels, all else being equal.

Aggregate Accident Trends

Table 8.1 presents information on the number of accidents, aircraft-miles, revenue departures, and accident rates per 100,000 departures for U. S. certificated air carriers' scheduled passenger and cargo operations, from 1955 through 1986. Both fatal and nonfatal accidents are included in total accidents (column 2); fatal accidents are also tabulated separately in column 3. There is a steady decline in both fatal and nonfatal accidents over time, a decline that continues after economic deregulation of the industry in 1978. The number of accidents declines even though the number of flights and aircraft miles increase sharply through time, implying even larger declines in the accident rates per 100,000 departures, as reported in columns 6 and 7 (and in accident rates per million miles, not shown). This improvement in safety is attributable primarily to the substantial improvements in aircraft and aviation technology over the last 40 years. These include the diffusion of radar technology, development of the jet aircraft (with enhanced power, range, and operating altitude), metallurgical and materials advances, the introduction and continued improvement of navigational and landing aids (such as automatic pilot, electronic glide slopes, ground proximity warning systems), more sophisticated simulators for pilot training, and the like.

Although accident rates continue to fall through the deregulation period, it is possible that deregulation has caused the rate of decline to deviate from its long-term trend. Figure 8.1 indicates that this is not the case. This figure plots actual total accidents per million departures and the accident rate that would be predicted by a logarithmic time trend predicted over the 1955–1977 period. The figure illustrates the sharp decline in accident

Table 8.1 Accident Rates, U.S. Certificated Air Carriers, Scheduled Passenger and Cargo Operations (FAA PART 121 Operations)

Year	Accidents Total	Accidents Fatal	Miles (millions)	Departures (100,000)	Total Accidents/ 10^5 Depart.	Fatal Accidents/ 10^5 Depart.
1955	61	11	779.93	32.76	1.86	.34
1956	70	7	869.31	35.03	2.00	.20
1957	73	7	976.17	37.69	1.94	.19
1958	67	8	972.99	36.33	1.84	.22
1959	78	14	1,030.25	39.12	1.99	.36
1960	72	12	996.92	38.56	1.87	.31
1961	66	6	969.66	37.50	1.76	.16
1962	47	6	1,009.68	36.60	1.28	.16
1963	54	6	1,094.52	37.88	1.43	.16
1964	59	11	1,189.14	39.54	1.49	.28
1965	65	8	1,353.50	41.97	1.55	.19
1966	56	5	1,482.27	43.73	1.28	.11
1967	54	8	1,833.56	49.46	1.09	.16
1968	56	13	2,146.04	53.00	1.06	.25
1969	51	8	2,385.08	53.77	0.95	.15
1970	43	4	2,417.55	51.00	0.84	.08
1971	43	7	2,380.66	49.99	0.86	.14
1972	46	7	2,347.96	49.66	0.93	.14
1973	36	8	2,448.11	51.34	0.70	.16
1974	43	7	2,258.14	47.26	0.91	.15
1975	29	2	2,240.51	47.05	0.62	.04
1976	21	2	2,320.00	48.33	0.43	.04
1977	19	3	2,418.65	49.37	0.38	.06
1978	20	5	2,520.17	50.16	0.40	.10
1979	23	4	2,791.12	54.00	0.43	.07
1980	15	0	2,816.30	53.53	0.28	.00
1981	25	4	2,703.22	52.12	0.48	.08
1982	15	3	2,698.93	49.64	0.30	.06
1983	22	4	2,808.57	50.34	0.44	.08
1984	14	1	3,113.57	54.48	0.26	.02
1985	18	4	3,331.88	57.24	0.31	.07
1986	22	1	3,723.00	64.35	0.34	.02

FAA PART 121, excludes air taxi and commuter airline operations.

SOURCES: 1955–1974 data from Civil Aeronautics Board (various years). 1975–1986 data from National Transportation Safety Board (various years).

rates: Accidents per million departures fell by a factor of 6 between 1955 and 1986, a rate of 6.6% per year. Moreover, the figure suggests that the post-1978 experience conforms quite closely to what one would have expected prior to deregulation. Accident rates after deregulation are all quite close to, or slightly below, the predicted trend line. Similar conclusions are suggested by graphs of the number of actual and predicted total accidents and fatal accidents.

A more formal way to structure this test is to model accidents as following a logarithmic decline over time and to allow for the possibility that the accident trend changes after deregulation. This suggests an equation of the form

$$\ln \text{(ACCIDENT RATE)} = \beta_0 + \beta_1 \cdot \text{TIME} + \beta_2 \cdot \text{TIME} \cdot \text{DEREG} \quad (8.1)$$

where ln denotes the natural logarithm, TIME is a linear time trend, and DEREG is a dummy variable equal to 1 for the years 1978–1986, 0 otherwise. β_2 measures the change in the time trend after deregulation. In 1980, fatal accidents are zero, and the logarithm of the fatal accident rate is undefined. I treat this by setting ln (fatal accidents) equal to 0 and introducing a dummy variable, ZERODUM, equal to 1 when the number of accidents is 0 and equal to 0 otherwise (see Pakes and Griliches, 1980, and Hausman, Hall, and Griliches, 1984, for a similar econometric treatment in a model of patents).

I estimate Equation 8.1 by ordinary least squares regression using three different measures of accident rates: the total number of accidents in a given year (TOTACC), the total number of accidents per 100,000 departures (TACCDEP), and the number of fatal accidents in a given year (FATACC). These results are reported in Table 8.2. The decline in accidents over time is quite strong, with the number of total and fatal accidents declining by 4.5% (standard error, 0.7%) to 4.7% (1.5%) per year throughout the period and the total accident rate per 100,000 departures declining by 6.5% (0.6%) per year. The equations for the two measures based on total accidents (columns 2 and 3) explain roughly 90% of the variance in accident rates over time. The third equation, for fatal acci-

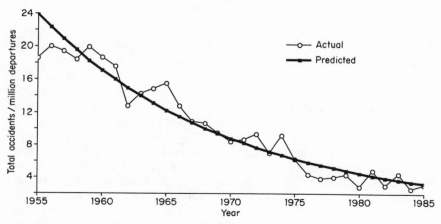

Figure 8.1 Actual Versus Predicted Total Accidents per Million Departures.

Table 8.2 Regression Analysis of Aggregate Accident Rates over Time

| | Dependent Variable (SE) | | | | | |
| | 1955–1986 (32 observations) | | | 1956–1984 (29 observations) | | |
	TOTACC	TACCDEP	FATACC	TOTACC	TACCDEP	FATACC
Constant (β_0)	4.447	0.915	2.465	4.404	0.925	2.292
	(0.092)	(0.081)	(0.202)	(0.122)	(0.115)	(0.286)
TIME (β_1)	−0.045	−0.065	−0.047	−0.048	−0.068	−0.043
	(0.007)	(0.006)	(0.015)	(0.007)	(0.006)	(0.016)
TIME·DEREG (β_2)	−0.008	−0.005	−0.005	−0.007	−0.005	−0.001
	(0.005)	(0.004)	(0.011)	(0.005)	(0.005)	(0.012)
OPMARG$_{t-1}$ (β_3)	—	—	—	1.637	0.685	2.226
				(1.094)	(1.029)	(2.561)
ZERODUM	—	—	−1.123	—	—	−1.165
			(0.496)			(0.496)
R squared	.86	.93	.61	.88	.93	.60

SE, standard error; TOTACC, total number of accidents in given year; TACCDEP, total number of accidents/100,000 departures; FATACC, number of fatal accidents in given year; DEREG, dummy variable; OPMARG, operating margin; ZERODUM, dummy variable. See text and Appendix 8.1 for explanation of variables.

dents, explains 60% of the variation in fatal accidents over the period. Moreover, in all three equations, accident levels decline *faster* after deregulation, although this effect is not statistically significant. They are qualitatively the same if deregulation is constrained to change the level of accident rates rather than the trend.

Aggregate Accidents and Industry Profitability

Aggregate data also can be used to explore the relationship between accidents and profitability at the industry level. I calculate the industry's average operating margin, OPMARG, defined as

$$1 - \left(\frac{\text{operating expenses}}{\text{operating revenues}}\right)$$

This measure is calculated before interest, lease, and tax payments, thereby treating all returns to capital—debt, equity, and capital leases—equivalently. To ensure that OPMARG is not contaminated by the costs of current accidents and to reflect the assumption that profits are likely to influence accident rates with a lag (if at all), I use OPMARG from the preceding period in the regression estimates. Similar results were obtained using unlagged values of OPMARG.

Table 8.2 presents the estimated coefficients for TIME, TIME·DEREG, and OPMARG using the three measures of accident rates described previously as dependent variables. Operating margin data were available only through 1983, so the equations are estimated over the 1956–1984 period. The TIME and TIME·DEREG coefficients are essentially unchanged from the earlier results. In all three regressions, OPMARG has a counterintuitive positive sign, implying that *lower* profit margins correspond to *lower* accident rates. In all the equations, however, the standard errors on the OP-MARG coefficients are substantial and the point estimates are statistically indistinguishable from zero. The small number of post-1977 observations and large standard errors on the OPMARG coefficients also make it impossible to tell whether the accident-profit relationship changes after deregulation. While the imprecision of the estimated OPMARG coefficients limits the power of these tests, there is no evidence of negative effects of average profitability on the level of aggregate accidents.

Conclusions from the Aggregate Accident Analysis

This analysis of aggregate accident data for the major domestic air carriers provides no support for the view that safety levels have deteriorated in the aftermath of deregulation, nor does it indicate any correlation of aggregate profitability with aggregate safety. The results here are consistent with McKenzie and Shughart's (1986) findings that deregulation had no discernible effect on aggregate airline fatalities, either directly or through its effects on increased air passenger miles, on the basis of data from 1973 through 1984. The results are consistent with Golbe's (1986) time series analysis of the influence of profitability on aggregate accident rates. Using data from 1952 through 1972 and controlling for the number of departures and general economic conditions, Golbe found that profitability tends to be associated with positive but statistically insignificant effects on accident rates.

Whereas the aggregate data indicate no adverse change in accident rates after 1978, they provide a weak test of the hypothesis that deregulation reduced safety. Interfirm differences in accident rates permit more precise and more powerful tests of deregulation's effects on accidents (see Barnett and Higgins, 1987). The next section analyzes firm-specific accident data.

ANALYSIS OF INDIVIDUAL AIR CARRIER ACCIDENT DATA

The argument that safety has declined since deregulation frequently is based on an implicit link between the financial health or profitability of a carrier and its investment in safety. Advocates of this view claim that more profitable carriers choose to spend more on maintenance and safety-enhancing procedures, while low-profit carriers "cut corners" on safety expenditures (see Chapter 13 in this volume). Deregulation, they argue, reduces overall profits and therefore safety. This argument rests on two links: a decline in average profitability or financial health as a result of deregulation and the existence of a relationship between safety and measures of firms' financial conditions. This section reports on a preliminary exploration of the determinants of accident rates for individual air carriers with particular attention to the relationship of accident rates to firms' profit margins. Note that this study addresses only the last part of a possible deregulation–financial condition—safety chain. Inferences about deregulation also require estimates of deregulation's effects on airline finances, which some sources suggest may be positive (Morrison and Winston, 1986).

Individual Air Carrier Data

The data set used to evaluate individual air carrier safety performance consists of information on 26 scheduled air passenger carriers certificated by the Civil Aeronautics Board (CAB) over the period 1954 through 1986. The carriers include domestic and international trunks (12), local service carriers (6), Alaskan and intra-Hawaiian airlines (5), and territorial carriers (3). The sample carriers and their dates of data availability are listed in Table 8.3. Not all carriers are observed over the entire period; a number exit the industry, primarily through mergers or acquisitions.

The sample omits five major classes of air carriers: commuter airlines and air taxis, intrastate air carriers, nonscheduled passenger carriers (charters), new jet carrier entrants, and cargo carriers. The first three groups were excluded primarily because of the unavailability or noncomparability of published data on their operations and financial conditions. Omission of commuter and nonscheduled air carriers also is supported on theoretical grounds: Most of these carriers differ substantially from scheduled jet carriers in terms of their services, technology, and scale and scope of operations, suggesting that statistical inferences based on pooled samples may be quite misleading (see Chapter 10 in this volume). New carriers are excluded due to possible noncomparability with existing carriers (see Barnett and Higgins, 1987, and Chapter 9 in this volume for an evaluation of new entrants). The deletion of all-cargo carriers is dictated by our focus on air passenger safety.

A variety of operating and financial information was collected for each carrier, including systemwide measures of accidents (TOTACC), revenue earning departures (DEPART), revenue earning aircraft miles (MILES), total operating expenses, and total operating revenues. A number of additional measures were constructed from these variables, including exposure to international operations (DINTL, equal to 1 if the airline has any international departures and to 0 otherwise), operating margins (OPMARG), average stage length (AVSTAGE, in thousands of miles), airline operating experience (measured in cumulative aircraft miles flown by the airline, EXPER, and the natural log of experience, LNEXPER), TIME (a linear time trend), and ZERODUM (equal to 1 when TOTACC equals 0, and to 0 otherwise). Several data series are missing information or are incompatible for part of the 1954–1986 time period (typically either the beginning or the end of the period); this may reduce the available sample size for any given application. Further details on the sources, construction, and availability of the data are provided in Appendix 8.1.

The statistical analysis focuses exclusively on total accidents rather than on fatal accidents. For individual carriers, fatal accidents are extremely rare events, even during the early part of the sample period; therefore it

Table 8.3 Sample Air Carriers

Carrier	Basic Data Availability
Trunk carriers	
American Airlines	1954–1986
Braniff Airways	1954–1981
Continental Air Lines	1954–1986
Delta Air Lines	1954–1986
Eastern Air Lines	1954–1986
National Airlines	1954–1979
Northeast Airlines	1954–1971
Northwest Airlines	1954–1986
Pan American World Airways	1954–1986
Trans World Airlines	1954–1986
United Air Lines	1954–1986
Western Air Lines	1954–1986
Local service carriers	
Frontier Airlines	1954–1986
Ozark Air Lines	1954–1986
Piedmont Aviation	1954–1986
Southern Airways/Republic[a]	1954–1986
Texas International	1954–1982
USAir/Allegheny	1954–1986
Alaskan and intra-Hawaiian	
Alaska Airlines	1954–1986
Aloha Airlines	1958–1974
Hawaiian Air	1954–1986
Northern Consolidated/Wien Consolidated[b]	1954–1967
Wien Air	1954–1986
Territorial carriers	
Caribbean Airlines	1954–1972
Pacific Northern	1954–1966
Pan American-Grace	1954–1966

[a]Data reflect Southern Airways operations prior to merger into Republic in 1979, and Republic operations thereafter.

[b]Data reflect Northern Consolidated operations prior to its merger with Wien Air in 1968, and combined operations thereafter.

would be difficult to draw any meaningful statistical inferences. Moreover, it is not clear that fatal accidents are a priori fundamentally different signals of underlying safety levels than are nonfatal accidents, even though they are ex post substantially worse outcomes.

Accident Probabilities across Carrier Groups

Before estimating models of the relationship between accidents and firms' operating characteristics, it is useful to ask whether different types of firms

Table 8.4 Accident Probabilities by Carrier by Time Period
(Accidents/1,000 Departures)

Carrier	1954–1960	1961–1965	1966–1970	1971–1975	1976–1980	1981–1986
American	0.024	0.014	0.008	0.009	0.006	0.004
Braniff	0.010	0.011	0.009	0.005	0.005	0.004
Continental	0.013	0.021	0.007	0.004	0.008	0.008
Delta	0.008	0.012	0.011	0.012	0.005	0.002
Eastern	0.015	0.015	0.007	0.008	0.003	0.004
National	0.020	0.010	0.009	0.010	0.000	—
Northeast	0.022	0.017	0.013	0.000	—	—
Northwest	0.030	0.017	0.006	0.012	0.000	0.002
Pan American	0.034	0.030	0.024	0.017	0.008	0.006
Trans World	0.025	0.022	0.008	0.015	0.003	0.002
United	0.011	0.013	0.013	0.005	0.002	0.005
Western	0.014	0.010	0.006	0.003	0.005	0.001
Frontier	0.009	0.012	0.007	0.010	0.002	0.003
Ozark	0.010	0.007	0.006	0.003	0.000	0.006
Piedmont	0.010	0.010	0.008	0.005	0.001	0.003
Southern/Republic	0.005	0.008	0.005	0.003	0.001	0.002
Texas International	0.010	0.002	0.008	0.009	0.008	0.006
USAir/Allegheny	0.012	0.008	0.007	0.006	0.005	0.004
Alaska	0.093	0.041	0.018	0.008	0.010	0.003
Aloha	0.000	0.010	0.015	0.009	—	—
Hawaiian	0.000	0.007	0.005	0.000	0.000	0.007
Northern Cons./Wien	0.172	0.031	0.036	0.011	0.003	0.004
Wien Air	0.152	0.090	0.043	—	—	—
Caribbean	0.000	0.007	0.027	0.000	—	—
Pacific Northern	0.069	0.000	0.000	—	—	—
Pan Amer-Grace	0.000	0.000	0.000	—	—	—
Total						
Accidents	344	214	204	176	80	86
Departures	19,344	15,212	21,107	21,876	22,053	25,527
Aggregate accident rate	0.018	0.014	0.010	0.008	0.004	0.003

appear to have different accident probabilities. Table 8.4 reports the accident rates per thousand departures for each sample carrier over six time periods beginning with 1954. Accident rates are calculated as N_{it}/D_{it}, the number of accidents for firm i during time t divided by the total number of departures (in thousands) for firm i during time t. The last three rows of the table report aggregate accidents, departures (in thousands), and the estimated aggregate accident rate for each period. The aggregate accident rate declines continuously through time, the 1981–1986 rate being one sixth the initial accident rate in 1954–1960. Likelihood ratio tests reject the hypothesis that sample aggregate accident rates in any two adjacent periods are the same, with the exception of the last two periods.

   The accident rates for individual carriers exhibit considerable variance, although some general patterns emerge. For example, Pan American has above-average accident rates throughout the period, as do the Alaskan carriers Alaska Airlines, Northern Consolidated/Wien Consolidated, and Wien Air. This illustrates the need for caution in interpreting the individual accident rates: Higher rates may reflect differences in the inherent risk of carriers' operations (such as potentially riskier conditions for international airports and Alaskan operations), rather than differences in carriers' safety per se.

   To test whether groups of carriers with different characteristics exhibit different accident probabilities, I group carriers on the basis of four characteristics: type of operations (trunk vs. others), profitability (operating margins), maintenance expenditures (as a fraction of total operating expenses), and size (passenger-miles). For the size, profit, and maintenance tests, carriers are separated into 'high" and "low" groups for each time period. The cutoff between high and low is based on natural break points in the data and is specific to each time period. This means that a carrier may be in the high group one period and in the low group the next period, and may be in different groups for different characteristics. The use of natural break points when these occur implies that the high and low groups may not be of equal size; the low group in each period and for each characteristic tends to be slightly larger than the high group. For the trunk/nontrunk test, trunk carriers are considered to be in the high group and all others are considered to be in the low group.

   I model the number of airline accidents per thousand departures as a Poisson random variable. The Poisson distribution is particularly suited to this type of problem, and has been employed widely in studies of accident probabilities (see Barnett, Abraham, and Schimmel, 1979; Barnett and Higgins, 1987; Golbe, 1986; and Appendix 8.2). Given this distribution, the accident probability per thousand departures, λ, can be estimated from the accident rate, N_j/D_j, for each group j. From these accident probabilities I construct likelihood ratio statistics to test the hypothesis that both high and low groups are drawn from a common accident probability distribution. The results of these tests are reported in Table 8.5. The table reports a " + " if the accident probability of the high group significantly exceeds the accident probability of the low group, a " − " if the accident probability of the high group is significantly below the accident probability of the low group, and a 0 if the likelihood ratio test fails to reject the hypothesis that the two rates are equal (using a 95% confidence interval). When the test rejects the equality hypothesis, the table also reports the ratio of the low groups's accident probability to the high group's accident probability. Note that these tests reflect only bivariate correlations, and say nothing about causal relationships.

   The results of these tests are mixed. There is weak evidence that trunks have higher accident probabilities than do other types of carriers. The estimated accident probability tends to be higher for trunks, although this

Table 8.5   Test of the Homogeneity of Accident Rates across Carrier Groups

Period	Characteristics on Which Carriers Are Grouped[a]			
	Domestic Trunks vs. All Other Carriers	Operating Margin	Maintenance Expense/Mile	Passenger-Miles
1954–1960	0	−(0.021/0.016[b])	0	0
1961–1965	+(0.011/0.015)	0	0	0
1966–1970	0	+(0.008/0.011)	0	0
1971–1975	+(0.006/0.009)	0	0	+(0.006/0.009)
1976–1980	0	0	−(0.005/0.002)	0
1981–1986	0	0	0	−(0.004/0.003)
Overall 1954–1986	+(0.007/0.010)	0	−(0.010/0.008)	+(0.008/0.009)

[a] "+" denotes statistical rejection of equality of accident rates, when the high group's accident rate is estimated as higher than the low group's accident rate. (See text for explanation of "high" and "low".) "−" denotes rejection of equality of accident rates, when the high group's accident rate is estimated as lower than the low group's accident rate. 0 denotes failure to reject the equality of accident rates.

[b] Numbers in parentheses for tests that reject equality are low group's accident rate/high group's accident rate. Trunks are considered high for the trunk/nontrunk test.

difference is statistically significant in only two of the six periods (1961–1965 and 1971–1975). Higher profit carriers are associated with lower accident rates in the 1954–1960 period, but this relation is reversed in the 1966–1970 period, and there is no statistically significant difference between the high and low profit groups over the rest of the sample. In only one period (1976–1980) is there a significant difference between the high and low maintenance expenditure carriers: Carriers that devoted a larger fraction of total expenses to maintenance had lower accident rates. Size (passenger-miles) does not appear to affect accident rates; there is a significant difference between accident rates for the high and low groups in 1971–1975 and 1981–1986, but the difference is of opposite signs in the two periods. It is unclear whether these null findings reflect the limitations of simple bivariate correlations or the presence of little real difference in accident rates across groups of carriers. The regression analysis, by estimating the effect of each characteristic holding all others constant, may shed light on this question.

Regression Analysis of Individual Carriers' Accident Rates

In this section I model the accident rates for individual carriers as a function of their traffic and financial characteristics. This permits an estimate of the correlation of a particular characteristic with accident rates, controlling for other factors that may affect accident levels. The model should be interpreted as a reduced form rather than as a structural model of the accident-generating process. The analysis is quite similar to studies by Graham and Bowes (1979) and by Golbe (1986), although the data and statistical techniques differ.

The basic specification of carrier i's accident rate in year t is

$$\ln(\text{TOTACC}_{it}/\text{DEPART}_{it}) = \beta_0 + \beta_1 \cdot \text{TIME}_{it} + \beta_2 \cdot \text{AVSTAGE}_{it}$$
$$+ \beta_3 \cdot \text{EXPER}_{it} + \beta_4 \cdot \text{OPMARG}_{it-1} + \beta_5 \cdot \text{DINTL}_{it} + \qquad (8.2)$$
$$\beta_6 \cdot \text{ALASKA}_{it} + \beta_7 \cdot \text{ZERODUM}_{it}$$

where ALASKA is a dummy variable equal to 1 for Alaskan carriers and to 0 otherwise and the other variables are as defined previously in this chapter and in Appendix A. A log-linear specification is used to ensure that the estimated accident rates satisfy the nonnegativity constraint.

The variables in Equation 8.2 have the following interpretations and predictions: First, the accident rate, (TOTACC/DEPART), can be interpreted as a measure of the underlying accident probability, λ. TIME allows for a logarithmic decline in accident rates over time (reflecting improvements in safety technology), where β_1 is expected to be negative. AVSTAGE measures the effect of longer flights (holding constant the number of flights and all other right-hand side variables) on accident rates. Its expected sign is positive. Cumulative flight experience (EXPER) is included to capture the notion that safety levels may rise with airline experience. Given this rationale, the expected sign of β_3 is negative. Equation 8.2 also was estimated using the log of experience (LNEXPER), as suggested by the literature on learning curves (see Joskow and Rose, 1985, and the references cites therein). OPMARG, the operating margin, is a measure of carrier profitability. If the argument that lower profits induce lower levels of safety is correct, β_4 should be negative. DINTL and ALASKA capture the effect of any higher risk associated with operations outside the United States or in Alaska. To the extent that these operations are more risky, both variables should have positive coefficients. ZERODUM is a correction factor for the treatment of zero accident years, and therefore is not reported in the tables.

Variations on this basic equation include measuring experience as the log of experience, LNEXPER (column 2 in Table 8.6); replacing TIME with a series of time-fixed effects (separate intercepts for each year) (columns 4 and 5); and including carrier-specific fixed effects (columns 3 and 5).

The estimation treats lagged profitability as an exogenous variable from the standpoint of the accident-generating process. Some empirical support for this was provided by Golbe (1986), whose data failed to reject the hypothesis of exogeneity for profitability measures. Using lagged values reduces potential simultaneity problems: Last period's profits should not be contaminated by costs that are incurred due to accidents this period (repair or replacement of aircraft, damage claims, higher insurance premiums, and the like). Lagged values may also be appropriate, since the impact of reduced profits on accidents is unlikely to be immediate. On a practical level, the exogeneity assumption is required by the dearth of reasonable instruments.

Table 8.6 Regression Analysis of Carrier Accident Rates, 1957–1986
(n of Observations = 663) [Dependent Variable: log(accidents/departures)]

Variable	Basic Model[a] (1)	Variations on Basic Model				Sample Mean (SD) (6)
		(2)	(3)	(4)	(5)	
Constant	4.010 (0.060[b])	−3.390 (0.112)	Fixed effects	−4.759 (0.174)	Fixed effects	1.00 (0.00)
TIME	−0.029 (0.003)	−0.018 (0.004)	−0.028 (0.005)	Fixed effects	Fixed effects	17.31 (8.51)
AVSTAGE	0.540 (0.125)	0.631 (0.129)	0.297 (0.276)	0.555 (0.121)	0.270 (0.266)	0.41 (0.30)
EXPER	−0.141 (0.027)	—[c]	−0.092 (0.034)	−0.151 (0.025)	−0.112 (0.032)	1.07 (1.67)
LNEXPER	—	−0.185 (0.031)	—	—	—	−1.27 (1.89)
OPMARG	−0.992 (0.292)	−0.418 (0.281)	−0.725 (0.314)	−1.154 (0.322)	−0.900 (0.319)	0.040 (0.079)
DINTL	−0.010 (0.060)	0.048 (0.065)	0.074 (0.111)	−0.018 (0.058)	0.019 (0.108)	0.481 (0.500)
ALASKA	0.758 (0.106)	0.532 (0.104)	—	0.752 (0.104)	—	0.104 (0.306)
Sum of squared residuals	211.26	212.59	188.76	202.99	180.93	

AVSTAGE, average stage length; EXPER, cumulative flight experience; OPMARG, operating margin; DINTL, exposure to international operations; ALASKA, dummy variable. See text and Appendix 8.1 for further explanation of variables; SD, standard deviation.

[a] See text for explanation of basic model and variations.

[b] White hetroskedastic–consistent standard errors in parentheses.

[c] Variables omitted from the regression.

Table 8.6 reports results for five variations of Equation 8.2 for the period 1957 through 1986 (over which complete data were available). Column 6 reports the sample mean and standard deviation for each variable. The variations reported in Table 8.6 are representative of a much broader range of specifications that have been estimated. Column 1 is the basic specification of Equation 8.2 Most variables have the expected signs, although some of the point estimates are not precisely identified. The primary coefficient of interest, OPMARG, is negative and statistically significant, implying that higher operating margins (profits) are associated with lower accident rates. A change in the operating margin by 10 percentage points reduces the accident rate by 9.92% with a standard error of 2.92. (From now on the standard error will be shown in parentheses after the point estimate.) For a carrier with the 1981–1986 sample aggregate accident rate (Table 8.4), a 10-percentage-point increase in its operating margin would reduce its expected accident rate from 0.003 to 0.0027.

Accident rates decline through TIME at the rate of 2.9% (0.3) per year. AVSTAGE has the expected positive effect on accidents: An increase in the average stage length from 500 to 1,000 miles would raise the expected accident rate by 2.7% (0.63). The coefficient on EXPER implies a pronounced learning effect: Increasing cumulative airline experience by 1 billion miles reduces the accident rate by 14.1% (2.7). The impact of experience may be more readily apparent from the LNEXPER specification in column 2, where the experience coefficient suggests that doubling cumulative experience reduces the accident rate by 18.5% (3.1). The experience coefficient may reflect in part nonlinear time effects (as experience trends strongly through time), but the persistence of the results across specifications replacing the time trend with individual year intercepts (see columns 4 and 5) suggests that experience is not a proxy for time effects. Finally, DINTL is estimated with an unexpected negative sign but is statistically indistinguishable from zero. ALASKA is large and positive, implying that Alaskan carriers have an accident rate double that of non-Alaskan carriers with the same characteristics.

The variations in columns 2 through 5 do not materially affect the qualitative conclusions. The coefficient on OPMARG is consistently negative, ranging from −0.418 (0.281) in the LNEXPER specification (column 2) to −1.154 (0.322) in the time-effects specification (column 4). Although its point estimate is fairly imprecise, we can reject the hypothesis that the coefficient is zero for all equations except the LNEXPER specification. The point estimates are about 1 standard deviation smaller in the specifications that allow for company-specific fixed effects (columns 3 and 5). These results provide weak evidence of a profitability-accident relationship.

The data in the full sample period seem to prefer the EXPER specification of experience over the LNEXPER specification, as evidenced by smaller sums of squared residuals for the former equations. Because of this, Table 8.6 reports primarily equations using EXPER. The coefficient on this variable falls by about one third when carrier-fixed effects are estimated, from −0.141 (0.027) to −0.092 (0.034). This may suggest that the experience measure in the regressions absorbs some effects of carrier size, which is correlated with experience in the cross section.

The coefficient on TIME is relatively constant at 2 to 3% in most specifications, with larger values estimated in equations that omit experience measures (not reported). The effect of AVSTAGE is fairly robust with regard to variations in time and experience measures (columns 1, 2, and 4), but both the size and precision of the coefficient decline substantially in the carrier-fixed-effect estimates (columns 3 and 5). DINTL varies between positive and negative but is never statistically significant and adds little to the regressions. The coefficient on ALASKA depends somewhat on the way experience is measured but is always large and significant. This effect is absorbed in the carrier-fixed effects in columns 3 and 5.

To explore the profit-accident relationship further, I estimated Equation 8.2 for various subsamples of firms and years. I first explored whether the results were sensitive to pooling different types of carriers, using divisions between trunk, local service, and territorial/Alaskan/Hawaiian (other) carriers. Full period regressions for these groups are reported in columns 1

Table 8.7 Sensivity Test of Basic Regression Results (Subsample Regressions) [Dependent Variable: log(accidents/departures)]

Variable	Trunk	Local	Other	1957–1965	1966–1977	1978–1986
N of obser.	336	176	151	232	266	165
Constant	−4.227	−4.439	−2.630	−3.972	−4.020	−4.926
	(0.101[a])	(0.083)	(0.132)	(0.117)	(0.190)	(0.344)
TIME	−0.035	−0.028	−0.038	−0.005	−0.028	−0.009
	(0.006)	(0.007)	(0.010)	(0.014)	(0.009)	(0.013)
AVSTAGE	0.879	2.888	0.006	0.594	0.555	0.579
	(0.148)	(0.564)	(0.198)	(0.204)	(0.221)	(0.176)
EXPER	−0.139	−1.081	2.592	−0.346	−0.205	−0.108
	(0.028)	(0.222)	(1.636)	(0.133)	(0.050)	(0.027)
OPMARG	−0.848	0.993	−0.646	−0.719	−0.966	−0.673
	(0.351)	(0.616)	(0.289)	(0.503)	(0.411)	(0.737)
DINTL	0.013	—	−0.037	−0.039	0.030	−0.051
	(0.089)		(0.076)	(0.091)	(0.096)	(0.096)
ALASKA	—	—	0.177	1.098	0.577	0.102
			(0.062)	(0.164)	(0.141)	(0.124)
Sum of squared residuals	97.49	21.77	19.71	61.97	70.18	35.71

AVSTAGE, average stage length; EXPER, cumulative flight experience; OPMARG, operating margin; DINTL, exposure to international operations; ALASKA, dummy variable. See text and Appendix 8.1 for further explanation of variables.

[a] White heteroskedastic-consistent standard errors in parentheses.

through 3 of Table 8.7. The results for trunks are similar to results for the whole sample (reflecting their dominance in the sample). The results are much less stable for local service and other airlines. OPMARG has a large negative and statistically significant effect for trunk and other carriers. It is, however, estimated with a large positive coefficient 0.993 (0.616) for the local service carriers. AVSTAGE is quite unstable, with coefficients ranging from 0.006 (0.198) for other carriers to 2.888 (0.564) for local service carriers, as is EXPER, with coefficients ranging from −0.139 (0.028) for trunk carriers to 2.592 (1.636) for other carriers. The carrier group results are not substantially affected by fixed-effects estimation, with the exception of the AVSTAGE coefficients.

I also examined the robustness of the relationship through time. The sample was divided into three times periods: 1957–1965, 1966–1977, and 1978–1986; the results are also reported in Table 8.7. The point estimates are broadly similar to the full-sample results in Table 8.6, although the coefficient estimates are much less precise for the individual periods. The coefficient on OPMARG, in particular, is estimated within the same range as in Table 8.6. This result is sensitive to specification, however, particularly for the 1966–1977 period. OPMARG is quite small and sometimes positive in specifications using LNEXPER or fixed effects. In general, when fixed-effects estimation is used (not reported), the standard errors on most coefficients are much larger (frequently larger than the estimated coefficient), and the point estimates are unstable.

These results provide mixed evidence on the stability of the accident-generating process estimated here. The sign of the relationship seems reasonably robust across variations on the basic specification and across a variety of subsamples of the data set. This provides suggestive evidence of a positive profits-safety relationship. However, there are enough aberrations to dictate further research into the robustness of the results.

CONCLUSION

The analysis provided in this study provides mixed implications for airline safety in a deregulated environment. On the one hand, the aggregate safety performance of the industry as measured by both fatal and nonfatal accidents is superb and shows no sign of deterioration. If there has been any change in accident performance relative to trends under regulation, it is a *reduction* in accident levels, not an increase. For three of the eight years under deregulation, for example, there have been no passenger fatalities, compared with no such years prior to deregulation. This provides strong evidence for the stability and soundness of the air safety system.

On the other hand, this study finds evidence that financial condition may be correlated with accident rates at the level of individual carriers. In the presence of controls for cumulative airline flight experience and other operating characteristics, higher operating margins appear to be correlated with lower accident rates. This finding, if it proves to be robust, may imply that more intense scrutiny of the safety practices and performance of financially marginal carriers is desirable. Note that the results do not, however, imply that deregulation is undesirable, even on safety grounds. In particular, deregulation appears to have increased overall industry profits (Morrison and Winston, 1986; this tendency is likely to be strengthened by the recent merger wave in the industry), leading to improved safety if profit-safety linkages exist.

APPENDIX 8–1: DATA DESCRIPTION AND SOURCES

Accident Data

An accident is defined by the National Transportation Safety Board (NTSB) as "an occurrence associated with the operation of an aircraft which takes place between the time any person boards the aircraft with the intention of flight until such time as all persons have disembarked, in which any person suffers death or serious injury as a result of being in or upon the aircraft or by direct contact with the aircraft or anything attached thereto, or in which the aircraft receives substantial damage."

Individual air carrier accident data are from the U. S. Civil Aeronautics Board (CAB), *Resume of Accidents, U.S. Air Carriers, Rotorcraft and Large General Aviation Aircraft* (annual, 1953–1959); U. S. CAB, *Statistical Review and Briefs of U. S. Air Carrier Accidents* (annual, 1960–1965); U. S. NTSB, *Annual Review of Aircraft Accident Data, U. S. Air Carrier Operations* (succeeds the CAB accident publications; 1966–1982); U. S. NTSB, *Preliminary Analysis of Aircraft Accident Data, U. S. Civil Aviation* (1979–1982), and U. S. NTSB Accident Briefs (unpublished computer printout, for 1983–1986). These data were collected from 1954 through 1986.

Traffic Data

Annual airline system revenue departures (DEPART) in thousands and aircraft miles completed (MILES) in millions for scheduled passenger service are from U. S. CAB, *Air Carrier Traffic Statistics* (various issues, 1954–1983) and U. S. Department of Transportation (DOT) *Air Carrier Traffic Statistics* (continues the CAB publication, various issues, 1983–1986). Average stage length (AVSTAGE) is computed as miles per departure and is measured in thousands of miles. These data were collected for 1954 through 1986.

Financial Data

Annual airline system operating revenues and operating expenses are from the U. S. CAB, *Air Carrier Financial Statistics* (various issues, 1954–1983) and U. S. DOT, *Air Carrier Financial Statistics* (continues CAB publication, various issues, 1983–1985). The system operating margin, OPMARG, is calculated as

$$1 - \left(\frac{\text{operating expenses}}{\text{operating revenues}}\right)$$

Experience

Airline experience (EXPER) in year t is calculated as the cumulative miles flown in billions from 1954 through year $t-1$.

APPENDIX 8–2: A PROBABILITY MODEL FOR ACCIDENTS

A natural stochastic specification for the number of air carrier accidents is based on the Poisson probability distribution. The Poisson distribution recognizes the infrequent and discrete natures of accidents, and has been applied extensively as a model of accident probabilities in a wide variety of contexts, including air carrier accidents (Barnett et al., 1979; Barnett and Higgins, 1987; Golbe, 1986). For our purposes it is most plausible to specify the number of accidents in a year as a function of an (unknown) accident rate per thousand departures and the number of departures (in thousands), rather than to specify an accident rate per year. (This is equivalent to assuming that each flight has some probability, p, of being involved in an accident, and that the number of flights is large, which takes advantage of the binomial distribution's convergence to the Poisson in the limit.)

Using the Poisson distribution, and denoting firm i's expected number of accidents as λ_i, its number of departures (in thousands) in a year as D_i, and its number of accidents in a year as n_i, we can express the probability that firm i experiences n_i accidents during the year as

$$\Pr(n = n_i) = \frac{[\exp(-\lambda_i D_i)](\lambda_i D_i)^{n_i}}{n_i!} \tag{8.3}$$

The maximum likelihood estimator (MLE) for the accident rate λ_i is $\lambda_i = n_i/D_i$.

The analysis of accidents across carrier groups relies on this simple parameterization of carrier's accident probabilities to investigate whether different types of air carriers have different underlying accident probabilities. As an illustration of this technique, consider a test of the equality of accident rates for domestic trunk carriers and all other carriers. Let N_1 be the total number of accidents and D_1 the total number of departures for group 1 (domestic trunk carriers), and let N_2 and D_2 be the corresponding totals for group 2 (all other carriers). We will impose the assumption of homogeneity within each group ($\lambda_i = \lambda_1$ for all i belonging to group 1; $\lambda_j = \lambda_2$ for all j belonging to group 2), and test for homogeneity across groups: $\lambda_1 = \lambda_2 = \lambda_0$. The MLEs for λ_1 and λ_2 are $\lambda_1 = N_1/D_1$ and $\lambda_2 = N/D_2$. The MLE for λ_0 is $\lambda_0 = (N_1 + N_2)/(D_1 + D_2)$. To test for homogeneity, we construct the ratio of the likelihood under the null hypothesis that groups 1 and 2 are drawn from the same probability distribution to the likelihood under the alternative hypothesis that $\lambda_1 \neq \lambda_2$. This likelihood ratio is

$$LR = \frac{\lambda_0^{(N_1 + N_2)} \exp[-\lambda_0(D_1 + D_2)]}{\lambda_1^{N_1} \cdot \lambda_2^{N_2} \exp(-\lambda_1 D_1 - \lambda_2 D_2)} \tag{8.4}$$

Substituting in the maximum likelihood estimates of λ_0, λ_1, and λ_2 and taking the log of LR yields a log-likelihood of

$$\text{Log } LR = (N_1 + N_2)\ln\left(\frac{N_1 + N_2}{D_1 + D_2}\right) - N_1\ln\left(\frac{N_1}{D_1}\right) - N_2\ln\left(\frac{N_2}{D_2}\right) \tag{8.5}$$

The test statistic $-2\log LR$ is distributed as a chi-square random variable.

In the regression analysis of individual carriers' accident rates we relax the assumption that all carriers within a particular group have identical accident rates. We parameterize λ_i as a function of a carrier's operating and financial characteristics. Denote these characteristics as the vector X_i.

We parameterize the accident rate per thousand flights as $\lambda_i = \exp(X_i\beta)$. This parameterization ensures that the estimated accident rates satisfy the nonnegativity restriction on λ (i.e., the expected number of accidents must be greater than or equal to zero). This parameterization of the accident rate is estimated by ordinary least squares (OLS) regression of the form $\ln(N_i/D_i) = X_i\beta$.

This is a simple linear equation in the log of the observed accident rate per thousand flights. The elements of the estimated coefficient vector β_{ols} have the interpretation that a one-unit change in the corresponding variable in X will lead to a $\beta_{ols} \cdot 100\%$ change in the accident probability. Note, however, that if accidents are Poisson distributions, OLS will no longer be an efficient estimator. In particular, the mean of a Poisson distribution (λ_i) is equal to its variance, which implies heteroskedasticity. The standard errors calculated under the OLS assumptions will be inconsistent estimates of the true standard errors. The standard errors reported in this chapter use White's (1980) estimator for heteroskedastic-consistent standard errors.

9
New Entrants and Safety

ADIB KANAFANI and THEODORE E. KEELER

Among those who were skeptical of the appropriateness of the Airline Deregulation Act of 1978, safety was an important concern. A prime concern was that if new firms were allowed to enter the interstate market, they might well be less safe than existing well-established firms. This concern persisted despite the fact that even under deregulation the Federal Aviation Administration (FAA) remained charged with ensuring that new entrants met the same strict standards of safety as established carriers. Skepticism was based on the claim that the Civil Aeronautics Board's (CAB) restrictions on entry allowed one additional check on the airworthiness of new carriers—since CAB regulations ensured that all carriers serving domestic markets had a reasonable chance of a good financial return, firms had no reason to skimp on maintenance, pilot training, and other variables that might affect a carrier's safety level.

Those who favored deregulation, on the other hand, noted that with continued vigilance toward all carriers by FAA, air safety under deregulation should not be a problem. As a double check, the Department of Transportation (DOT) would continue to monitor entrance with a "fitness to serve" criterion, based on standards of financial and organizational capability to provide service.

The nation has had over eight years of experience under deregulation, and it is now possible to begin evaluating whether potential fears regarding the safety performance of new entrants have been realized. We begin our investigation of this subject by discussing earlier theory and evidence relating to new entrants—in other words, whether a priori reasoning would suggest that new entrants should have better or worse safety records than existing carriers.

We then undertake empirical investigation of the various dimensions of the safety records of the carriers entering interstate jet service markets since 1978. Large structural econometric models would be difficult to specify with certainty, and extensive data would be needed to estimate them. Therefore, the main focus of this chapter is more modest. We will analyze, in a basic way, data about the relative safety records of the new entrants and of established airlines.

Because large, fatal airline crashes are relatively rare events, the amount of evidence concerning them is rather limited. Therefore, for much of our evidence we shall rely on indirect but, we believe, still relevant and reliable data on safety. Airline accidents can be traced to two different causes: (1) failure to maintain and service equipment properly and (2) traffic and piloting problems. Given the statistical rarity of airline crashes, one way to measure a firm's basic degree of safety is through evidence on accident proneness from these two potential causes of accidents. We thus present statistical analyses of differences in maintenance expenditures between new entrants and established carriers; of data on near midair collisions, as reported by the FAA; and of data on actual observed accidents.

EXPECTATIONS ON THE SAFETY OF NEW-ENTRANT CARRIERS

A priori arguments can be made both ways regarding the safety of new-entrant airlines relative to their established rivals. One argument is that the relative inexperience of new carriers could make them less safe. New entrants could be controlled by less experienced managers, their aircraft could be flown by pilots less familiar with the airport environs they serve, and their maintenance personnel could have less experience and fewer resources available to them. Another argument might be that new entrants have weaker financial status than established carriers and thus have an incentive to spend less on maintenance and on quality control.

On the other hand, one can argue that new entrants have a stronger incentive to invest in preventing accidents than established carriers. This is what we would call the "reputation effect." An established carrier can more readily survive the effects of a serious crash than can a new entrant. Thus, when American Airlines had a crash in 1979, the fact that it was fined for failure to maintain its DC-10 aircraft properly did not put it out of business or even seriously divert traffic from it. On the other hand, Air Florida's crash in 1982 probably played a critical role in its financial downfall. The same was true of California Central Airlines, a 1950s intrastate airline entrant—shortly after a crash it was driven to bankruptcy by lack of patronage. The reputation-effect argument suggests that an established carrier with a reputation can withstand the effects of a crash more than a new entrant.

Such arguments remain by and large conjectural. Little empirical evidence has been brought to bear on the question of the safety of new entrants after airline deregulation. The only work we are aware of that indirectly touches on the issue relates to the relationship between financial performance and safety. Golbe (1986) documented clearly that, at least over the years immediately prior to deregulation, the financial strength of carriers did not have an effect on their propensity for accidents. However, more recent work by Advanced Technology (1986) found a clear statistical correlation between the financial status of a carrier and its safety level as measured by performance on the FAA's National Air Transportation Inspection (NATI) Program. It should be noted that new entrants were not specifically identified as the carriers beset by such a profitability-inspection rating problem. In fact, as we shall show in this chapter, new entrants did not fare differently from established carriers in the NATI inspections.

MAINTENANCE EXPENDITURES

For our analysis of maintenance expenditures we selected 22 airline firms (Table 9.1). The sample includes practically all the large major carriers for which data were available over the period of study. When mergers occurred, we aggregated the figures for unmerged firms in earlier years. In

Table 9.1 List of Airlines in Datasets for Maintenance Expenditures, Inspections, and Near Midair Collisions

Airline	New Entrant (= 1)	% Satisfactory Inspection	Crew Cost/ Plane Hour ($)	Maintenance Expenditure/ Plane Hour ($)
Air Atlanta	1	93.9	197.43	330.88
America West	1	90.2	98.45	96.68
American	0	94.4	541.20	379.32
Braniff	1	95.7	157.84	320.79
Continental	1	87.8	256.31	220.65
Delta	0	93.3	573.28	295.52
Eastern	0	93.8	487.22	354.11
Florida Express	1	96.4	94.36	177.58
Frontier	0	99.1	305.85	299.73
Midway	1	92.3	140.03	279.59
New York	1	93.2	104.62	334.23
Northwest	0	95.8	485.44	399.35
Pan American	0	82.6	522.03	350.04
People Express	1	87.6	116.48	356.37
Piedmont	0	95.3	363.66	252.09
Pacific Southwest	0	94.0	412.31	286.16
Republic	0	94.9	334.44	254.97
Southwest	1	92.9	197.63	187.19
Trans World	0	87.3	581.58	493.45
United	0	91.1	592.64	418.69
USAir	0	96.1	476.80	310.58
Western	0	85.8	471.09	331.03

SOURCES: Total flight hours 1983–1985: Civil Aeronautics Board (various years); average crew cost for 1983–1984: U.S. Department of Transportation (various years); % Satisfactory inspections, 1984: Federal Aviation Administration; maintenance expenditure 1983–1984: U.S. Department of Transportation (various years).

addition, we included in our data a large sampling of the new-entrant carriers using jet aircraft. Criteria for inclusion were mainly related to availability of data for a significant number of years. Admittedly, the sample we used may be somewhat limited, but since it includes practically all important carriers, both established and newly entered, we believe that our results are relevant to a broad range of circumstances.

The maintenance data we used for established carriers extended from 1971 to 1985. That time span allows for a large data set but is recent enough that jet aircraft technology can be expected to have changed relatively little. For new-entrant carriers we incorporated data starting with their first year of operation, ranging from 1979 for Midway Airlines to 1981 for America West Airlines.

Two carriers, Braniff Airways and Continental Air Lines, deserve special attention here. Although they are both old, well-established carriers, they went bankrupt after deregulation and were reorganized with the low wages and streamlined operating procedures that characterize new entrants. We thus reclassified Braniff and Continental as new entrants after their reorganizations. However, we assigned them a special dummy variable post-reorganization. The extent that this dummy variable differs from zero in the estimated statistical equations indicates the difference of Braniff and Continental from other new entrants.

Two other carriers in our sample also posed problems for classification, Pacific Southwest Airlines and Southwest Airlines. Both were intrastate carriers in unregulated environments before deregulation and "new" entrants to interstate markets after the Deregulation Act. We classified Pacific Southwest as an established carrier, because it has operated since 1948 and its wage structure and level of unionization have risen to levels more like those of established carriers than of new entrants. We classified Southwest as a new entrant, because it was founded only in 1970 and has maintained wages and operating procedures consistent with those of the newest entrants.

Finally, in our estimation we desired to separate potential effects stemming from bigness on the one hand and from being a new firm on the other. We thus included a variable for size, in this case total airline flight hours. Use of revenue passenger-miles as an alternative size variable did not change the results.

The Analysis

The equations estimated for maintenance expenditures took the following form:

$$\text{Maintex}_{it} = a_1 + a_2 D_{1972} + \ldots + a_{15} D_{1985} + b \ \text{NEWENT} + c \ \text{BCDUM} + e \ \text{FLTHRS} \tag{9.1}$$

The constant a_1 measures the initial value in 1971. The variables D_{1972} through D_{1985} are dummy variables, varying between 1 during the respective years in their labels, and 0 during other years. The coefficients of these dummy variables represent the change in maintenance for each year beyond 1971. The variable NEWENT takes the value 1 if the relevant carrier is a new entrant and is 0 otherwise. The variable BCDUM takes the value 1 if the carrier is Braniff or Continental after each carrier was reorganized following bankruptcy (1983 onward for Continental and 1984 onward for Braniff) and is 0 otherwise. Both carriers in those years also have values of 1 in the NEWENT variable, as explained previously. Finally, the variable FLTHRS, total fight hours per year for the carrier, is a measure of the scale of the carrier, to see if bigness per se has an impact on maintenance expenditures. All data for the maintenance equations came from U. S. DOT (formerly CAB) Form 41 data.

The left side of the equation is the various measures of maintenance expenditures for firm i at time t. These measures require some care in calculation. Clearly, with costs varying over the time period of the sample, undeflated maintenance expenditures would be inappropriate. Indeed, because new-entrant carriers have much lower factor prices, especially wages, than established carriers, direct comparison of maintenance expenditures in any one year may also have major problems. Unfortunately, no available cost indices can accurately correct these maintenance expenditure data over time and across airlines.

In our statistical analysis we used three alternative measures of maintenance expenditures: (1) total expenditures on maintenance of all sorts, (2) (more closely connected to issues of safety) maintenance cost connected with flight equipment, and (3) (more closely connected with the aircraft themselves) direct maintenance of aircraft. All are measured as a percent of total cost, to normalize these expenditures across airlines.

Results

The results of our analysis, done by least squares regression, are shown in Table 9.2. They are striking on several counts. First, they show that new-entrant carriers do not, in fact, spend a significantly lower fraction of their total costs on maintenance than their established peers. Indeed, as regards maintenance of flight equipment specifically, they actually spend a higher portion of their total expenditures on maintenance. The reorganized carriers, Braniff and Continental, do seem to spend less on maintenance than other new entrants, but the difference is quite insignificant, so no conclusions can be drawn here.

There may be good reasons for the new-entrant result. New entrants would typically operate older aircraft, with less efficient maintenance programs, than established carriers. They may contract out their maintenance disproportionately to higher cost established carriers, which would mean that relative to their other costs, maintenance would be higher than for established carriers. Also, if established carriers have newer aircraft, they will to some degree have the benefit of warranty programs not enjoyed by the new entrants.

Table 9.2 Estimation Results for Maintenance Equations

Independent Variable	Dependent Variables (SE)			
	PCMAINT	PCFLT	PCDIR	PCDEP
Constant	0.154184	0.097748	0.101177	0.102972
	(0.005820)	(0.004640)	(0.004646)	0.007156
New-entrant dummy	0.003933	0.017975	0.014521	−0.009664
	(0.004923)	(0.003925)	(0.003930)	(0.006053)
Braniff-Continental dummy	−0.008143	0.000709	0.000207	0.011302
	(0.009223)	(0.007657)	(0.007363)	(0.011340)
Flight hrs.	0.002194	0.031390	0.023200	0.000918
(millions)	(0.005995)	(0.004780)	(0.004786)	(0.007371)
R squared	0.064035	0.556741	0.553170	0.298520
Degrees of freedom	208	208	208	208

SE, standard error; PCMAINT, total maintenance expenditures; PCFLT, expenditures for maintenance of flight equipment; PCDIR, expenditures for direct maintenance of aircraft; PCDEP, depreciation costs; all of above as percent of total costs; year specific dummy variable results are not shown.

We can shed some light on the older aircraft effect without detailed data on aircraft age. We do so by analyzing airlines' depreciation expenditure as a fraction of their total costs (PCDEP, Table 9.2). The results indicate that new entrants have lower depreciation expenses than established carriers, suggesting that the new entrants have older aircraft. But this result is doubtful because it is of marginal statistical significance.

In any case, while these factors could possibly explain the fact that new entrants spend more of their costs on maintenance, the important point is that there is no evidence that the new entrants observed spend less on maintenance than the established carriers. As regards small versus large carriers, the results present no clear associations. Large firms spend more of their resources on maintenance of flight equipment, but less on direct maintenance. There is thus some evidence for scale economics in maintenance.

The results for the year specific dummy variables, which are not shown in Table 9.2, indicate something of a secular downward trend in expenditures on maintenance as a fraction of total expenditures over the period since 1971 and for all carriers in the sample. This too is a weak trend, however, without much statistical significance.

In sum, then, the evidence presented in this section indicates that new-entrant carriers have tended to spend a significantly higher fraction of their costs on aircraft maintenance than have established carriers. It is possible that some of this difference is due to older aircraft on the part of new entrants, but the statistical support for this interpretation is not strong. It is also possible that the higher fraction of total cost devoted to aircraft maintenance by new entrants comes about because the other components of operating cost are lower for them. There is some evidence that this may be true of labor costs and of indirect operating costs.

FAA SAFETY INSPECTION RESULTS

In an effort to improve aviation safety, the federal government instituted the NATI program to increase FAA safety inspection and surveillance of all certificated carriers. This one-year program was conducted in 1984 and involved 13,467 inspections of over 300 air carriers. The inspection results provide another important indicator of the safety level of a carrier. Safety can be measured in a number of ways, for example by the percent of inspections deemed satisfactory, or by the severity ratings used by the inspectors. In the NATI program, inspectors judged each inspection as either satisfactory if no violation of safety rules and procedures was found, or unsatisfatory if there was such a violation. For the unsatisfactory inspections, a severity rating was assigned depending on a set of criteria that had to do with the extent of the violation and its relation to the integrity of the aircraft. For simplicity we will look at the percent satisfactory results as an indicator of the safety level and then look for differences in inspection success between the established and the new-entrant carriers.

Table 9.3 shows the results of regression between the percent satisfactory inspections and the classification of the carrier as a new or incumbent carrier. Two separate equations were used. The new-entrant dummy was used in the first, and expenditures on maintenance as a fraction of the total cost was used in the second. In addition, both equations contained variables to reflect overall carrier size (total flight-hours) and dummies to separate Braniff and Continental from more normal new entrants. The statistics clearly speak for themselves. There is no evidence that new entrants have a different safety record than established carriers. The evidence also

Table 9.3   Results of Regression on Percent Satisfactory Inspections

Independent Variable[a]	Coefficient (SE)
Constant	83.8219
	(4.6036)
New-entrant dummy	3.2072
	(2.3083)
Braniff-Continental dummy	−1.4753
	(3.2130)
% Expenses on maintenance	0.6386
	(0.4822)
Flight-hours	5.7820
(millions)	(4.1430)
R squared	0.2308
Degrees of freedom	17

SE, standard error.

[a] See text for explanation of variables.

indicates that Braniff and Continental do not differ significantly from more normal new entrants.

NEAR MIDAIR COLLISIONS

Given the rarity of airline accidents, near midair collisions (NMACs) can be used as a measure of the accident proneness of an airline. It might be argued that the flight crews of new-entrant carriers are less experienced or not as well trained as the crews of more established carriers. If after correcting for the effects of other important factors the incidence of near collisions is higher for the new entrants than for the established carriers, there would be grounds for concluding that the new entrants are in fact more prone to traffic problems.

The FAA keeps a record of all near midair collisions reported by the pilots of the aircraft involved, but, as noted below, there are weaknesses in the data. First, such data are not available on a consistent basis before 1983. This reduces the sample, and in that sense is a drawback. However, it is important to not go too far back in time with this variable, because of inherently different levels of air congestion at different times in the past and dramatic changes in the air traffic control system after President Reagan's firing of striking air traffic controllers in 1981. Therefore, restriction of the sample to the more recent period of 1983 through 1985 has some advantage in making the estimates of proneness to midair collisions more accurate and more relevant to the present time period. Furthermore, since we are interested at this point only in inter-airline comparisons and specifically in whether new entrants have a different collision risk than established carriers, a cross-sectional data set may be more appropriate.

A second problem with the near midair collision data is that it is difficult to control for all the factors other than new-entrant status that affects NMACs. For example, we are not able at this point to correct for differences in airport locations frequented by the different groups of carriers, nor in the complexity of airspace they fly in. We do not know whether one group of airlines or another flies in airspace more actively used by general aviation aircraft.

Finally, and most important, we do not know about the accuracy of NMAC reporting, which is to a large degree voluntary. We suspect that the data on NMACs is biased in favor of reporting when the report is made by the nonviolating party. We also suspect that the recent increase in NMACs is due to increased vigilance in reporting and in keeping statistics. A connected problem is that in some cases the offending aircraft is not identified. However, the number of near midair collisions in which the aircraft were large jets and the carriers were unknown is quite small.

Our statistical analysis of near midair collisions is based on a sample of 50 air carriers over the 1983–1985 period for which we have NMAC data. To control for the amount of exposure to collisions among airlines,

Table 9.4 Results of Regressions on Square Root of Near Midair Collisions Divided by Square Root of Flight Hours

Independent Variable[a]	Coefficient (SE)	With Crew Costs (SE)
Constant	0.004030 (0.000603)	0.004347 (0.000665)
Flight hours (million)	−0.001979 (0.001072)	−0.008911 (0.005169)
Braniff-Continental dummy	0.001598 (0.000940)	0.002507 (0.001142)
New entrant dummy	−0.002061 (0.000714)	−0.002943 (0.000834)
Crew cost/million hours		1.257000 (0.870800)
D_{1984}	0.000980 (0.000585)	0.001195 (0.000572)
D_{1985}	0.001155 (0.000586)	
R squared	0.226012	0.404786
Degrees of freedom	44	27

SE, standard errors.

[a]See text for explanation of variables.

we used flight-hours as an explanatory variable. As in the case of accidents, reported below, near midair collisions can be thought of as random events, and thus, as Golbe (1986) has argued, can be expected to have a Poisson distribution. As a result, to correct for the problem of heteroskedasticity, we made the dependent variable the square root of near midair collisions. The independent variables in the equation were flight hours and the new-entrant dummy variable multiplied by flight-hours. The argument here is that it is not total near midair collisions per firm per year, but rather total near midair collisions per flight hour that might be expected a priori to vary between new entrants and established firms. As in the maintenance equations, dummy variables are included to allow for system changes over time.

The equations and results are shown in Table 9.4. If anything, new entrants seem have a lower propensity for NMACs than incumbent carriers, although the result is not statistically significant. There is also an indication of a secular increase over time in NMACs, although it is not clear that the trend is a significant one.

Another possible hypothesis relating to near midair collisions is that lower crew wages will result in less experienced, less skilled crews and create more potential for traffic-related accidents. For the years 1983 and 1984, we were able to obtain data on crew wages (pilots, co-pilots, but not flight attendants) per plane hour. This resulted in 33 observations for analysis, with crew wage costs included in the equation. The data were available from the DOT's *Operating Cost and Performance Report*.

With crew wages added as a variable, the new-entrant variable was significant. Results indicate that new entrants had fewer near midair collisions than established carriers. Also, the model indicates that carriers with higher crew costs encounter a higher incidence of near midair collisions than those with lower wages. These results are curious. One could look for explanations such as that higher wage crews tend to fly in the larger, more congested hubs, but such explanations are fairly weak. In any event these results lead us to conclude that generally new entrants are not more prone to near midair collisions than more established carriers. Neither new entrants nor low-cost, low-wage carriers appear to have risks different from those of other carriers. Braniff and Continental do appear to be involved in more near midair collisions than typical new entrants, but the difference is not significant.

ACCIDENTS

Naturally, the most direct evidence on air safety relates to accidents themselves. These are defined as occurrences involving injury to people or damage to property. Although major fatal airline crashes are rare occurrences, accidents as just defined above do occur with a considerably higher frequency than is generally perceived. Evidence from the FAA suggests that the number of U. S. air carrier accidents has averaged about 30 annually during the last decade. Most of these accidents are relatively minor, but some do involve injury or significant damage to property. Typical of these accidents are landing with gears unextended, injury during unexpected severe turbulence, or minor collisions between aircraft and ground equipment at an airport. While these events are not as serious as large, fatal crashes, which are also numbered in these statistics, they nonetheless give a fair indication of a carrier's safety performance.

Accident data are compiled in the United States by the National Transportation Safety Board. The specific data we have worked with include both accidents and "incidents," the latter being of milder significance than accidents. For example, a collision between an aircraft and a baggage cart is reported as an incident, but we believe it is still reflective of the safety performance of a carrier. Our data on accidents and incidents are as reported in the aviation journal *Flight International* for each respective year. For our analysis of accidents we used data for all carriers operating jets in U. S. domestic service from 1982 through 1985. This comprehensive sample, listed in Table 9.5, provided 151 observations over the period. It included all new-entrant carriers using jet aircraft, and thus analysis of a larger number of carriers than in our maintenance expenditures analysis was possible.

Before we discuss statistical evidence on accidents, it is worth considering some figures relating to accidents for different carrier types. To compare accidents between carriers, it is appropriate to divide by departures, since 80% of accidents occur during takeoffs and landings (Golbe, 1986).

Table 9.5 Carriers in Accident Sample

Carrier	Years in Sample	Accidents	Departures	Accidents/ Million Departures
Established carriers				
Air California	1982–1985	1	267,736	3.7
Alaska	1982–1985	0	203,700	0
Aloha	1982–1985	0	148,229	0
American	1982–1985	20	1,458,925	13.7
Delta	1982–1985	12	2,039,556	5.9
Eastern	1982–1985	20	1,985,355	10.1
Frontier	1982–1985	4	549,676	7.3
Hawaiian	1982–1985	0	176,364	0
Northwest	1982–1985	6	610,440	9.8
Ozark	1982–1985	2	437,898	4.6
Pacific Southwest	1982–1985	0	428,339	0
Pan American	1982–1985	5	259,878	19.2
Piedmont	1982–1985	7	1,007,687	6.9
Republic	1982–1985	11	1,684,702	6.5
Texas International	1982	0	73,919	0
Trans World	1982–1985	9	752,757	12.0
United	1982–1985	25	1,873,708	13.3
USAir	1982–1985	7	1,297,482	5.4
Western	1982–1985	2	608,963	3.3
Wien	1982–1985	0	192,543	0
New Entrants				
Air Atlanta	1984–1985	0	13,094	0
Air Florida	1982–1984	1	74,951	13.3
Air One	1984	0	6,005	0
America West	1983–1985	1	124,405	8.0
American International	1982–1984	0	14,552	0
Braniff	1982–1985	1	109,270	9.2
Capitol	1982–1984	1	14,642	68.3
Continental	1982–1985	9	587,645	15.3
Emerald	1982–1984	0	22,149	0
Empire	1982–1985	0	179,730	0
Florida Express	1984–1985	0	27,429	0
Frontier Horizon	1984–1985	0	7,906	0
Hawaii Express	1982–1983	0	1,237	0
Jet America	1982–1985	0	20,581	0
Jet Charter	1985	0	15	0
Midway	1982–1985	1	137,063	7.3
Midwest Express	1984–1985	1	6,734	148.5
Muse	1982–1985	1	113,964	8.8
New York Air	1982–1985	1	165,274	6.1
Northeastern	1982–1985	1	27,176	36.8
Pacific Express	1983–1985	0	40,130	0
People Express	1982–1985	4	371,000	10.8
Southwest	1982–1985	2	724,154	2.8

Table 9.5   Carriers in Accident Sample (cont.)

Carrier	Years in Sample	Accidents	Departures	Accidents/ Million Departures
Sunworld	1984–1985	0	19,232	0
World	1982–1985	1	28,568	35.0
Totals				
All firms		156	18,909,663	8.2
New Entrants		25	2,851,806	8.8
Established carriers		131	16,057,857	8.2

SOURCES: Accidents: National Transportation Safety Board (various years); departures: U.S. Department of Transportation (various years).

Evidence on accidents per million departures indicates that it is difficult to distinguish between new entrants and incumbent carriers on this count. The evidence shows an average of 8.15 accidents per million departures for established carriers versus a figure of 8.76 for new entrants. On this basis, the new entrants in our sample had a record almost equal to that of the established carriers over the 1982–1985 period. However, a more complete statistical analysis of this issue needs to take account of the fact that established carriers typically fly longer stage lengths than new entrants, which could easily result in an inflated accident rate per departure for established carriers relative to new entrants. A more complete analysis also needs to account for whether the difference between new entrants and established carriers is statistically significant.

The Analysis

We estimated the following equation, based on our comprehensive panel of firms operating jets over the period 1982 through 1985:

$$\mathrm{SQRT}\!\left(\frac{\mathrm{ACC}}{\mathrm{DEP}}\right) = a_1 + a_3\,\frac{\mathrm{FLTHRS}}{\mathrm{DEP}} + a_4\,\mathrm{FLTHRS} + a_5\,\mathrm{NEWENT}$$
$$+ a_6\,\mathrm{BCDUM} + a_7\,\mathrm{D}_{1983} + a_8\,\mathrm{D}_{1984} + a_9\,\mathrm{D}_{1985} \qquad (9.2)$$

where SQRT(ACC/DEP) = square root of total accidents divided by DEP (millions of departures)
FLTHRS = hundreds of thousands of flight hours
NEWENT = dummy variable for new entrants (described earlier)
BDDUM = dummy variable for Braniff and Continental (described earlier)

As with NMACs, the accident variable is transformed to a square root to compensate for heteroskedasticity, which is likely to occur. Accidents are divided by departures because from carrier to carrier, accidents are likely to be randomly distributed only when correction is made for the fact that some carriers have more departures than others. Flight-hours per departure are included as a variable to allow for the fact that longer flights can increase the probability of accident per departure, and flight-hours alone is included to allow for the possibility of scale effects in accidents, that is, bigness alone could make an airline more or less prone to accidents. As in the other equations, year dummies are included to allow for effects of changes in operating environments and technology.

Results

Results are shown in Table 9.6. Although the accident rate appears to be higher for new entrants than for established carriers, the difference is not statistically significant. It is difficult, if not impossible, to infer that the new entrants in our sample have a safety record that is inferior to that of the incumbents.

The analysis of the accident data indicates some other interesting results regarding carrier safety. First, it appears that, all other things equal, larger carriers are more accident prone than smaller ones, a rather surprising and counterintuitive result that would be worth more research. Second, it appears that Continental Air Lines, but not Braniff, tended to be somewhat more accident prone than the typical new-entrant airline, although the difference did not prove to be statistically significant.

Table 9.6 Results of Equations on Square Root of Accidents Divided by Departures

Independent Variable	Coefficient (SE)
Constant	0.000280 (0.000626)
Braniff-Continental dummy	0.001579 (0.001289)
Flight-hours/departure	0.000165 (0.000249)
Flight-hours (millions)	0.003442 (0.001222)
New-entrant dummy	0.000600 (0.000573)
D_{1983}	0.000916 (0.000644)
D_{1984}	−0.000138 (0.000628)
D_{1985}	0.000159 (0.000649)
R squared	0.095857
Degrees of freedom	143

SE, standard error.

CONCLUSIONS

Our evidence consistently suggests that there is no difference in safety performance between the established carriers and new entrants who joined the market after airline deregulation. Neither aircraft safety nor traffic safety appear to be any different for these two groups of carriers. New entrants, if anything, appear to be spending more of their resources on maintenance than the large established carriers.

The evidence we have brought to bear on this question can be considered preliminary. A number of issues remain that should be looked at when additional data permit. A thorough analysis of fleet age characteristics, as well as an in-depth analysis of the recent changes in maintenance productivity, would be needed to complement our work on maintenance expenditures. More detailed analysis of the FAA's NATI inspection program results might also shed additional light on the question of relative safety levels. Network effects, and the geographic distribution of the incidence of near midair collisions, would have to be corrected for to ascertain whether any air carrier group, new entrant or otherwise, is at a higher risk of these events because of the specific structure of the network it serves. Finally, analysis of service difficulty statistics, which reports such data as engine failures in flight, may be useful, as this is a wealthy source of information.

Preliminary as it may be, our study seems to point to one major conclusion: If the safety of air transportation is any different after deregulation than it was before, the cause is not due to new entrants. It must be found elsewhere.

10
Is It Still Safe to Fly?

CLINTON V. OSTER, JR. and C. KURT ZORN

The structure and operations of the domestic airline industry have changed markedly since passage of the Airline Deregulation Act of 1978. Deregulation liberalized entry and exit restrictions, making it easier for jet carriers to abandon low-density routes and redeploy equipment to more profitable higher density routes, and for both established and newly formed carriers to enter new markets. These liberalizations have ushered in a period of intense service and fare competition.

Deregulation's impact on safety was a concern even in the prederegulation congressional debates. Recently, however, concern has grown, in part perhaps because 1985 was widely hailed as the worst year in history for world wide aviation accidents. Although the most publicized of 1985's accidents were not in U. S. domestic scheduled service, public sensitivity about aviation safety increased during that year. Many passengers were then further alarmed by announcements of record fines assessed against large, well-established carriers for maintenance and safety violations. For example, American Airlines was fined $1.5 million, Pan American World Airways $1.95 million, and most recently Eastern Air Lines agreed to pay a fine of $9.5 million. While American Airlines has prospered in the postderegulation period, both Eastern and Pan American have struggled financially which leads to the question whether deregulation's economic pressures or, perhaps in the case of American, rapid growth has interfered with maintenance. Additional concerns emerged with the August 1986 midair collision of an Aeromexico jet and a small private aircraft in the Los Angeles area. This accident raised the question whether the air traffic control system had recovered adequately from the dismissal of striking air traffic controllers in 1981, particularly in light of increased activity at major airports in the wake of deregulation.

While the most highly publicized airline accidents of 1985 and 1986 involved foreign or charter carriers or seemed clearly related to weather, and thus did not appear to be caused by U. S. airline deregulation, there is nevertheless reason to question whether the airline industry's adaptation to deregulation might have an effect on safety. As will be discussed here, some changes in response to deregulation have the potential to erode the excellent safety performance of the airline industry, although there are also strong safeguards against such deterioration. This potential gives rise to four basic questions, which are the focus of this chapter:

1. Has the substitution of commuter airlines for the former trunk and local service airlines in small community service increased risk for travelers to and from those communities?
2. Have the economic pressures accompanying deregulation led to an increase in accidents that might likely result from shortcomings in maintenance procedures?
3. Has increased pressure on the air traffic control system from the combination of deregulation-induced changes in routes and schedules and the dismissal of striking controllers reduced the safety of the system?
4. Has the growth of new entrant and former intrastate jet carriers degraded the safety of the industry?

AIRLINE INDUSTRY CHANGES SINCE DEREGULATION

Deregulation has brought forth a wide array of changes in the domestic airline industry (Bailey, Graham, and Kaplan, 1985; Meyer and Oster, 1981). From the standpoint of airline safety, however, some changes would seem to have the potential for far greater impact than others. Specifically, changes in carrier route strategies and scheduling as well as the impact on carrier finances would seem of greatest importance.

Table 10.1 shows the average annual growth in domestic scheduled enplanements for trunk airlines, local service airlines, other jet airlines, and commuter airlines for the period 1970 through 1985. As can be seen in the table, growth in the postderegulation period varied considerably across these carrier groups. At the end of this period, in 1985, trunks accounted for 59.2% of domestic scheduled enplanements, local service airlines accounted for 18.5%, other jet carriers accounted for 15.9%, and commuters accounted for 6.5%. Because the average flight lengths of the trunks were much longer than those of the other segments of the industry, trunks

Table 10.1 Percent Growth in Enplanements by Industry Segment, Domestic Scheduled Service 1970–1985 (Compound Average Annual Growth)

Period	Trunk Airlines	Local Service Airlines	Other Jet Airlines	Commuter Airlines
1970–1978	6.3	9.0	n.a.	14.1
1978–1985	1.6	5.5	21.6	17.9
1970–1985	4.0	7.0	n.a.	14.6

n.a., not available due to data limitations.

SOURCE: Derived from U.S. Civil Aeronautics Board, Forms 41 and 298; and Regional Airline Association, Annual Reports, various years.

accounted for a larger share of passenger miles. As discussed in this chapter, however, such distance-based measures are not appropriate for calculating accident rates.

Although the established jet carriers, the trunk and local service airlines, had relatively modest postderegulation growth rates, their operations were far from static. The most important feature of postderegulation route strategies has been the development and strengthening of hub-and-spoke route networks by all of the established jet carriers (Meyer and Oster, 1987; Morrison and Winston, 1986). Hub-and-spoke route networks have long been a dominant feature of domestic airline route strategies, but prior to 1978 Civil Aeronautics Board (CAB) regulation severely hampered the rational development of such networks. The basic notion of a hub-and-spoke system is that flights from many different cities converge on a single airport—the hub—at approximately the same time, and after passengers are given sufficient time to make connections, all then leave the hub airport bound for different cities. Such a convergence of flights on a hub is often called a "connecting complex" or "connecting bank," and a passenger on any of the incoming flights of a bank has a convenient connecting opportunity to any of the cities served on the outgoing flights of the same bank.

The main advantage of such a system is that by collecting passengers bound for all the spoke cities, less dense markets can be served more frequently, and because of higher load factors usually more cheaply, than if all city-pair markets were served only by direct flights. A disadvantage to passengers is that hub-and-spoke service is usually connecting, or in a few cases one-stop, service instead of the preferred nonstop service. The more spokes in a hub-and-spoke system, the more connecting opportunities; therefore large systems have service advantages over small systems, at least to the point where the hub airport's landing and takeoff capacity becomes saturated, with accompanying rise in delays.

It is precisely this tendency toward large connecting banks that gives rise to safety concerns. The concentration of operations during connecting banks can put intense pressure on the air traffic control system. Moreover, reliance on connections increases difficulties for passengers and the airline if a flight is late and thus may increase pressure on airlines to maintain schedules.

A second important feature of postderegulation route strategies has been the transfer of much low-density, small community service from local service airlines (and to a lesser extent trunks) to commuter carriers (Meyer and Oster, 1984). To extend air service to the more remote parts of the country and to smaller communities, under regulation the CAB had evolved a program of providing subsidies to local service airlines. By the mid-1970s most of this service was provided in jet aircraft not particularly well suited to such markets. Before deregulation, local service carriers typically used their jets in multistop service from one small community to another during late evening or midafternoon hours with the main economic incentive being the collection of a government subsidy (Meyer and Oster, 1981). These jets were thus kept available for more remunerative service in higher density markets elsewhere during peak business hours. With deregulation and route entry freedoms, these jets could be deployed far more profitably in unsubsidized markets at all times of the day, usually in markets just a bit denser in terms of traffic and a bit further apart than those on subsidy.

The reassignment of local service jets opened up replacement service opportunities for commuter airlines and helps explain the commuters' rapid postderegulation growth seen in Table 10.1. The commuter airlines' market niche even before deregulation had been to provide service to smaller communities whenever and wherever the larger jet carriers pulled out in search of better opportunities elsewhere. Deregulation simply enhanced those opportunities and hastened this "rationalization" of small community airline service.

The accelerated transfer of small community service from trunk and local service airlines to commuters gave rise to the concern that travelers to and from small communities might face undue risks from the poorer safety record of commuters. Widely cited statistics based on fatalities per passenger mile gave credence to the perception that commuters were less safe, commuter airlines often being portrayed as from 10 to 30 times more dangerous than the jet carriers they replaced. Although, as will be discussed below, distance-based measures are wholly inappropriate for comparing commuters and jets, such perceptions were furthered by public comment such as that by a former chief administrator of the Federal Aviation Administration (FAA) Langhorne Bond, who told a gathering of commuter operators and other airline officials that, "no matter how you cook or juggle the statistics on commuter accidents, they add up to a safety record that is unacceptable" (Federal Aviation Administration, 1980). Although the differences between the risk faced by passengers on commuters and passengers on jet carriers have been shown to be not nearly as great as distance-based measures would imply, the safety of commuter airlines remains a concern of many travelers (Oster and Zorn, 1983).

The recent development of marketing alliances in the commuter industry also has the potential to influence safety. Whereas close cooperation between commuters and jet carriers has always been an important feature in the airline industry, such ties have been strengthened by formal alliances between commuters and large jet carriers (Oster and Pickrell, 1986b). Such alliances, which typically have the common feature of the smaller airline sharing the larger airline's two-letter designator code in the Official Airline Guide, in computer reservation systems, and on tickets, spread rapidly from late 1984 through early 1986, until by the end of 1986 virtually all the largest 50 commuter airlines belonged to one or more such alliances. In many respects, the prototype for such alliances was the Allegheny Commuter System, which began in 1967 (Oster and Pickrell, 1986a). The Allegheny Commuter System, operating under close scrutiny of Allegheny Airlines (now USAir) and using carefully selected carriers, achieved a safety record during the 1970–1980 period that rivaled or exceeded that of the jet carrier industry, giving rise to some hope that the new alliances might help improve the safety record of the commuter industry.

The third important postderegulation development with a possible impact on safety is the rapid growth of jet carriers other than the trunks and local service airlines. These carriers are or were either entirely new companies, such as Midway or People Express, or emerged from the ranks of intrastate carriers previously confined by CAB regulation to operations within a single state or from the ranks of supplemental carriers confined by CAB regulation to nonscheduled charter operations (Meyer and Oster, 1984). With deregulation entry and fare freedoms, such carriers expanded rapidly, as seen in Table 10.1 giving rise to concern whether maintenance and training procedures could keep up with growth.

These postderegulation developments—increased hub-and-spoke route operations, replacement of jet carriers by commuters in small community service, and rapid growth of new entrant jet carriers—give rise to the basic questions with which this chapter is concerned. In the course of addressing these questions, the roles of such factors as general aviation, weather, pilot error, and ground crew error will also be examined.

CHARACTERISTICS OF THE ANALYSIS

The diversity of the domestic airline industry presents the risk that highly aggregated safety statistics can mask important differences and thus provide misleading comparisons of air travel safety relative to other modes or even among segments of the airline industry. Without adjusting for varying stage lengths, different mixes of equipment, the variety of airports served, varied maintenance requirements and practices, and differing training and operation procedures, measures of air travel safety suitable for comparison cannot be developed. The basic approach taken in this analysis recognizes the differences among types of carriers in the airline industry and examines the variation in safety rates among these segments of the industry. Disaggregation of the industry into subsets of similar operating characteristics allows examination of some of the factors thought to affect safety within the industry. A comparison of relative safety rates across industry segments can then help shed light on some underlying determinants of airline safety.

Segmenting the Industry

The airline industry falls easily into two distinct segments—jet carriers and commuters. For the purposes of this study, jet carriers are those airlines with aircraft fleets predominantly powered by jet engines, whose operations are certificated, and who are subject to FAA Part 121 operating regulations.

Despite the obvious similarities of aircraft type and operating regulations, there remain sufficiently important differences among jet carriers to suggest further disaggregation. The jet carrier segment is divided into established carriers and new-entrant carriers, and the established carriers are further divided into trunks and local service airlines. The "new entrants" are composed of former intrastate jet carriers that expanded into interstate operations in the wake of deregulation, former charter carriers that began scheduled service, and entirely new airline companies. Not all the carriers included in this category may be considered new-entrant airlines in the manner of carriers such as Air Florida, Midway Airlines, or People Express. World Airways, for example, was new to scheduled passenger service but had considerable experience in charter service with the type of aircraft involved in their postderegulation accident in Boston. Similarly, carriers such as Southwest Airlines and Pacific Southwest Airlines (PSA) had many years experience flying aircraft in intrastate service prior to deregulation.

Trunk carriers are those that evolved from the 16 carriers that were certificated when economic regulation of the industry began in 1938. Local service carriers were certificated on an experimental basis in 1944 to extend air service to small communities and became permanently certificated in 1955. Over the years, some local service airlines evolved and came to resemble small trunk operations, but throughout most of the 1970–1985 period their route structures and operating policies still differed from the trunks in important ways (Meyer and Oster, 1981). Although the local service airlines have operated some propeller equipment, they are nevertheless considered jet carriers for this analysis.

Commuter airlines, as defined in this study, are carriers with predominantly commuter-type operations, that is, operations using propellor-driven aircraft, covered by FAA Part 135 regulations, involving relatively short-haul flights on low-density routes, and seating 60 passengers or less (Oster and Zorn, 1984). Carriers operating fleets comprising solely small propellor-driven aircraft obviously fit into the commuter segment of the industry. In addition, those carriers operating fleets with a preponderance of propellor-driven aircraft are also considered commuters even if they acquired some jet aircraft after deregulation or operate under FAA Part 121 regulations. This definition of commuters parallels the definition of regional airlines used by the Regional Airline Association, formerly the Commuter Airline Association. Thus, the terms *commuter* and *regional* are interchangeable in this analysis.

By no means is the commuter segment of the airline industry homogeneous. Operators vary widely in terms of size, experience, managerial sophistication, fleet composition, financial condition, and routes served. Failure to recognize and account for these differences among commuter carriers can mask important systematic differences in safety performance among different segments of the industry. Earlier research clearly demonstrated this point. Comparing the safety records of the top 20 commuters (based on passenger enplanements) with the rest of the industry (Oster and Zorn, 1983), during the 1970–1980 time period, the top 20 commuter carriers were found to have a safety record approximately five times better than that of the rest of the industry.

Another clearly identifiable segment of the commuter industry, the Allegheny Commuters, had a much safer record than other commuter operators during the 1970–1980 period, suggesting that more stringent operations and maintenance standards imposed by Allegheny (now USAir) support safer operations (Oster and Zorn, 1983). Due to the similarities between the Allegheny Commuter System and newly formed marketing alliances, commuter marketing alliances, essentially those involving the top 50 commuters, get special attention in the analyses that follow.

Measures of Safety

The first step in constructing appropriate measures of safety is to understand exactly what the measures are intended to describe. This study is concerned with the likelihood that an individual passenger will be killed or seriously injured while taking an airline flight, how and why that likelihood varies across segments of the industry, and how and why that likelihood may have changed over time. Keeping these goals in mind, passenger fatalities, passenger serious injuries, accidents resulting in passenger fatalities, and accidents resulting in serious injuries to passengers are the basic measures of the incidents that pose a risk to passenger safety.

An additional useful element to include in the analysis is accidents resulting in minor injuries or no injuries to passengers. By definition, an accident is an unintended happening. A robust safety measure would encompass all unintended happenings, regardless of their severity. While fatal and serious-injury accidents are dramatic, newsworthy, and cause for immediate concern, frequent noninjury accidents may be a portent of problems that could lead to more severe accidents in the future. Moreover, the difference between an accident that kills many passengers and one in which the passengers escape unharmed is often very small.

Fortunately an airline accident is an extremely rare event. By focusing on all accidents, the analysis can be expanded. For example, the recent spate of well-publicized FAA fines of trunk carriers would seem to lend credence to concerns about maintenance shortcuts in the wake of deregulation. However, these fines may also reflect changes in FAA inspection procedures. If only fatalities and fatal accidents were considered, the limited number of observations would make it hard to determine whether the fines indicate increasingly dangerous maintenance practice or increasingly stringent inspections. But, by considering all accidents, it is easier to deduce whether the fines are a result of the former or the latter.

The second step in constructing a safety measure involves the choice of the unit of observation of exposure to risk—the denominator. Transportation safety measures often are based on either the distance traveled or the number of trips. In the case of passenger air travel, distance-based measures are often expressed in terms of passenger-miles or aircraft hours flown, but such measures are biased against carriers carrying small passenger loads, and more important, flying short flight stages. Such measures fail to take into account the fact that a typical commuter must take off and land many more times than a typical jet carrier to amass the same number of passenger-miles. For example, for a passenger to travel as many miles on typical commuter flights as on a flight from New York to Los Angeles would require over 20 such commuter flights. Because most accidents occur during the takeoff and landing phases of flight, the risk of being killed, injured, or involved in an accident would seem to be strongly related to the number of takeoffs and landings (Oster and Zorn, 1982). Thus, long-haul carriers can accumulate many passenger-miles while exposing passengers to relatively little takeoff and landing risk.

Departure-based measures are far more useful for comparing different segments of the industry, since every landing is associated with a prior takeoff and these measures take into account the increased risk associated with these phases of flight. These measures must be used carefully in evaluating passenger fatalities and serious injuries. A passenger fatality or injury per aircraft departure measure, for example, reverses the bias in favor of small-passenger-load air carriers. Thus, commuters generally will be favored relative to jet carriers by the use of this statistic, and carriers using similar equipment but with different passenger loads will receive different treatment.

A preferred measure would be passenger departures rather than aircraft departures. Unfortunately, data on passenger departures are unavailable. Enplanement data can act as a reasonable proxy for passenger departures, with a few qualifications. A passenger is counted as an enplanement each time he or she boards a flight, but if the flight involves multiple stops, a passenger is counted as one enplanement regardless of the number of times the plane takes off and lands with the passenger on board. Thus, for flights with intermediate stops, enplanements and passenger departures are not equal. Few data are available to assess the magnitude of this divergence. There is, however, no reason to believe that serious systematic biases for or against any particular segment of the airline industry would be introduced by using enplanements for the denominator.

Data

National Transportation Safety Board (NTSB) Accident Briefs on Part 121 and Part 135 operations provide a rich source of accident data. These data include the number of fatalities and injuries for both passengers and crew, a description of the aircraft involved, details concerning the circumstances surrounding the accident, and probable cause and contributing factors. Data regarding passenger fatalities, serious injuries, fatal accidents, serious-injury accidents, and other accidents were derived from this source. Although data on pilot and crew fatalities and injuries were also available, the risk to pilots and crew in the airline industry was not considered because the focus of the analysis was on the safety of air travel for passengers.

Operations data for jet carriers are also fairly complete. Jet carrier operation statistics, including aircraft departures and passenger enplanements, were obtained from CAB Form 41. The U.S. Department of Transportation has continued collection of these data after the sunset of the CAB. Commuter operations data are more limited than jet carrier operations data. Passenger enplanements were obtained from CAB Form 298 for the 1970–1980 period. Data for the 1981–1985 period were obtained from the annual reports of the Regional Airline Association. For those carriers operating under Part 121 but still classified as commuter carriers for purposes of this study, enplanement data were obtained from CAB Form 41 and Regional Airline Association annual reports.

Unfortunately, there is no complete source of aircraft departure data for commuters. As a result, the safety measures based on aircraft departures constructed for the jet carriers could not be constructed for commuter carriers. Measures of the risk of serious injury or death to passengers in both the jet and commuter segments were constructed: passenger fatalities per 1 million enplanements and passenger serious injuries per 1 million enplanements. The lack of departure data for commuters, however, means that the risk of an accident can be measured only for jet carriers.

Because the focus of the analysis is on the effects of deregulation, the analysis is limited to domestic scheduled passenger service. All intra-Alaska, intra-Hawaii, international, caribbean, and charter operations and accidents are excluded. There is better comparability across segments of the industry when operations in the noncontinental states are omitted—in other areas commuters, in particular, serve a much different role and frequently face a much harsher operating environment (Oster and Zorn, 1982).

AIRLINE SAFETY SINCE DEREGULATION

In the wake of deregulation, competition among carriers, enhanced flexibility in setting route strategies, and shifts in service responsibilities among segments of the industry have kindled rising public concern over safety within the industry. Table 10.2 suggests that, in the aggregate, this concern about worsening safety as a result of deregulation may well be unwarranted. During the postregulation period (1979–1985), commuter and local service carrier safety, as measured by passenger fatalities per 1 million enplanements, improved substantially while trunk carrier safety was essentially unchanged.

Part of the improvement in the safety record of commuter carriers can be attributed, at least in part, to revisions in commuter safety regulations that occurred in 1978. In response to concerns about commuter airline

Table 10.2   Passenger Fatality Rates, Domestic Scheduled Service
1970–1985 (Fatalities/1 Million Enplanements)

Industry Segment	1970–1978	1979–1985	1970–1985
Commuters	2.65	1.27[a]	1.66
Trunks	0.36	0.38	0.37
Local service	0.65	0.01[b]	0.29

[a]The 1979–1985 rate is lower than the 1970–1978 rate at the 90% confidence level.

[b]The 1979–1985 rate is lower than the 1970–1978 rate at the 95% confidence level.

SOURCE: Derived from computer printout of Part 121 and 135 operation accident briefs provided by the National Transportation Safety Board; U.S. Civil Aeronautics Board, Forms 41 and 298; and Regional Airline Association, Annual Reports, various years.

safety, the FAA tightened regulations in three areas. First, pilot qualifications were upgraded to parallel those for established jet carriers. Second, maintenance requirements were improved to include more detailed and extensive procedures for all types of classes of aircraft used by Part 135 operators. Third, training program requirements of Part 135 operators were increased (Oster and Zorn, 1984). These changes are discussed in more detail later in the chapter.

Commuter Substitution in Small Community Service

As commuter carriers have replaced withdrawing jet carriers, there has been concern that small community residents have undergone degradation in the safety of their air transportation. Table 10.2 suggests at first glance that such substitutions may indeed increase risk to travelers to and from small communities. During the entire 1970–1985 period, passengers were more than four times as likely to be killed when boarding a commuter flight than one operated by trunk or local service carriers. More important, in the postderegulation period, commuter air travel remained more risky than jet air travel by more than a factor of 3.

If commuters were a homogeneous segment of the air transportation industry, the analysis might well end here, supporting the claim of serious degradation of safety from these substitutions. However, the commuter industry is not homogeneous but rather is composed of many carriers of greatly varying sizes, fleet mixes, and route structures. Moreover, the widely varying safety records in the commuter industry have been shown to be correlated with carrier size (Oster and Zorn, 1983).

Table 10.3 clearly demonstrates that safety rates within the commuter industry are inversely related to carrier size. In the 1970–1978 period, the top 20 carriers were more than 4.5 times safer than the rest of the top 50 carriers (carriers 21 through 50, size ranking) in terms of passenger fatalities per 1 million enplanements, and more than 19 times safer than the rest of the industry, the relatively small carriers who make up the non-top-50 category. In terms of passenger serious injuries per 1 million enplanements, the top 20 carriers were almost 2.5 times safer than the rest of the industry.

Just as the commuter industry as a whole experienced an improvement in safety in terms of both fatalities and serious injuries, each of these segments of the industry also experienced improvements. The safest portion of the industry in the preregulation period, the top 20 commuters, experienced the smallest (and not statistically significant) improvement in their safety records, most likely because these largest commuter carriers had previously instituted many of the procedures that the FAA mandated in 1978 and thus the safety revisions had little additional effect. On the other hand, the dramatic improvement in safety records for the rest of the top 50 and the rest of the industry is most likely the direct result of the 1978 safety revisions.

The levels of statistical significance reported in this and following tables were calculated in two ways. For the accident rates presented in Table 10.7 through 10.10, the tests were based on the binomial distribution, with each aircraft departure as the unit of observation. For the passenger

Table 10.3 Commuter Carriers Passenger Fatality and Serious Injury Rates, Domestic Scheduled Service 1970–1978 versus 1979–1985

Group	1970–1978	1979–1985
Total industry:		
Passenger fatalities/1 million enplanements	2.65	1.27[a]
Passenger serious injuries/1 million enplanements	1.62	0.66[a]
Top 20 carriers		
Passenger fatalities/1 million enplanements	0.69	0.67
Passenger serious injuries/1 million enplanements	0.93	0.21
Rest of the top 50		
Passenger fatalities/1 million enplanements	3.27	1.21[b]
Passenger serious injuries/1 million enplanements	2.27	0.48
Rest of the industry		
Passenger fatalities/1 million enplanements	13.32	4.08[b]
Passenger serious injuries/1 million enplanements	4.30	3.00

[a]The 1979–1985 rate is lower than the 1970–1978 rate at the 90% confidence level.

[b]The 1979–1985 rate is lower than the 1970–1978 rate at the 95% confidence level.

SOURCE: Derived from computer printout of Part 121 and 135 operation accident briefs provided by the National Transportation Safety Board; U.S. Civil Aeronautics Board, Forms 41 and 298; and Regional Airline Association, Annual Reports, various years.

Table 10.4 Transition from Jet Service to Commuter Service: 60 City-Pair Markets Where Commuters Replaced Jets, 1978 versus 1986

Mix of Carriers	Average Weekly Departures	Average Intermediate Stops
1978		
15 Trunks	2.88	0.59
45 Local service		
1986		
49 Code-sharing	6.29	0.30
6 Top 20 independents		
4 Other independents		
1 Service loss		

SOURCE: U.S. Civil Aeronautics Board Staff Study. *Report on Airline Service, Fares, Traffic, Load Factors and Market Shares, Service Status on September 1, 1984*. Washington, D.C.: U.S. Government Printing Office, December 1984; and Official Airline Guide, North American Edition, July 1, 1978, 1984, and 1986 editions.

fatality rates presented in Tables 10.2, 10.3, 10.5, 10.6, and 10.7, however, the assumption required for the binomial distribution—that each observation (passenger fatality in this case) was an independent event—was not tenable. For these tests, the annual fatality rates were considered the unit of observation and the means were tested using a normal distribution.

After viewing the data in Table 10.3, a conclusion that the substitution of commuter carriers for jet carriers in small community service has degraded safety must be drawn with more caution. Recognizing that differences in safety records among segments of the commuter industry exist, however, implies that the degradation may not be as bad as the aggregate commuter safety measure originally suggested. If the replacement service were provided by the larger commuter carriers, the impact on safety would be less severe. Indeed, the first column of Table 10.4 shows that most of the replacement service did involve the very largest carriers, largely because they had the personnel, equipment, and expertise to move most easily into markets abandoned by the jet carriers.

An important consideration thus far overlooked in the discussion of safety implications of commuter replacement stems from the improved service typically offered to small communities by commuters. Table 10.4 summarizes the experience between 1978 and 1986 in the 60 communities that lost jet carrier service between June 1978 and June 1984 and where commuter carriers provided replacement service in 1986. Commuter replacement service into these communities resulted in an increase in frequency of weekday flights to these cities, improving this aspect of air transportation services. In addition, the average number of intermediate stops on flights between a small community and its nearest major hub decreased by almost half after the transition to commuter service. Thus on average, service improved in these communities where commuters provided substitute service for jets, because passengers have more direct connections to hubs where they can link up with jet carrier service.

The data in Table 10.4 not only indicate improved service when commuters substitute for jets in small communities, but also suggest that a straight comparison between jet carrier and commuter safety measures is not appropriate. If the service improvement includes fewer intermediate stops on flights to and from these small communities on commuter flights, the relative safety rates of commuters and jet carriers must be adjusted to reflect the additional risk from extra takeoffs and landings on the jet flights. According to Table 10.4, the average number of takeoffs and landings associated with jet service to these small communities was 1.59 (the original takeoff and subsequent landing and 0.59 takeoff and landing at intermediate stops), but only 1.30 for commuters serving the same cities. Thus, to get comparable safety measures that take into account this improvement in service and its impact on safety, the jet carrier rate should be inflated by 22.3%. So while there may be some increased risk associated with the service provided by commuters relative to that provided by jet carriers, there is not the dramatic effect suggested by a simple comparison of aggregate safety measures.

A final consideration must be added to complete this safety comparison. As a result of the increased frequency of service offered in small communities in the wake of the exit of jet carriers, ridership is generally found to increase (Meyer and Oster, 1981). The added convenience offered by commuters attracts passengers who would have previously opted for the automobile. Assuming that the average commuter flight length of between 120 and 130 miles substitutes for an auto trip to the same hub airport of about 150 miles and further, assuming that average auto occupancy for such a trip is about 1.5 (slightly below the average for all intercity auto trips), the fatality rate associated with motor vehicle trips that would be comparable to the commuter rate is somewhere between 1.9 and 2.3 passenger fatalities per 1 million passenger trips. Comparing this motor vehicle fatality rate with fatality rate of the top 50 commuters indicates that travel is safer for those passengers switching from auto to commuter as a result of improved commuter service. Thus commuter service to small communities may actually improve the transportation safety record relative to the prior service by jet carriers, once these adjustments have been made.

Economic Pressures and Airline Maintenance

The record high fines levied against such airlines as American, Pan American, and Eastern, coupled with several postderegulation commuter accidents in which equipment failure was a clear contributing factor, have combined to raise the question of whether the increased competitive pressures accompanying deregulation have caused the airlines to cut corners in maintenance practices in an attempt to lower costs. Were strong evidence found that deregulation had intensified competitive pressures to the point where safety was compromised by maintenance cost cutting, then the efficacy of deregulation itself might come into question. However, the large fines could also be the product of an admitted change in inspection and fine policy by the FAA. Moreover, accidents caused by equipment failures are regrettably not a new experience in the airline industry.

One approach to assessing the impact of deregulation on maintenance practices is to examine the frequency of accidents in which inadequate maintenance might have been a contributing factor. Thus, accidents for both commuter airlines and jet carriers were classified according to their "primary contributing factor." Each accident was put into one of eight categories according to its primary contributing factor:

1. equipment failure;
2. seatbelt not fastened;
3. weather;
4. pilot error;
5. air traffic control;
6. ground crew error;
7. general aviation;
8. other.

Many, indeed most, accidents are the result of several factors coming together at the same time and thus have multiple causes. Determining the primary contributing factor from the information contained in NTSB Accident Briefs for multiple-cause accidents is a difficult task requiring admittedly subjective judgments in which there is room for disagreement. The purpose, however, was not to determine the "main" cause, but rather to apply a consistent set of criteria that would allow a longitudinal analysis of the relative frequency of types of accidents. Thus it is important to be clear about what sorts of accidents were included in each category before results are presented.

Equipment failure

If the events that culminated in the accident were precipitated by some sort of mechanical or electrical malfunction in the aircraft, then the accident was considered as equipment failure even if there were other important factors. In some cases, such as improper installation of a part or failure to detect cracks or corrosion, more meticulous maintenance might have prevented the accident. In other cases, such as failure of a tire that did not show excessive wear, maintenance practices could not reasonably be blamed. No attempt was made to distinguish whether inadequate maintenance was at fault. Rather, all equipment-failure accidents were treated the same on the assumption that those where maintenance was not a factor would be randomly distributed over time so that any time trend in equipment failure would be the result of changing maintenance practices or improved aircraft design.

Seatbelt not fastened

A surprisingly common source of serious passenger injury, but fortunately not death, is failure of passengers to fasten their seatbelt when turbulence is encountered despite the illuminated seatbelt sign in the cabin and requests to do so from the cockpit and cabin crew. Such injuries include broken ankles, broken legs (in one case both legs), head, and neck injuries. If the seatbelt sign was illuminated in sufficient time for passengers to return to their seat and fasten their seatbelt prior to the injury, the accident was placed in this category. If, however, the turbulence was unexpected and the seatbelt sign was not on, the accident was placed in the "weather" category discussed next.

Weather

Weather is a factor in many airline accidents but is frequently not regarded by the NTSB as the cause of the accident. Rather, the NTSB often determines the cause to be the cockpit crew not responding properly to weather conditions. The approach taken for this study differs from the NTSB approach in that accidents were rarely classified as "pilot error" (discussed later) if weather was a significant contributing factor. Thus an accident was usually classified as weather even if the pilot, taking precisely the right action at precisely the right time, could have prevented the accident.

There were, however, some notable exceptions to this approach. If an aircraft took off under weather conditions that led to a takeoff accident, as was the case with an Air Florida accident in January 1982 at Washington National Airport and a commuter airlines accident in March 1970 in Binghamton, New York, the accident was considered pilot error. If an aircraft was unable to stop after landing on a slick runway, it was considered a weather accident, unless the pilot landed excessively long on the runway, as was the case with a Sunbird Airlines accident in Hickory, North Carolina, in May 1984, in which case it was considered pilot error. As a final example, if a pilot attempted to land when the weather at the airport was below minimum accepted standards, and an alternate airport was available, the accident was considered pilot error. The distinction between weather and pilot error was certainly the most subjective, but considerable effort was made to achieve consistency in the classifications over the 16-year study period.

Pilot error

In addition to the weather-related examples just cited, an accident was classified as pilot error only in those cases in which the error appeared undeniable, for example when a pilot attempted a landing without lowering the landing gear or taxied into a stationary object. Another example of pilot error was running out of fuel because of failure to refuel or failure to switch fuel tanks during flight.

Air traffic control

An accident was classified as air traffic control (ATC) when normal action by a controller could have prevented the accident, such as a USAir accident in June 1976 in Philadelphia in which ATC failed to advise the crew of unsafe weather.

Ground crew

Accidents attributed to ground crew error included such cases as the American Airlines accident at Chicago in October 1978, in which a service truck collided with a parked aircraft, and the Tampa Air accident in Tampa, Florida in January 1972, in which a ground crew member walked into a propellor while delivering a message to the pilot of an aircraft about to depart.

General aviation

General aviation accidents included those where the accident would not have occurred had general aviation aircraft not been operating in the area, such as the PSA collision with a general aviation aircraft during approach into San Diego in 1977, or the Air U. S. collision with a general aviation aircraft carrying skydivers in April 1981 over Loveland, Colorado. This determination does not necessarily imply that the general aviation aircraft was at fault, just that its presence contributed to the accident.

Other

Accidents not falling into one of the above categories were classified as "other." Such accidents included a wide array of causes, ranging from a passenger tripping over a baby bottle in the aisle to injuries sustained during an evacuation due to a bomb threat. Two accidents in late 1984 and seven accidents in 1985 were classified as other because NTSB accident investigations had not yet released findings as to their causes.

Commuters Before and After Deregulation

Table 10.5 breaks down the passenger fatality rate for the entire commuter industry into these eight categories for both the 1970–1978 preregulation period and the 1979–1985 postderegulation period. As the table indicates, most of the improvement in fatality rate between these two periods comes from a reduction in fatalities due to three reasons: equipment failure, pilot error, and weather (although only the equipment failure rate was significantly lower in the later period).

Such improvement was probably to be expected. As mentioned earlier, coincidental with the onset of deregulation, commuter safety regulations underwent a major revision by the FAA in 1978. The revisions evolved from a review of safety regulations that was prompted in part by a 1972 NTSB report (National Transportation Safety Board, 1972, 1980). The report pointed out several inadequacies in FAA regulations of commuter

Table 10.5   Commuter Carriers: Passenger Fatality Rate
by Principal Contributing Factor, Domestic Scheduled
Service 1970–1978 versus 1979–1985

| Contributing Factor | Fatalities/1 Million Enplanements | |
	1970–1978	1979–1985
Equipment failure	1.07	0.35[a]
Seatbelt not fastened	0.00	0.00
Weather	0.61	0.27
Pilot error	0.46	0.05
Air traffic control	0.04	0.00
Ground crew error	0.00	0.01
General aviation	0.22	0.20
Other	0.24	0.39
Total	2.65	1.27[a]

[a]The 1979–1985 rate is lower than the 1970–1978 rate at the 90% confidence level.

SOURCE: Derived from computer printout of Parts 121 and 135 operation accident briefs provided by the National Transportation Safety Board; U.S. Civil Aeronautics Board, Forms 41 and 298; and Regional Airline Association, Annual Reports, various years.

airlines, including (1) requirements for maintenance and training programs for crew members, (2) pilot qualification requirements, and (3) rules concerning which items of onboard equipment may be inoperative yet the aircraft be allowed to fly, the "minimum equipment list." Among the FAA's 1978 revisions was a provision requiring the pilot in command of a commuter aircraft seating 10 or more passengers to hold an airline transport pilot certificate rather than simply a commercial pilot certificate, and a provision requiring a pilot in command to have made three takeoffs and three landings in the same type of aircraft within 90 days preceding a scheduled flight. Similarly, initial and recurrent training programs became a basic requirement in 1978. Thus commuter pilot requirements were brought more in line with those for jet airline pilots. Revised maintenance requirements included more detailed and extensive procedures for all types and classes of aircraft used by commuter operators as well as additional maintenance record keeping. Minimum equipment lists for commuters were also established for the first time in the 1978 FAA revisions and addressed such items as ground proximity warning systems and fire extinguishers as well as radio and navigational equipment for instrument flight rule (IFR) operations.

Given the thrust of many of the 1978 revisions, the pattern of fatality rate reduction in Table 10.5 might have been expected. The tightened maintenance procedures have probably contributed to the reduction in accidents involving equipment failure. The added pilot certification requirements coupled with added recurrent training have likely contributed to both reduced pilot error and reduced weather-related fatalities. Weather-

Table 10.6 Passenger Fatality Rate by Principal Contributing
Factor, 1979–1985 Domestic Scheduled Service, Top 20 versus
Rest of Top 50 versus Rest of Industry

| | Fatalities/1 Million Enplanements | | |
Contributing Factor	Top 20	Rest of Top 50	Rest of Industry
Equipment failure	0.11	0.52	1.15[b]
Seatbelt not fastened	0.00	0.00	0.00
Weather	0.15	0.03	1.21[b]
Pilot error	0.07	0.00	0.06
Air traffic control	0.00	0.00	0.00
Ground crew error	0.01	0.00	0.00
General aviation	0.00	0.45[a]	0.64[b]
Other	0.32	0.21	1.02[b]
Total	0.67	1.21	4.08[b]

[a] The rate is higher than the rate for the top 20 carriers at the 90% confidence level.

[b] The rate is higher than the rate for the top 20 carriers at the 95% confidence level.

SOURCE: Derived from computer printout of Parts 121 and 135 operation accident briefs provided by the National Transportation Safety Board; U.S. Civil Aeronautics Board, Forms 41 and 298; and Regional Airline Association, Annual Reports, various years.

related fatalities may also have been reduced by tightened regulations regarding navigational equipment for IFR flights.

Table 10.6 shows the 1979–1985 passenger fatality rate by type of accident for three segments of the commuter industry: the top 20 carriers, the rest of the top 50 (those carriers ranked in size from 21 through 50), and the rest of the commuter industry (those carriers ranked 51 and higher). The most striking pattern in the table is the markedly poorer safety record of the carriers outside of the top 50. The pattern suggests that even with the tightened FAA regulations, the larger carriers may still have more effective operating and maintenance practices than the smallest carriers have been able to achieve.

One hypothesis as to why carrier size might contribute to lower rates of equipment failure is that a larger carrier can afford greater specialization in maintenance. A larger carrier with more aircraft may also find it easier to maintain the necessary inventory of spare parts, both financially and in terms of space and organization. Although size and depth of inventory should not matter too much for scheduled maintenance overhauls, it may matter, on occasion, for marginal unexpected repairs; if a part shows wear or slightly diminished performance, it might be replaced quickly if the part is on hand, while replacement might be postponed if the part has to be ordered.

Another possible reason for the higher rate of equipment failure among the smallest carriers is that they are least likely to operate a fleet of all turbine-engine aircraft and more likely to operate small piston-engine aircraft seating nine or fewer passengers. Earlier analysis has shown that small carriers operating either piston-engine aircraft only or a combined fleet of piston- and turbine-engine aircraft experience a much higher rate of passenger fatalities than either larger carriers operating any type of fleet or smaller carriers operating turbine-engine only fleets (Oster and Zorn, 1982). Piston engines have far more moving parts and are more complex to maintain, particularly, it appears, for small carriers. A mixed fleet of both piston and turbine engines aggravates the maintenance specialization and spare parts inventory, again particularly for smaller carriers. Moreover, many of the 1978 FAA revisions did not apply to operators of aircraft seating nine of fewer passengers, so that these aircraft may not be subject to some of the same tighter maintenance requirements (General Accounting Office, 1984). Indeed, even aircraft seating 10 to 19 passengers had somewhat less extensive requirements than larger aircraft, a difference that may have contributed to some accidents (National Transportation Safety Board, 1986a).

It is interesting that the smallest carriers also experience a substantially higher rate of fatalities from weather-related accidents. Again, it may be that the smaller piston-engine aircraft usually operated by these carriers are more difficult to operate in adverse weather than the larger aircraft usually flown by the largest carriers. Also, these smaller aircraft may not be equipped with the same navigational equipment for instrument flights.

Jet Carriers Before and After Deregulation

While the impact of deregulation on commuter safety is difficult (indeed probably impossible) to separate from the impacts of the FAA safety regulation revisions, there was no such dramatic change in safety regulations for the jet carriers. Table 10.7 contains several measures of airline safety for the combined trunk and local service airlines for both the prederegu-

Table 10.7   Combined Trunk and Local Service Carriers: Passenger Fatality and Aircraft Accident Rates, Domestic Scheduled Service 1970–1978 versus 1979–1985

Measure	1970–1978	1979–1985
Passenger fatalities/1 million enplanements	0.42	0.30
Passenger serious injuries/1 million enplanements	0.25	0.03 [a]
Fatal accident/1 million aircraft departures	0.46	0.22 [a]
Serious injury accidents/1 million aircraft departures	1.92	0.83 [a]
Minor accidents/1 million aircraft departures	2.90	1.37 [a]

[a] The 1979–1985 rate is lower than the 1970–1978 rate at the 95% confidence level.

SOURCE: Derived from computer printout of Parts 121 and 135 operation accident briefs provided by the National Transportation Safety Board; U.S. Civil Aeronautics Board, Forms 41 and 298; and Regional Airline Association, Annual Reports, various years.

Table 10.8 Combined Trunk and Local Service
Carriers: Total Accident Rate by Principal Contributing
Factor, Domestic Scheduled Service 1970–1978 versus
1979–1985

	Accidents/1 Million Aircraft Departures	
Contributing Factor	1970–1978	1979–1985
Equipment failure	1.49	0.43[a]
Seatbelt not fastened	1.49	0.68[a]
Weather	0.82	0.33[a]
Pilot error	0.54	0.21[a]
Air traffic control	0.26	0.11[b]
Ground crew error	0.23	0.11
General aviation	0.10	0.04
Other	0.39	0.50
Total	5.28	2.42[a]

[a] The 1979–1985 rate is lower than the 1970–1978 rate at the 95% confidence level.

[b] The 1979–1985 rate is lower than the 1970–1978 rate at the 90% confidence level.

SOURCE: Derived from computer printout of Parts 121 and 135 operation accident briefs provided by the National Transportation Safety Board; U.S. Civil Aeronautics Board, Forms 41 and 298; and Regional Airline Association, Annual Reports, various years.

lation 1970–1978 period and the postderegulation 1979–1985 period. All five measures presented in the table were sharply lower in the postderegulation period.

As mentioned earlier, consistent aircraft departure data are available for jet carriers throughout the period, so that in addition to passenger fatality and passenger injury rates, accident rates can be examined. As Table 10.7 illustrates, all of the various rates declined between the two periods, so the general conclusions about airline safety do not seem particularly sensitive to the choice of appropriate measure. By focusing on accident rates for the jet carriers, accidents that do not result in passenger fatalities or serious injuries but may still represent other potentially serious situations can be included in the analyses.

Table 10.8 shows the combined accident rates for trunks and local service airlines broken down among the eight principal contributing factors used earlier in Tables 10.5 and 10.6. The accident rates for the jet carriers declined for seven of the eight types of accidents, and in five cases the decline was significant.

Perhaps the most striking observation is in the table's first line, where the rate for equipment failure-related accidents in the postderegulation period is seen to be less than one third of the prederegulation rate. If deregulation had indeed induced shortcuts in aircraft maintenance, the rate of equipment failures would be expected to have increased. That the rate decreased sharply suggests that, at least through 1985, deregulation has not led to widespread maintenance deficiencies.

The accident rates are also down for both air traffic control and for general aviation, although these rates are low both before and after deregulation. Part of the postderegulation period was prior to the air traffic controllers' strike (in August 1981), but much of the period was after the striking controllers had been dismissed. In terms of accidents, there is no evidence that the air traffic control system functioned less safely after deregulation than before. Similarly, the rate of general aviation-related accidents has not increased since deregulation. Thus, despite the added pressure on the air traffic control system and on the air-space surrounding large airports because intensified hub-and-spoke operations cause a bunching of flights, the rates for the kinds of accidents that might be expected to have increased have, in fact, gone down.

Deregulation has also stimulated increased pressure on airline labor for less restrictive work rules. Pilots and cabin attendants are flying more hours per month and ground crews are performing a wider variety of tasks than before deregulation (Meyer and Oster, 1987). Despite these pressures, the rates for pilot error and ground crew error have both declined. The rate for accidents related to unfastened seatbelts has also declined, suggesting that, if anything, cabin attendants have been more rather than less effective in making sure that passengers fasten their seatbelts when turbulence is expected. Again, the sorts of accidents that might have been feared to have increased because of labor practice changes induced by deregulation have been less frequent following deregulation than they were before.

The rapid growth and route expansion of the new-entrant and former intrastate jet carriers has been one of the major developments following deregulation and has given rise to concerns about these carriers' impacts on safety. One concern was that it would be difficult to achieve adequate training and maintenance practices in the face of rapid growth; a second, brought about by the Air Florida accident in Washington in 1982, was that expansion might confront these carriers with weather and operating conditions for which they were not fully prepared.

Table 10.9 contains the passenger fatality, passenger serious injury, and various accident rates for trunks, local service airlines, and other carriers for the 1979–1985 period. As the table indicates, the other jet carriers have a safety record very similar to that for other segments of the jet carrier industry, although comparisons are somewhat limited by the comparatively small number of operations by these other jet carriers. Of the five measures presented in the table, only the rate of fatal accidents per 1 million departures appears higher than the comparable rate for trunk or local service airlines. In essence, the other jet carriers experienced three such accidents rather than the one that might have been expected based on the trunk and local service airlines' experience.

Table 10.9   1979–1985 Passenger Fatality and Aircraft Accident Rates, Domestic Scheduled Service, Trunks versus Local Service versus Other Jet Carriers

Measures	Trunks	Local Service	Other Jet Carriers
Passenger fatalities/1 million enplanements	0.38	0.01 [a]	0.42
Passenger serious injuries/1 million enplanements	0.03	0.04	0.05
Fatal accidents/1 million aircraft departures	0.22	0.22	0.90 [a]
Serious injury accidents/1 million aircraft departures	0.86	0.77	0.90
Minor accidents/1 million aircraft departures	1.30	1.53	1.51

[a] The rate is different from the trunk airline rate at the 95% confidence level.

SOURCE: Derived from computer printout of Parts 121 and 135 operation accident briefs provided by the National Transportation Safety Board; U.S. Civil Aeronautics Board, Forms 41 and 298; and Regional Airline Association, Annual Reports, various years.

Table 10.10   Total Accident Rate by Principal Contributing Factor, 1979–1985 Domestic Scheduled Service, Trunk versus Local Service versus Other Jet Carrier

Contributing Factor	Accidents/1 Million Aircraft Departures		
	Trunks	Local Service	Other Jet Carriers
Equipment failure	0.59	0.11 [a]	0.30
Seatbelt not fastened	0.70	0.66	0.60
Weather	0.32	0.33	0.60
Pilot error	0.16	0.33	0.60 [b]
Air traffic control	0.05	0.22	0.60 [a]
Ground crew error	0.10	0.11	0.30
General aviation	0.00	0.11 [b]	0.00
Other	0.43	0.66	0.30
Total	2.38	2.52	3.31

[a] The rate is different from the trunk airline rate at the 95% confidence level.

[b] The rate is different from the trunk airline rate at the 90% confidence level.

SOURCE: Derived from computer printout of Parts 121 and 135 operation accident briefs provided by the National Transportation Safety Board; U.S. Civil Aeronautics Board, Forms 41 and 298; and Regional Airline Association, Annual Reports, various years.

Table 10.10 breaks down the accident rates by primary contributing factor for all three segments of the jet industry. As the table indicates, the other jet carriers' experience again is quite similar to that of trunks and local service airlines. The accidents are fairly uniformly distributed among contributing factors, with no factors appearing disproportionately represented. In sum, while these other jet carriers may have experienced a somewhat higher rate of fatality-causing accidents, their overall safety experience does not appear to differ markedly from the other segments of the jet industry. However, it should be noted that there are a limited number of observation points.

SUMMARY AND CONCLUSIONS

The analyses reported in the preceding section shed considerable light on the four basic questions posed in the introduction.

Has the substitution of commuter airlines for the former trunk and local service airlines in small community service increased risk for travelers to and from those communities?

While the commuter industry as a whole has a safety record clearly inferior to that of the established jet carriers, three factors suggest that there has not been a substantial reduction in safety for travelers to and from small communities as a result of the transition to commuter service. First, the larger commuter carriers have a safety record nearly comparable to that of the local service jet carriers before deregulation, and these larger carriers have provided most of the replacement service. Second, commuter flights between small communities and large hubs typically operate with fewer intermediate stops than the jet flights they replace, thus exposing passengers to fewer takeoffs and landings. Third, many commuter passengers had previously used autos for their trips because of the relatively low frequency jet flights; deregulation has shifted those passengers to a safer mode of travel.

Have the economic pressures accompanying deregulation led to an increase in accidents that might likely result from shortcomings in maintenance procedures?

For both commuters and jet carriers, the rate of accidents related to equipment failure is substantially lower in the postderegulation period, suggesting that at least through 1985 there is no evidence of increased accidents from worsening maintenance practices.

Has increased pressure on the air traffic control system from the combination of deregulation-induced changes in routes and schedules and the

dismissal of the striking air traffic controllers reduced the safety of the system?

Despite the increased pressure on the air traffic control system because of increased emphasis on hub-and-spoke operations and reduction in the number of air traffic controllers, the rate of air traffic control-related accidents is lower in the postderegulation period than prior to deregulation.

Has the growth of new-entrant and former intrastate jet carriers degraded the safety of the industry?

Although the accident rates, based on limited operations, appear slightly higher for the new-entrant and former intrastate jet carriers than for the established trunk and local service carriers, the differences are small and do not suggest a serious degradation of safety.

11

Can Truth in Airline Scheduling Alleviate the Congestion and Delay Problem?

ELIZABETH E. BAILEY and DAVID M. KIRSTEIN

Flight delays and airport congestion have become a common experience for virtually every air traveler. To ameliorate this problem in the short run, we advocated in the spring of 1987 a policy of airline truth in scheduling (Bailey and Kirstein, 1987). Our idea was to require airlines to publish schedules based on realistic, rather than optimistic, flight times. The truth-in-scheduling approach is premised on the belief that if a flight regularly has lengthy delays, whatever their cause, the public has a right to know. Because people value their own time and seek to avoid delay, they would use the information to select times of day, airlines, and airports that minimize travel time.

Policies to give consumers true information on flight delays have since been adopted by a number of air carriers, by the Department of Transportation (DOT), and by Congress. Evidence from recent months suggests that these policies, along with the introduction of new technologies and procedures for improved traffic flow by the Federal Aviation Administration (FAA), are contributing to substantial improvement in delay statistics.

There is, however, a longer term problem. Demand for air travel has exploded since deregulation, and air transportation shows every promise of continuing to be a preferred mode in the decades ahead. Capacity limitations are already causing stress and constraint, and the relief of these constraints, particularly to the extent that they involve more airports and runways and/or a fully modernized and automated en route control system, takes many years. Long-run solutions to congestion and delay will require institutional arrangements and policies that spread the peak, make better use of existing capacity, and provide revenues for capacity expansion and for the introduction of new and better technologies.

DELAY AS A PROBLEM FOR CONSUMERS

Airline delays have become a major problem for consumers. The FAA defines delay to mean arrival time 15 minutes or more after that published in an airline schedule. Figure 11.1 shows the number of airline delays for the first nine months of 1986 and 1987. During the spring and summer of 1986, there were around 40,000 delays per month. Evidence from the first few months of 1987 showed that the problem was significant even in the off-peak winter season, at 25,000 delays per month, and was not improving compared with 1986 levels. Consumers were clearly frustrated and concern was deepening in Congress.

To show the degree of the problem from a consumer's viewpoint, it is useful to convert the figures to percentage of flights delayed. Table 11.1 shows that delays affected between 30% and 50% of flights in the March 1986–March 1987 time frame. Even a carrier with a good record of on-

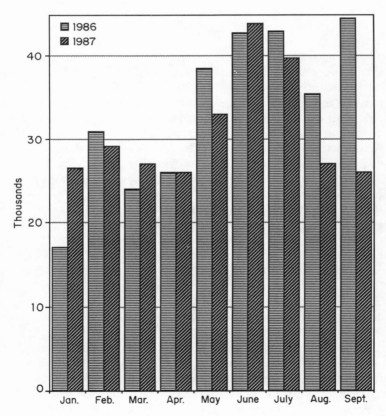

Figure 11.1   Airline Delays: First Nine Months 1986–1987. SOURCE: Federal Aviation Administration.

Table 11.1   Percent of On-Time[a] Arrivals for Certain Airlines

Airline	March 1986–March 1987	March 1987
Piedmont Aviation	70.2	77.4
Continental Air Lines	67.4	56.9
USAir	65.9	68.1
Eastern Air Lines	62.3	56.6
American Airlines	59.9	61.7
Northwest Airlines[b]	55.3	50.5

[a] Arrived within 15 minutes of schedule.

[b] Excludes September and October 1986 because of merger with Republic.

SOURCE: Federal Aviation Administration.

time performance, such as Piedmont Aviation, had only 70.2% of its flights arriving within 15 minutes of schedule in the 13-month period. A poorer performer, Northwest Airlines, had only 55.3% of its flights meeting this on-time standard over the same 13-month period.

Figure 11.2 displays causes of delays by categories, such as weather and inability to handle volume, for the period January–April 1987. Contrary to popular conception, the "weather" category, and not demand or air carrier practices, is the prime cause of delays. However, it should be recognized that current FAA policies do affect the weather component. For example, if bad weather is expected at a destination airport, the FAA currently delays takeoff of the plane. Thus, if the weather clears, the plane is not in a position to land since it has not been circling.

Attribution of blame for the "inability to handle volume" category should

Figure 11.2   Causes of Air Traffic Control System Delays, January–April 1987. SOURCE: Federal Aviation Administration.

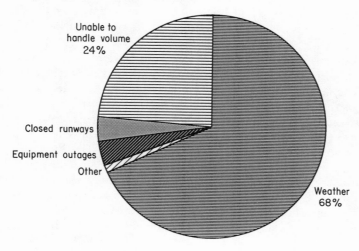

be divided among consumers, airlines, and the government. Consumers bear responsibility because delays arise most often during early morning and early evening rush hours. Business travelers show a strong preference for travel at these times, and thereby strain existing airport and en route capacity.

Air carriers also must assume a share of the blame. Since deregulation, they have moved to adopt hub-and-spoke systems. Bailey and Williams (1987) argued that such systems contribute to stability of profit for carriers. Carlton, Landes, and Posner (1980) pointed out that these systems are also valued by consumers. Consumers who must change planes in an airport strongly prefer single-carrier service because such service involves savings in trip time and reductions in uncertainty and other transaction costs. Moreover, Morrison and Winston (1985) argued that hub-and-spoke operations allow frequent jet service in many city-pair markets whose traffic would otherwise be too sparse to support it. At the hub airport, however, there are now banks of flights queued up to land and then take off again within narrow time frames. Thus, planes are stacked up at hub airports during peak periods, contributing to saturation of the airspace and to concerns about overloading traffic controllers.

Another factor on the airline side that compounds delays relates to flight scheduling practices. During the time frame depicted in Table 11.1, March 1986–March 1987, air carriers declared flight schedules that were unrealistic. Airlines realized that passengers generally try to book the most convenient flight, that is, the one with the shortest elapsed time. The computer reservation systems used by all scheduled airlines and travel agents displayed such flights prominently. The result was a computer full of flight times for every airline between major U. S. cities that represented an industrywide "wish list." Individual air carriers were fearful of dropping optimistic scheduling because of the advantage such a decision might give competitors. Another scheduling practice that contributed to the delay problem was that airlines tended to schedule flights at the most requested time slots, generally the half hour or the hour, and as a group, they often scheduled more arrivals and departures at an airport during peak periods than there was capacity to handle them.

Government resource allocation decisions have also contributed to delays. The Administration dismissed 11,500 striking air traffic controllers in 1981. The new controller work force is still below 1980 levels, despite the strong increase in air traffic over the period. Moreover, many of the newly trained controllers have not yet reached full performance status. Of the additional 6,311 fully qualified controllers authorized, only 3,802 actually are on board. Some of the slowness in training is attributed to limited radar simulation capabilities. Some is due to pay structures that do not accommodate high-cost living areas, such as Chicago and New York.

In any event, as a result of the strike, the FAA changed many of its procedures to reflect the reduced number and experience level of controllers. These changes have substantially increased delays. For example, to

Table 11.2 U.S. Scheduled
Service Flights 1980–1986

Year	Flights (in Millions)
1980	568
1981	395
1982	311
1983	475
1984	589
1985	758
1986	839

SOURCE: Federal Aviation Administration.

ensure safety, given the reduced work force and the fact that many of the new controllers have not yet reached fully qualified status, the FAA spreads flights and discourages circling of aircraft. If there is likely to be congestion or bad weather at a destination airport, the plane is kept on the ground at the originating airport. The FAA also has increased the separation of aircraft in flight from 5 miles to 20 miles in the period 1981–1986.

The administration has also been criticized for slowness in upgrading what many view as antiquated air traffic control systems. A comprehensive National Airspace System plan was published by the FAA in 1981. A key element of this plan was an upgrading of air traffic control with modern, automated systems. All parts of the program are behind schedule, and the advanced automation system is five years behind. It is also felt that the FAA has lagged in introducing a number of technological and procedural enhancements that offer delay relief. One positive sign is that some of these are now being implemented.

Other constraints center on airport and runway capacity. As Table 11.2 shows, flights increased from 311 million during the 1982 recession to some 839 million in 1986. Yet the last major commercial airport built in the United States was Dallas/Fort Worth, which was completed in 1973. Many feel that construction of new runways and terminal facility expansion during the 1978–1987 period substantially lagged the rapid increase in demand of the flying public since deregulation. As of 1986, for example, only about nine new runways at commercial airports are planned for completion by the year 2000.

LINK BETWEEN DELAYS AND SAFETY

Before turning to some solution concepts for the short- and long-run congestion problem, it is well to consider briefly the link between delays and safety. Evidence suggests that despite the recent delays and congestion, aviation is a highly safe method of travel. Moreover, the level of safety has improved, rather than deteriorated, in the period since deregulation. Accidents per aircraft departure for major U. S. air carriers have shown a 35% improvement during the past decade.

However, there is a concern about near collisions between aircraft. These incidents have recently increased at a much faster rate than the increase in the number of flights. The National Transportation Safety Board has cited this as evidence of a deteriorating margin of safety. However, it should be noted that actual midair collisions are extremely rare, and that the connection between reported near midair collisions and actual collisions is controversial. Many of the reports appear to be of "no hazard" incidents.

An intuitive argument is that congested airports and airspace are characterized by an increased probability of collision compared with uncongested areas, and that delays can be seen as symptomatic of highly congested airports. However, to a great extent the delays witnessed today ensure rather than compromise safety. The increased spacing of aircraft and the reduction in numbers of aircraft circling airfields through the policy of holding them on the ground at their originating point may increase delays but reduce the probability of collision. This is not to say that there is no link between congestion and safety concerns. Currently, however, delays are induced in order to ensure safety.

BETTER INFORMATION AS AN AID TO CONSUMERS

Clearly, travelers who have access to carrier-based delay information could benefit by using this information in making flight choices. A passenger wishing to choose between Piedmont and Continental Air Lines on a flight between Norfolk and Newark might well select Piedmont because of its overall on-time record in recent months (77.4% in March 1987 contrasted with 56.9% at Continental) (see Table 11.1).

However, consumers would benefit more from more explicit information. They might wish to know that Piedmont's 6:45 a.m. Flight 16, Norfolk to Newark, was delayed only 3% of the time in March 1987, whereas flights later in the day by either Continental or Piedmont were more likely to experience delays. Or they might wish to learn that Piedmont's late morning Flight 20, Orlando to Dayton, was late 61% of the time during March 1987, while Piedmont flights at other times of the day on this route had better on-time performances.

Indeed, consumers would benefit most by full truth-in-scheduling on a flight-by-flight basis. In March 1987, for example, United Air Lines Flight 1, scheduled to depart from Newark, New Jersey at 7:30 a.m. and to arrive at Chicago O'Hare at 8:52 a.m., had an average delay of 31.8 minutes in good weather and 66 minutes in poor weather. But United Flight 951 from the satellite White Plains (New York) airport leaving at 7:50 a.m. actually arrived in Chicago at 9:11 a.m. with an average delay of just 2.2 minutes in good or poor weather. In other words, the informed consumer who lived roughly equidistant from the two airports could leave later and arrive earlier by selecting the White Plains airport rather than Newark.

Trade practices in 1986 made it virtually impossible for the ordinary consumer to gain such information about scheduling. The only exception was a frequent traveler with a pattern of travel that was both predictable and involved the same city-pairs, for example, travel from Pittsburgh to New York two weekday mornings and evenings each week. For other consumers, some form of information on flight delays would offer the dual advantages of correcting a misleading trade practice and of giving them the ability to minimize travel delay.

We recognize that information is costly and that people get overloaded from too much information. Transaction costs rise as more information needs to be conveyed. Yet, in Bailey and Kirstein (1987), we argued that a requirement for truth in scheduling would not further overload people nor significantly raise transaction costs in the airline industry. Rather, this approach might provide consumers the means to improve their welfare. As Levine (1987) has pointed out, in the air transportation industry information is vital. There is an enormous amount of it, and it changes day by day. Information about fares and schedules must be produced, inventoried, and transmitted to a large group of travel agents for use by an even larger group of consumers. Individual consumers use only a tiny fraction of the information being produced and on only an occasional basis.

The marketplace for information currently provides information about price alternatives in an accurate, if sometimes hard to elicit, manner. The consumer searching for the lowest fare must ask a number of probing questions, book well in advance, meet restrictions as to travel times, and often be willing to pay a cancellation penalty to receive this fare. Similarly, the consumer who cannot meet the restrictions for the lowest possible fare can, with effort, locate the least costly alternative still available. The marketplace also provides nonprice information in some dimensions, including scheduled departure times of flights, availability of first-class seating, and so forth.

However, in spring 1987 accurate information on another nonprice dimension, expected arrival time, was not available to the typical consumer. Instead, the marketplace quoted the same quality of service in terms of total travel time for all flights between an origin and destination city, no matter what the time of day or which particular carrier provided the service. This suppression of true information on the differential delays actually experienced by different flights meant that consumers were being misled. They were unable to adjust their choices on the basis of a dimension of competition that would shift demand for service across time of day, carriers, and airports. As Spence (1976) has argued, efficiency and performance are therefore less than they would be if accurate information on this important quality dimension were provided.

The provision of accurate delay information is well within current computer capabilities, and indeed airlines currently use details of flight-by-flight performance in their internal planning. The missing element was the provision of the true, rather than optimistic, arrival time information to consumers. We believed that for both fairness and efficiency, such information should be reflected in the schedules travel agents and airline representatives give consumers at the time they make their reservations. By receiving accurate estimates about expected travel time, consumers could improve their flight choices. They could make a better guess about the time needed to ensure on-time arrival for a meeting at a distant city. They could schedule more realistic connections. They would make travel choices that minimized their travel time, other things being equal. Specifically, information can be used to achieve the economic benefit of spreading the peak: If delays are longer in peak than off-peak periods, accurate delay information will tend to spread the peak.

This proposition is readily established. Suppose, for simplicity, that travel time during the peak period, t^*, equals a minimum flight time, f, as well as a delay, d. Off-peak travel of duration t incurs no delay; that is, for A^*, A, peak and off-peak arrival times, respectively, and D^*, D, peak and off-peak departure times, respectively:

$$A^* = D^* + t^* = D^* + f + d \qquad (11.1)$$

$$A = D + t = D + f \qquad (11.2)$$

Since consumers care about actual arrival times, they will shift their flights to off-peak travel whenever the experienced arrival time of the off-peak flight is no worse than that of the peak flight:

$$A \leq A^* \qquad (11.3)$$

or, using equations 11.1 and 11.2, so long as the off-peak departure is scheduled to leave by the end of the anticipated delay of the peak flight:

$$D \leq D^* + d \qquad (11.4)$$

Indeed, since consumers value their total travel time, there would be a willingness to switch even for scheduled departure times somewhat later than the delay period d, since the total travel time is then reduced from $f + d$ to f. Thus, we see that two aspects of consumer behavior would tend to spread the peak. First, consumers derive utility from accurate knowledge of actual arrival times, and second, consumers prefer less travel time.

The preceding argument holds as well for other delay causes. If instead of peak and off-peak arrival times A^* and A, we had asserted that one carrier's flights incurred delay d, as in Equation 11.1, while another carrier's flights incurred no delay, as in Equation 11.2; then, by the same line of reasoning as just outlined, consumers would shift to the carrier with no delay for flight schedules departing within the delay period d. Similarly, if the delays were airport related, consumers would switch to an alternate airport if such an airport choice were available to them. Thus, the policy avoids the need to attribute causality to airline delays. Under the truth-in-scheduling approach, the total travel time builds in the average delay regardless of where the delay occurs.

The extent to which information will shift consumer behavior is, of course, largely empirical. Truth-in-scheduling would certainly induce short-run changes in behavior, which could be studied using techniques such as those described in Mohring, Schroeter, and Wiboonchatikula (1987). The economist would expect there to be a transition to a new equilibrium as consumers and carriers adjusted their travel schedules as a result of the truth-in-scheduling requirement.

IMPROVED TRUTH-IN-SCHEDULING

Since May 1987, a number of steps have been taken to implement a truth-in-scheduling policy. A move toward true scheduling that reflects delays that airlines can control, but not those due to problems in air traffic control, was adopted voluntarily by American Airlines. In full-page advertisements, American Airlines described its decision to quote realistic, rather than overly optimistic, flight times to improve on-time dependability. It rescheduled 1,537 of its 1,600 daily departures in an effort to keep on schedule. It also changed its arrival and departure times to ease congestion during peak flying hours. In total, it added more than 150 hours a day of flight time to its schedules. American's new policies have contributed significantly to the solution of the delay problem for its customers. Indeed, by the fall of 1987, it became the carrier with the best on-time arrival figures (see Figure 11.3).

Other carriers began to follow American's lead. As of July 31, 1987, seven major airlines had agreed to revamp a procedure that leads to inaccurate listings of estimated flight times in computer reservations systems. DOT had asked the airlines to make the changes voluntarily, amid widespread criticism from Congress and consumers, and with the threat that stronger measures would otherwise be enacted into law. By late August, six carriers signed a consent decree with DOT promising to end unrealistic flight practices. Beginning November 1, 1987, they adjusted schedules so that their flights operate within 30 minutes of published schedules at least half the time. By April 1, 1988, 75% of flights should operate within 30 minutes of schedule.

Another short-term policy to mitigate delay involves schedule adjustments. Basically, DOT waives antitrust laws that prevent the airlines from discussing business practices among themselves. Airlines then negotiate new

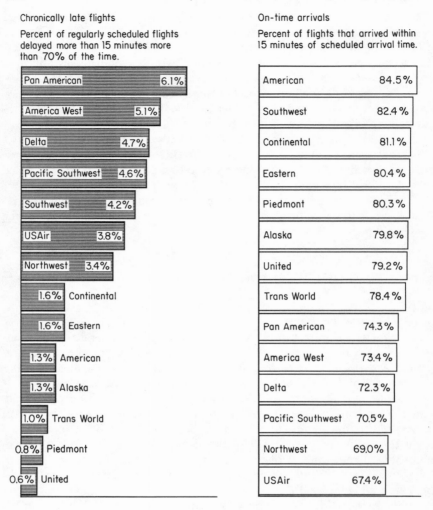

Figure 11.3 First U.S. Rating of Airline Promptness. SOURCE: Department of Transportation.

schedules that serve to better spread peaks. Such schedule adjustment discussions have been standard for a number of years at the highest density airports: New York LaGuardia, Washington National, Chicago O'Hare, and New York Kennedy. In anticipation of the summer 1987 peak season, over 444 takeoffs and landings were changed through such negotiations, affecting a broad group of airports. The largest numbers of changes took place at Atlanta Hartsfield, Dallas/Fort Worth International, and Chicago O'Hare. Most of the schedule changes involved a few minutes to 30 minutes. These agreements did not eliminate delays, but did serve to spread the peak.

An entrepreneurial approach to informing consumers about delays also became available in 1987. A software program devised by St. Louis air traffic controller Wayne Dimmic, called the *Peak Delay Guide*, could be purchased by large firms. The program is proprietary and costs more than $3,000 per year to each of its subscribers. It identifies and displays peak congestion at major airports and calculates the probability of delay and the estimated delay time. It incorporates data that include the scheduling plans of all airlines at a particular airport, the availability of traffic controllers and runway capability, and the likely weather patterns at the airport. The program then computes the strategic points of congestive buildups and letdowns. Each period of the day is assigned a rating that depicts the probability of delay and its severity, expressed in minutes, compared with other periods. The program has found that in some cases a five-minute scheduling difference can make an overwhelming difference in on-time performance.

Throughout this period, Congress has also been active. A comprehensive truth-in-scheduling plan was proposed by Senators Wendell Ford and Nancy Kassebaum in July 1987. In their bill, computerized airline reservations systems would be required:

1. to display an elapsed time for each flight which is not less than a minimum realistic time based on a formula on the distance between the airports, the standard cruise speed for the involved type of aircraft, the typical taxi, landing, and takeoff times for the type of aircraft involved and, where the administrator determines it to be appropriate, for the airport and for meteorological factors;

2. to provide information for each flight which embodies the actual times of arrival for such flight during the previous month.

This bill would thus mandate a base of realistic flight times for all carriers and would provide consumers with more specific information about the immediate history of the particular flight choices they are considering. The threat of this legislation no doubt served as a spur to DOT and helped to stimulate the voluntary actions by carriers to lessen delay.

Another important milestone for improved truth in scheduling has been a DOT rule to require airlines to provide information on flight delays. The first monthly reports, on September 1987 flights, were made available to the public in early November (Figure 11.3 is an example). Major newspaper coverage outlined the performance of all major carriers.

This flurry of activity in the latter part of 1987 meant that detailed information began to make its way to consumers. As Figure 11.1 shows, the number of delays is decreasing. As true delay information becomes incorporated into computerized reservations systems, there should be continued improvement.

Other extremely helpful short-term policies contributing to the reduction in delays portrayed in Figure 11.1 come from recent FAA efforts to change its procedures and introduce improved technologies. For example, there has been a major restructuring of en route control along the East

Coast. New routing rules were put in place in the New York area, permitting an increase from 17 to 27 departure routes. The new plan enables better coordination among facilities and realigns paths permitted in the corridor. This has been highly beneficial to the Boston and New York area airports.

Also, the FAA has introduced a new display system for all U. S. en route control. The new system can provide on a single screen, for the first time, a radar display of every airplane from coast to coast that is flying under air traffic control. Traffic managers can focus on any section of airspace that is threatened with traffic congestion. With greater precision than was previously possible, they can forestall traffic jams. Since September 1987, this new system permits controllers to predict two hours in advance when one of the nation's 649 airspace sectors will be overburdened and to alleviate such traffic congestion by slowing or rerouting individual flights or by ordering longer waits between takeoffs.

Other FAA policies to alleviate congestion have involved changes in procedures to reduce minimum aircraft separation for some airport runways and to ease the rules of separation of aircraft in flight from 20 miles back to 10 miles. Congressional and press-related pressures also caused DOT to agree in summer 1987 to spend an additional $51.5 million and hire 955 more air traffic controllers, supervisors, and managers to meet the next year's traffic growth.

LONGER RUN SOLUTIONS TO CONGESTION

For the longer run, we must look for solutions that truly add capacity to meet the demand for airspace services. When faced with capacity constraints, economists advocate surcharges. The argument is that in deciding whether to schedule another flight out of Chicago O'Hare during rush hour, each carrier should be charged the incremental delay this flight would impose on all flights using the airport during this period. Such tolls would be fed back to airport managers and FAA to increase capacity until the marginal cost of increased capacity would be just equal to the marginal cost of delay (see Chapter 12 in this volume).

In lay terms, the peak-load surcharge has been described in Jones' report (1982, page 3) as follows:

The management of air traffic to produce smoother workloads should be considered as a permanent policy. It would be preferable to persuade user groups to accomplish this objective with a form of market pricing rather than relying on FAA-mandated systems of flow control and general aviation reservations. Smoother traffic produces substantial benefits to FAA work skills, people planning, and working relationships; and greater efficiency in the use of airplanes, airways, airports, control equipment facilities, and human resources through the aviation community.

We have not proposed the use of surcharges in the short run, mainly because this policy would tend to negate consumer gains from deregulation through lower prices, as described in Bailey (1986), without alleviating the causes of stress and constraint. We would, of course, support the use of surcharges if a way could be found to channel the money so collected toward its intended purpose of expanding capacity of the airport and airways system.

To tie fees collected to capacity expansion, there would need to be a change in the current way the system is funded. At present, monies are collected from a variety of sources, mainly an 8% ticket tax and taxes on aviation fuel, and placed in an Aviation Trust Fund. For a variety of reasons, this fund has been politicized and now contains some $5.7 billion in unspent funds. Simultaneously, however, the government has supported FAA needs through a regular appropriations process.

A system is needed to match cost of and demand for airspace services with supply of these services. The beneficiaries of air transportation should pay the costs they impose. There should be an efficient structure linking capacity to demand. Moreover, the overall level of charges should be high enough to give incentives for the development and deployment of new technologies. Basically, the public sector should adopt these technologies with much the same speed as if infrastructure supply was in the hands of private industry. With this in mind, it has been suggested that the FAA be spun off from DOT as a quasi-independent federal corporation. An alternative, supported by the trade association, the Air Transport Association (Bolger, 1988), provides for the reestablishment of the FAA as an independent entity within the federal government. The idea is to bring together airport and airway enhancement programs in a single administrative structure that can match its expenditures against its user fee revenue, that has some incentive to do long-range planning, and that is managed by an administrator who serves for a reasonably long term.

CONCLUSION

In sum, we are pleased that a number of information-based remedies to mitigate airline delays in the short run have been adopted. We especially applaud the move toward truth in scheduling as a short-term contribution to ameliorating the delay problem. The use of new displays and improved procedures by the FAA also accomplish the goal of easing congestion without intrusive intervention of the government into airline scheduling. Both the better information to consumers and the better procedures for controllers serve the same public purpose. There is an easing of congestion and an enhancement of safety without the need for detailed regulation of schedules. In addition, both the improved procedures and the truth-in-scheduling policy avoid placing blame for delays. They are oriented to solving the current congestion problems using visible, powerful displays for traffic controllers and invisible, but equally powerful, forces to inform consumer choice.

Over the longer term, we hope that there will be a better structural solution to congestion and safety concerns. We must ensure that an improved administrative structure is established to modernize the U. S. airport and air traffic control system and that payment for provision of these services by the traveling public is efficiently linked to capacity needs.

12

Congestion Pricing to Improve Air Travel Safety

RICHARD J. ARNOTT and JOSEPH E. STIGLITZ

The potential efficiency gains from congestion pricing have long been appreciated by economists (Pigou, 1920). In this chapter we make the simple but important point that in the context of air travel, *congestion pricing* also *improves safety*—congestion pricing reduces congestion, and reduced congestion improves safety.

THE THEORY OF OPTIMAL PRICING AND CAPACITY OF CONGESTABLE FACILITIES

Drawing on Mohring and Harwitz (1962), Strotz (1965), in a classic paper, derived the remarkable result that

a stretch of road should be taxed, subsidized, or exactly self-financing depending on whether returns to scale of the congestion function are adverse, favorable, or constant; this rule applies whether we have a single road, a multiplicity of roads, a road subject to variation in the flow of traffic; whether the purpose be pleasure or work; whether residential space be considered explicitly; whether the road itself takes up space; or whether there are external economies to the concentration of productive activity.

The results extend to any congestable facility, including the air travel network.

In this section, we rederive the Strotz result.

Social Optimum

In a planned economy, the planner would choose user trip frequencies and capacity to maximize the social surplus (SS) ($\equiv$social benefits, SB $-$ social

costs, SC from travel, which include travel and capacity construction costs).
Thus

$$SS = SB - SC = \sum_i n_i \, B_i[(q_i^t)] - \sum_i \sum_t n_i \, q_i^t \, c_i \, (N^t, I) - K(I) \quad (12.1)$$

The first-order conditions are

$$q_i^t: n_i\left[\left(\frac{\partial B_i}{\partial q_i^t}\right) - c_i^t - \sum_i N_i^t\left(\frac{\partial c_i^t}{\partial N^t}\right)\right] = 0 \quad (12.2)$$

$$I: -\sum_i \sum_t N_i^t\left(\frac{\partial c_i^t}{\partial I}\right) - K' = 0 \quad (12.3)$$

where

n_i = number of (identical) users of type i of the transport facility

q_i^t = trip frequency of each user of type i in time interval t

$B_i[(q_i^t)]$ = gross social benefit of each user of type i over all intervals from trips $\{q_i^t\}$

$N_i^t \equiv n_i \, q_i^t$

$N^t \equiv \sum_i n_i \, q_i^t$ total trip frequency in interval t

I = capacity of public infrastructure

$c_i^t \equiv c_i \, (N^t, I)$ (exclusive of any toll) travel cost of a user of type i in interval t, a function of the capacity of public infrastructure and total trip frequency in the interval[1]

$K(I)$ = amortized construction costs of public infrastructure

K' = derivative of capacity construction cost function

Time intervals are parts of a period, such as hours within a day. The travel cost functions $c_i(\cdot)$ characterize the congestion technology, and have the characteristics that $\partial c_i/\partial N > 0$ and $\partial c_i/\partial I < 0$.

Equation 12.2 states that within each interval each user should take trips to the point where the marginal social benefit $(\partial B_i/\partial q_i^t)$ from a trip equals the marginal social cost $[c_i^t + \sum_i N_i^t (\partial c_i^t/\partial N^t)]$. There are two types of cost associated with an extra trip: the cost of the trip itself, c_i^t and the cost of the extra congestion caused by the trip. The extra trip causes the cost of each trip by a user of type i in interval t to increase by $\partial c_i^t/\partial N^t$, and such users take N_i^t trips in the interval. In standard terminology, c_i^t is the marginal private cost of a trip (MPC_i^t) and $\sum_i N_i^t(\partial c_i^t/\partial N^t)$ is the marginal congestion externality (MCE^t); the sum of the two is the marginal social cost (MSC_i^t). Equation 12.3 states that the capacity of public infrastructure should be expanded to the point where marginal construction cost, K', equals marginal social benefit, $-\sum_i\sum_t N_i^t (\partial c_i^t/\partial I)$. A unit increase in capability reduces the trip cost of a type i user in interval t by $\partial c_i^t/\partial I$, and the cost of all trips by a type i user by $\sum_t q_i^t(\partial c_i^t/\partial I)$, and so on.

Market Equilibrium With an Optimal Congestion Toll

To simplify, we assume, in our discussion of market equilibrium, that the only relevant distortion is the congestion externality. Other distortions in the context of air travel safety are discussed by Stiglitz and Arnott (1988) and in Chapter 4 in this volume. How the analysis can be modified to incorporate them is discussed in Drèze and Stern (1987) and in the appendix to Stiglitz and Arnott (1988).

In a market setting, the government chooses the public infrastructure, but users decide on the number of trips to take in each interval. The government sets the level of the toll in a given interval so that users choose to take the socially optimal number of trips (conditional on capacity) in that interval. In absence of a toll, users face only marginal private cost; in deciding on how many trips to take in an interval, they neglect the congestion externality the trip imposes on others. They will choose to take the optimal number of trips in an interval if they face the marginal social cost of each trip in that interval. The government can get them to face this cost if it sets the toll in an interval equal to that interval's marginal congestion externality (evaluated at optimal traffic, conditional on capacity). Thus, the optimal (denoted by τ_*) congestion toll in interval t conditional on capacity I is

$$\tau_*^t(I) = \sum_i N_i^t \left(\frac{\partial c_i}{\partial N^t} \right) \qquad (12.4)$$

Euler's Theorem

Let $x_1,...,x_m$ denote a set of m inputs and let Q denote the output. They are related by the formula $Q = f(x_1,...x_m)$. Euler's theorem states that if $f(\cdot)$ is homogeneous of degree[2] h, then

$$hQ = \sum_j \left(\frac{\partial f}{\partial x_j} \right) x_j \qquad (12.5)$$

The Financing of Optimal Capacity

Let h^c denote the degree of homogeneity of the congestion cost functions (assumed the same for each group i). If a doubling of trips and the infrastructure capacity leave per-trip travel costs unchanged, then $[c_i(\cdot)]$ is homogeneous of degree zero. If the doubling increases per-trip travel costs, then $h^c > 0$, and so on. Let h^K denote the degree of homogeneity of the construction cost function. If a doubling of capacity more than doubles cost, then $h^K > 1$, and so on.

Applying Equation 12.5 to $c_i(\cdot)$ and $K(\cdot)$, respectively, yields

$$h^K K = IK' \qquad (12.6)$$

$$h^c \, c_i^t = \left(\frac{\partial c_i^t}{\partial N^t}\right) N^t + \left(\frac{\partial c_i^t}{\partial I}\right) I \qquad (12.7)$$

Substitution of Equations 12.6 and 12.7 into Equation 12.3 gives

$$\frac{-\sum_t \sum_i N_i^t\left[h^c c_i^t - \frac{\partial c_i^t}{\partial N^t}\cdot N^t\right] - h^K K}{I} = 0$$

which upon rearrangement gives

$$\sum_t N^t\left[\sum_i N_i^t\left(\frac{\partial c_i^t}{\partial N^t}\right)\right] = K + [(h^K - 1)K + h^c \sum_t \sum_i N_i^t \, c_i^t] \qquad (12.8)$$

Substitution of Equation 12.4 into Equation 12.8 gives

$$\sum_t N^t \, \tau_*^t(I_*) = K + [(h^K - 1)K + h^c \sum_t \sum_i N_i^t \, c_i^t)] \qquad (12.9)$$

In the above formulation, a distinction was made between the physical quantity of capacity, I, and expenditure on capacity, K. Suppressing this distinction, travel costs in time may be taken to be a function of the number of trips taken in that period and *expenditure* on capacity, that is, $c_i^t = \tilde{c}_i(N^t, K)$. To distinguish between $c_i(\cdot)$ and $\tilde{c}_i(\cdot)$, we refer to the former as the travel cost function and to the latter as the congestion function. If $h^{\tilde{c}}$ is taken to be the homogeneity of the congestion function, then Equation 12.9 simplifies to

$$\sum_t N^t \tau_*^t(I_*) = K + h^{\tilde{c}} \sum_t \sum_i N_i^t \, c_i^t \qquad (12.9a)$$

Letting C denote total travel costs over the entire period ($\equiv \sum_t \sum_i N_i^t \, c_i^t$) and R denote the total toll revenue collected over the entire period, Equation 12.9a reduces to

$$R = K + h^{\tilde{c}} C \qquad (12.9b)$$

Since $h^{\tilde{c}} = 0$ corresponds to constant returns to scale in transportation, $h^{\tilde{c}} > 0$ to decreasing returns, and $h^{\tilde{c}} < 0$ to increasing returns, we have:

Proposition 1. With optimal capacity and optimal congestion tolling, toll revenues (a) just cover construction costs if there are constant returns to scale in transportation, (b) more than cover costs if there are decreasing returns, and (c) less than cover costs if there are increasing returns.

Proposition 1 is remarkable because it indicates that the revenue raised from the optimal peak-load pricing of a congestable facility of optimal size that exhibits constant returns should just cover the costs of constructing and operating the facility—the facility should be self-financing. In the context of air travel, one expects the airports of a large city taken together to exhibit decreasing returns or increasing average costs; because of the increased crowding of airspace over the city, a doubling of traffic and capacity should increase the level of congestion. Thus, if metropolitan airports are of optimal scale and peak-load price efficiency, they should raise more than enough revenue to cover the costs of capacity and operation. Due to the fixed costs of building a runway, operating a control tower, and so on, one expects airports in small cities and rural areas to operate in the region of increasing returns to scale. The optimal pricing of such rural airports, if it is of optimal size, should raise insufficient revenue to cover construction and operation costs.

In off-peak hours, the marginal traveler causes virtually no additional congestion. As a result, in the off-peak period the congestion toll is very low. In peak periods, however, since the marginal traveler imposes a significant congestion externality on other travelers, the congestion toll may be substantial. Thus, the imposition of optimal peak-load pricing encourages passengers to travel at less crowded times, which results in an efficient utilization of airport capacity.

Financing with Nonoptimal Pricing and Capacity

A related set of results may be obtained when pricing and/or capacity are nonoptimal:

	$I < I_*$	$I = I_*$	$I > I_*$
$\tau > \tau^*(I)$	$R > K + h^{\tilde{c}}C$	$R > K + h^{\tilde{c}}C$	?
$\tau = \tau^*(I)$	$R > K + h^{\tilde{c}}C$	$R = K + h^{\tilde{c}}C$	$R < K + h^{\tilde{c}}C$
$\tau < \tau^*(I)$	?	$R < K + h^{\tilde{c}}C$	$R < K + h^{\tilde{c}}C$

The results accord with intuition. For example, if optimal congestion tolling is applied to an airport of below-optimal capacity, more revenue will be raised and less spent on capacity than if capacity were optimal; consequently, $R > K + h^{\tilde{c}}C$.

Diagrammatic Presentation

A diagrammatic presentation clarifies these results. To simplify, we assume that demand is uniform over the period and ignore the heterogeneity of travelers. Figure 12.1a shows a plot of long-run average social cost (LRASC) and the corresponding marginal social cost (LRMSC), assuming that there are decreasing costs, which corresponds to $h^{\tilde{c}}$ being negative. The demand curve coincides with the marginal social benefit (MSB) curve, since there

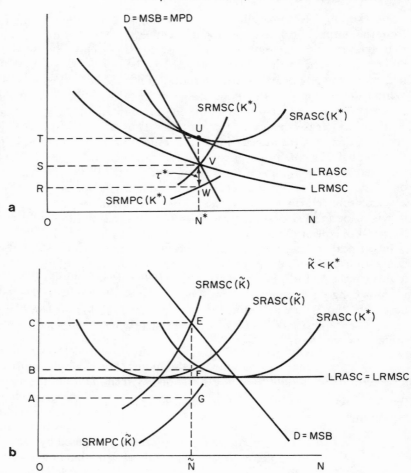

Figure 12.1 Plot of Social Costs (Long-run, LRASC; Short-run, SRASC) and Corresponding Marginal Social Costs (LRMSC, SRMSC) (a) Assuming Costs Are Decreasing and (b) Assuming Costs Are Constant. With Decreasing Costs, the Revenue Raised from Optimal Pricing Is Less than the Costs of Constructing Capacity of Optimal Size. With Constant Costs, the Revenue Raised From Optimal Pricing Exceeds the Costs of Constructing Capacity of Suboptimal Size. See Text for Further Discussion and Explanation of Symbols.

are no other distortions. The optimum occurs where the LRMSC and MSB curves intersect, with N^* trips. Corresponding to this optimum is a level of capacity K^*. The corresponding short-run average social cost curve [SRASC(K^*)] is tangent to LRASC at N^*. The corresponding short-run marginal social cost curve [SRMSC(K^*)] must both pass through the point of minimum SRASC and also (by the envelope theorem) intersect LRMSC and MSB at N^*. Finally, the figure shows the short-run marginal private cost curve [SRMPC(K^*)]. The optimal toll equals the marginal congestion externality evaluated at the social optimum, that is, $\tau^* = [\text{SRMPC}(K^*) - \text{SRMPC}(K^*)]_{N^*}$. Toll revenues are $N^*\tau^* = $ area RSVW. Total costs at the optimum are $[(\text{LRASC})N]_{N^*} = $ area OTUN*, while total travel costs are $[\text{SRMPC})N]_{N^*} = $ area ORWN*. Since total capacity construction costs equal total costs minus total travel costs, total capacity construction costs are given by the area RTUW. Hence, $N^*\tau^* < K^*$, which accords with Equation 12.9b, since $h^{\bar{c}}$ is negative. From the diagram it is evident that the Strotz result (Equation 12.9b) is a particular case of the general proposition that in the presence of decreasing (constant, increasing) costs, marginal cost pricing results in operation at a loss (zero profit, positive profit).

Figure 12.1b illustrates one of the inequality results, that with constant costs ($h^{\bar{c}} = 0$), the revenue raised from the optimal congestion toll exceeds the cost of constructing capacity of suboptimal size. In the diagram, K^* is the optimal expenditure on capacity, whereas $\tilde{K} < K^*$ is the actual expenditure. Conditional on capacity $\tilde{K}$, the optimal number of travelers is $\tilde{N}$, for which marginal social benefit equals short-run marginal social cost. The optimal congestion toll equals the marginal congestion externality with this volume of travelers $= [\text{SRMSC}(\tilde{K}) - \text{SRMPC}(\tilde{K})]_{\tilde{N}}$, and the revenue raised from the optimal toll equals area ACEG. Construction costs equal total costs (area OBF$\tilde{N}$) minus total travel costs (area OAG$\tilde{N}$) – area ABFG, which is smaller than area ACEG.

The actual technology of air travel congestion is considerably more complex than the technology assumed in the model. For example, most of the congestion at airports is due to congestion between *planes* in the air, landing, taxiing to the hangar, and taking off. If the congestion technology were such that the congestion in each of these operations were related to the *number* of planes, optimal pricing would entail a time-varying toll on planes set equal to the marginal congestion externality caused by each plane, independent of the plane's size, weight, takeoff and landing speed, and number of passengers. Imposing the congestion toll on passengers would be inefficient, since it would not give carriers the appropriate incentives; for example, small planes and large planes with few passengers would pay a lower toll than large planes with many passengers, even though all the planes caused the same congestion. It can be shown, however, that the results just developed continue to apply with more realistic descriptions of the congestion technology. Later in this chapter we shall discuss some of the practical aspects of efficient congestion tolling in the context of air travel.

EXTENSION OF THE THEORY TO TREAT SAFETY

There are sufficiently few air travel accidents and the technology of air travel safety is sufficiently complex that our empirical knowledge of the "air travel safety production function" is very poor. Nevertheless, it is safe to say that, all other things being equal, an increase in the level of congestion increases the probability of an air travel accident.

To simplify the notation, we shall assume in this section that passengers are identical, as are time intervals. To extend the theory of the previous section to treat the congestion-related safety costs of air travel, we employ the following additional notation:

s = level of private safety
S = level of public safety
$d(N,I,S,s)$ = expected social damage or accident costs per trip
$c(N,I,S,s)$ = social travel costs per trip, excluding accident costs
$K(I,S)$ = social cost of public infrastructure

Depending on context, S and s will have different interpretations. S may refer to the level of some public safety *activity* (e.g., the frequency of inspection of aircraft safety procedures), to a set of regulations (e.g., regarding quality of runway surface), or to the safety characteristics of public infrastructure (e.g. air traffic control equipment). Likewise, s may refer to the level of some private safety activity (e.g., the frequency of safety inspections of aircraft by the carrier), to a set of regulations (e.g., safety procedures set by the company), or to expenditure on safety equipment (e.g., safer seats in aircraft). I is supposed to capture the *quantity* of public infrastructure (e.g., floor space of an airport or area devoted to runways), whereas S is supposed to capture the *safety quality* of that infrastructure. The distinction between the two is not completely clear-cut.

The Planning Problem

The planning problem is essentially the same as that considered earlier, except for the addition of the safety variables. The planner wishes to maximize

$$SS = nB(q) - nqd(nq,I,S,s) - nqc(nq,I,S,s) - K(I,S) \qquad (12.10)$$

The first-order conditions are

$$q: nB' - n(d + Nd_N) - n(c + Nc_N) = 0 \qquad (12.11)$$

$$I: -Nd_I - Nc_I - K_I = 0 \qquad (12.12)$$

$$S: -nqd_S - nqc_S - K_S = 0 \qquad (12.13)$$

$$s: -nqd_s - nqc_s = 0 \qquad (12.14)$$

where a prime denotes a derivative and a subscript denotes a partial derivative. All are readily interpretable. Equation 12.11 states that the optimal number of trips (per person or per carrier, depending on context) is determined by the condition that the marginal social cost of a trip equals marginal social accident costs plus marginal social travel costs. Equation 12.12 states that expenditure on public infrastructure should be carried to the point where the marginal social benefit, the reduction in travel and accident costs, of infrastructure equals marginal social cost. Equations 12.13 and 12.14 state that the levels of public and private safety should be chosen such that marginal social benefit equals marginal social cost.

Optimal Financing of Capacity

Let $\hat{h}^c$ denote the degree of homogeneity of $c(\cdot)$ *in N and I*. If, for instance, $\hat{h}^c = 0$, then a doubling of traffic and public infrastructure, *holding the levels of private and public safety fixed*, has no effect on the social travel costs of a trip. Define $\hat{h}^d$ and $\hat{h}^K$ accordingly. Then, using Euler's theorem, Equation 12.11 may be rewritten

$$N(c_N N + d_N N) = K + [(\hat{h}^K - 1)K + N\hat{h}^c c + N\hat{h}^d d] \qquad (12.15)$$

The optimal toll should be set equal to the marginal travel cost congestion externality plus the marginal accident cost congestion externality, that is

$$\tau^* (I,S,s) = N(c_N + d_N) \qquad (12.16)$$

travelers should be charged not only for the travel cost congestion externality they impose on others, but also for the expected increase in accident costs to others their trips cause.

Substitution of Equation 12.16 into Equation 12.15 yields

$$R = K + [(\hat{h}^K - 1) \ K + \hat{h}^c C + \hat{h}^d D] \qquad (12.17)$$

where D is total expected accident damage costs. Thus

Proposition 2. If the travel cost and accident cost functions are homogeneous of degree zero in the number of trips and infrastructure size, with levels of private and public safety fixed, and if, in addition, there are constant costs to capacity expansion, with level of public safety fixed, then revenues from the optimal toll just cover the costs of optimal capacity.

We shall omit the interpretation of Equation 12.17 for the other cases, and also the presentation of the inequality version of this equation for nonoptimal capacity and toll size, since it should be evident that the results of the model in the previous section extend straightforwardly to treat the accident costs of increased congestion. For reasons analogous to those given earlier, we should expect optimal toll revenues to more than cover the costs of airport construction, operation, and safety for optimum-sized airports in metropolitan areas, whereas for optimum-sized airports in less densely populated areas, we should expect a revenue shortfall. Thus, for an optimal system of airports, there is no strong a priori case for either aggregate taxation or aggregate subsidization.

PRACTICAL APPLICATION

Several very good papers deal with the technology of air travel congestion. Levine (1969) provided a readable and quite comprehensive qualitative discussion of the technological determinants of air travel congestion, and Walters (1978) gave a useful survey of the literature up to 1976 on the economics of airports and air travel. Empirical studies of air travel congestion are presented in Carlin and Park (1970), Likens (1976), and Morrison (1983), while Dorfman (1983) provides a model of an unregulated airline market.

We know of no papers, however, that discuss the relationship between congestion and air travel safety. As we noted earlier, the reason is that the technology of air travel safety is not sufficiently well understood to establish a priori the quantitative relationship between accident probabilities and various elements of safety, while there are too many variables, compared with the number of observations of accidents, to establish quantitative relationships empirically. In the absence of such knowledge, it would appear sensible to price on the basis of congestion, on the implicit assumption that the marginal safety externality due to congestion is directly proportional to the corresponding marginal travel time externality.

The Current Pricing of Airport Usage

In the United States almost all airports are owned and operated by local authorities on a cost-recovery basis. There is a complex system of federal grants for investment in capacity. Airport concessions are typically charged monopoly rents. And whatever additional revenue is needed to cover costs is generated by airplane-weight-based landing fees, which are independent of the time of day. Congestion is regulated by the nonprice rationing of arrival and departure slots for commercial carriers. The system is similar in other countries, except that abroad passenger fees (arrival and/or departure fees) are typically charged, and some airports use crude forms of peak-load pricing (for example, about 1972, Heathrow Airport in London imposed a flat £20 surcharge on all planes arriving and departing in the peak period [Little & McLeod, 1972]). Even though airlines do not pay higher peak landing fees, this is not to say that passengers at peak times do not currently pay higher average fares than in the off-peak period. Airlines price discriminate among their passengers to maximize revenues, and this dictates that deep discount fares are usually restricted or unavailable in the peak. The deficiencies of the current pricing system will be discussed later in this chapter.

Optimal Pricing of Airport Use

A general principle of decentralized resource allocation is that agents will make socially efficient decisions if they derive the social benefit of their actions and pay the social costs. An airport authority will design runaways efficiently if it faces the full social costs and benefits of any design of runway system. Air carriers should be charged for the full social cost generated by each of their arriving and departing planes. By pricing efficiently, they in turn will cause each passenger to face the full social cost of each airplane trip.

While the general principle is clear, its practical application entails a host of complicating considerations. There are three general sources of these complications. First, there are second-best considerations. The most obvious of these stems from the fact that, due to the impracticality of charging cars for the congestion they create, at least on city streets (but see Vickrey, 1963), automobile congestion is underpriced. If the price of air travel is reduced, this will generate additional car trips to and from the airport (traffic creation) but will also cause some auto travelers to switch to air travel (traffic diversion). Thus, the level of congestion tolls on planes affects the efficiency loss due to underpriced auto congestion. Another possibly important second-best consideration is equity. If the income tax is inequitable,[3] it may be desirable to improve equity by adjusting prices. Since the average air traveler is wealthier than the average citizen, this consideration may provide an argument for setting air fares above what would otherwise be optimal.

Second, airline companies may act as monopolistic competitors rather than as perfect or contestable competitors, and may therefore charge inefficient fares. The form of competition between airlines depends, inter alia, on the form of fixed and sunk costs, as well as on the nature of dynamic competition on networks, which has received little attention in the literature because of its complexity.[4] Optimal congestion tolling would again entail second-best considerations.

Third, stochasticity-unanticipated mechanical difficulties, unpredictable variations in traffic, and variability in weather conditions, which lead to delays in departure and arrival and reduced safety, complicate efficient pricing.

For the moment, we put these complications aside, assuming no distortions in the rest of the economy, competitive pricing by carriers, and no uncertainty. In such an ideal economy, what would be the characteristics of optimal pricing and facility usage? Under the assumed conditions, the envelope theorem applies. The marginal social cost of a flight could then be computed by imagining removing the flight, with all other flights fixed, and calculating the reduction in total social cost.[5]

Based on a quick perusal of the literature, we put forward three tentative stylized facts concerning the cost technology of air travel:

1. It is most efficient for aircraft landing speeds to be the same and for takeoff speeds to be the same.

2. The interval between a large plane and a small plane in takeoff is greater than that between two large planes, since the small plane has to wait until the drag created by the large plane has subsided.
3. The runway damage caused by a plane is based on "footprint pressure," not on weight, and some heavier planes have less footprint pressure than lighter planes because of larger tires and lower tire pressure.

To the extent that these stylized facts are correct, it is evident that the current system of airplane-weight-based landing fees is highly inefficient. Not only does this pricing scheme provide no incentives for planes to take off and land in off-peak hours, but it also provides no incentives for airplane manufacturers to design planes with lower footprint pressure (to reduce runway damage) or with similar takeoff and landing speeds. Furthermore, it greatly undercharges small planes.

The optimal pricing scheme would cause planes to face the social cost of their landing, taxiing, sitting, and takeoff. Since we have for the moment assumed no uncertainty, there would be no queues in takeoff and landing. Planes would therefore be charged on the basis of footprint pressure (to cover runway damage) and the slot time bought, both time and duration (to allocate scarce slot space). The slot duration would reflect the safe time-distance between the aircraft in question and the aircraft ahead of it and behind it.[6] Such pricing would provide air carriers with the incentive to purchase aircraft with socially desirable characteristics, to charge socially efficient fares, and to schedule their departures at socially efficient times. Aircraft manufacturers would have the incentive to produce planes with economically efficient performance characteristics. Local airport authorities would incorporate socially correct landing fees in cost-benefit analyses of infrastructure investments. Also, the revenue raised from such pricing would go a large part of the way or perhaps all the way to financing construction and operation of the optimal air traffic system.

We now briefly consider a couple of aspects of how the optimal pricing scheme should be modified in light of the complications discussed earlier. The modification of prices to reflect distortions and inequities in the rest of the economy would probably entail more trouble than it is worth. Not only would the data requirements to compute second-best prices be formidable, but also a host of conceptually contentious issues are involved.

Several authors (e.g., Vickrey, (1971) have advocated an auction for slot times. This particular form of congestion tolling is sometimes referred to as marketable landing rights. As noted previously, models have not yet been developed to determine the efficiency of such a procedure with dynamic competition on a network. Since there is no general agreement on the characteristics of air carrier competition, we support adoption of an auction for slot times without any modifications to reflect possible imperfections.

The problems stemming from stochasticity, such as delays in departure and arrival due to unanticipated weather conditions, mechanical breakdowns, and so on, and the apparent substantially increased probability of accident from flying in inclement weather, are, however, too important to be ignored in a pricing scheme. At one extreme is completely ex ante pricing, whereby planes and passengers are charged on the basis of expected cost without reference to the actual state of congestion. The equilibrating variables are delay and flight cancellations. At the other extreme is ex post pricing, which charges on the basis of realized cost. Intermediate pricing schemes are preferable to either extreme. We shall discuss only one possible feature of such a scheme: priority classification of flights.

One of the disadvantages of complete ex ante pricing is that in bad weather, travelers with a high shadow value of time are delayed as long as those with a low shadow value. This situation can be improved by having a flight priority system similar to that employed in computer batch processing, whereby those flights with a higher priority would move up the takeoff queue faster. The local authority could auction off x high-priority slots per hour, subsequently y medium priority slots, and finally the residual slots at low priority. Carriers would pass on the increased cost of a high-priority slot in the form of higher fares to passengers. Those passengers with tight schedules, primarily business people, would choose to pay the premium to reduce expected delays. The typical vacationer, meanwhile, would choose a low-priority flight. With bad weather, the order of flight cancellation could also be based on this priority system. Passengers would then be able to telephone the airport prior to leaving home to ascertain which priority flights had been cancelled. A similar system could be organized for arrivals. Carriers would purchase scheduled arrival time slots, higher priority flights would move up more quickly in the landing queue, and low-priority flights would be the first to be rerouted to land at other airports.

The pricing system just outlined—auctioning of priority slots—would have numerous advantages over the current system. Demand would be smoothed over the day, which would result in a more efficient use of existing capacity, reduced delay, and reduced (optimal) capacity requirements. Air traffic, especially smaller planes, would be induced to switch from more to less crowded airports. Business people would be able to avoid delays, while vacationers would be able to travel more cheaply by choosing low-priority flights at off-peak hours. Safety would presumably be improved. And the air traffic system would be closer to self-financing.

Optimal Financing of Air Traffic Facilities

Should all the revenue collected at a specific airport be spent on the construction and maintenance of facilities at that airport—*specific* earmarking? Instead, should the revenue raised from all airports together be pooled and allocated across airports according to their operating expenses and need for capacity expansion—*general* earmarking? If funds are perfectly fungible and properly allocated, the answer to both questions is no. The revenues should be put into the general revenue fund and spent on whatever creates the greatest social benefit. However, because of the sometimes perverse incentives facing politicians and bureaucrats, resulting from problems in institutional design, funds are not always spent on their best use, in which case some earmarking may be desirable. As we have seen, specific earmarking would generally be undesirable because it would result in excess capacity at metropolitan airports and inefficient capacity at small city and rural airports. One means of circumventing this problem would be for the federal government to tax metropolitan airports (either in lump-sum fashion or on the basis of revenue collected) and to subsidize airports in small cities and rural areas.

Why Hasn't Congestion Tolling Been Implemented?

If congestion tolling can be so beneficial, why has it not been implemented? This question entails two subquestions. Are there legitimate arguments against congestion tolling? And accepting for the sake of argument that the social gains from congestion pricing are substantial, why might it nonetheless not be implemented?

A commonly heard argument against congestion tolling is that customers would have difficulty in understanding it and adjusting to it. This argument has some validity in the context of, say, household electricity pricing. For instance, the time allocation process would be made considerably more difficult if householders had to take electricity pricing into account in deciding when to launder the clothes or operate the dishwasher. But the complexity of the air traveler's decision would be little affected by congestion tolling. As it is, the system of airline pricing is sufficiently complex that typical travelers simply ask their travel agents for a list of travel options, along with prices, and make their decisions. The process would be no more complex with the congestion tolling scheme advocated earlier.

Another argument is that congestion tolling hurts the poor. This is true in some contexts. For example, blue-collar workers typically have less time flexibility than white-collar workers (see Small, 1982). They generally have to perform their routine activities at peak periods and would therefore pay heavily for the congestion tolling of urban transportation as well as of electricity and water consumption. The argument carries little force, however, in the context of air travel, where poorer travelers who are typically on vacation have more time flexibility than richer travelers who are typically on business. In this context, congestion pricing would probably benefit all classes of travelers. The benefit vacationers would receive from lower fares would more than offset the disutility of traveling in off-peak periods, and the benefit business persons would receive from reduced delays in peak periods would more than offset the higher fares they would have to pay.

Why has a system of congestion tolls not been implemented? There are a number of possible reasons, all stemming from the nexus of incentives, political economy, and institutional structure. First, there is a marked revealed preference by bureaucracies for regulation over pricing, for direct control over indirect control. While a set of regulations can under ideal conditions perform as well as pricing, in practice pricing is typically more effective. For one thing pricing is flexible, whereas regulations are rigid: Prices respond automatically and instantaneously to shocks and policy changes, whereas changing regulations is a slow and cumbersome discretionary process. For another, pricing is informationally more efficient than regulation. As an example, under a system of slot allocation a bureaucrat would have to collect a lot of information and go through complex computations to determine the efficient allocation, whereas under an ideal slot auction the efficient allocation would be achieved automatically. Relatedly, under a system of regulation the bureaucrat receives little information to indicate how effective the regulations are, and if there are problems, how to correct them.

A second possible reason that congestion tolls have not been implemented stems from the discretionary nature of the regulatory process. Almost any change results in winners and losers. Potential losers will fight hard, through the political process, to maintain the status quo. Airline companies are likely to oppose congestion pricing at airports, perceiving, perhaps incorrectly, that such pricing would lower their profits. Business people, too, may oppose such pricing, failing to acknowledge the trade-off between the level of fares and the length of delays. Inefficient regulation is more likely to persist when the gains from regulation are more concentrated and perceptible than the losses. In the context of air travel, the principal losers are taxpayers whose tax dollars go toward capacity expansion, which should be financed largely or completely from congestion toll revenue, but these losses are so diffuse and indirect that taxpayers have little incentive to lobby for congestion pricing at airports, even if, as is unlikely, they perceive the benefits from doing so.

A third hypothesis is that the agents who make the pricing decisions face incentives, deriving from the bureaucratic, institutional, and political structures, as well as from tax and subsidy policy, which discourage them from adopting congestion pricing. As an example, suppose that the federal government had the sole responsibility for congestion pricing and capacity expansion at all the nation's airports. Because of the complexity of determining efficient congestion prices and the appropriate pattern of capacity expansions over time, pork-barrel politics might interfere with the efficient allocation of resources, and even when it did not, the federal government might be accused of political favoritism.

Suppose, alternatively, that local airport authorities had the power to set congestion fees, but that all the revenues were sent to the federal government, which allocated these revenues to expand capacity at the most congested airports as well as to cover airport operating expenses. A Bertrand-Nash airport authority might reason that by lowering its fees below those of other airport authorities, not only would it generate more traffic, which would be good for local business, but also, by inducing increased congestion, it would generate additional funds from the federal government for airport expansion. This institutional structure would therefore likely result in excessively low fees.

Finally, suppose that the local authority had the power to set fees and to allocate slots, or to auction off slots, and that its fee revenue (after transfers by the federal government from metropolitan to rural and small-town airports) was earmarked for capacity expansion and operating expenses at the local airport. Under perfect competition between airports, this institutional structure would be efficient.[7] However, the monopoly status of local airport authorities and the friction of space combine to give local airport authorities some monopoly power. The outcome would then depend on the objective of the local authorities and the nature of the game played between them. If their aim were to maximize revenue and capacity, fees would be too high. Alternatively, if they destructively competed with one another to increase traffic or to attract a carrier to use their airport as a hub, fees would be too low.

Although not necessarily the best possible, the policy we outlined earlier, compulsory slot auctioning with general earmarking of the revenue, would likely be preferable to any of the previous three institutional structures.

Other Policies

While congestion pricing would go a long way toward solving the current congestion problems at airports and toward improving air travel safety, it is not a panacea. Stiglitz and Arnott (1988) and Panzar and Savage (Chapter 4 in this volume), have discussed various sources of market failure vis-à-vis air travel safety, most of which are best addressed via safety regulation. For example: There are several reasons to expect that, in the absence of regulation, the agents who make safety decisions are unlikely to face the correct incentives. First, the notion of legal liability differs substantially from the economist's notion of marginal responsibility. For instance, a carrier may be found liable because of operator error. But the operator might not have made the error if there had been less congestion at the airport, or if the government safety procedures had been better researched, or if the airport had provided better runway lighting, and so on. Thus, charging for accidents on the basis of legal liability, even in the absence of insurance, is unlikely to provide appropriate incentives for safety. Second, the provision of insurance blunts incentives to take care. In some contexts, this moral hazard problem can be substantially mitigated by means of experience rating, but airline accidents are sufficiently infrequent to render the experience rating of insurance ineffective. Third, fly-by-night operators might, in the absence of regulation, neglect safety, reasoning that they could declare bankruptcy should one of their planes have a serious accident. Fourth, consumers are ignorant of safety technology and the safety record of air carriers and cannot therefore properly trade off safety against price in deciding between carriers.

Whereas safety regulations are manifestly desirable and necessary, the same is not true of fare, route, and entry regulations. Prior to deregulation, fares were set too high, load factors were too low (Morrison & Winston, 1986), and slots were allocated inefficiently (Koran & Ogur, 1985). Morrison and Winston (1986) estimated that the welfare gains from deregulation to consumers have exceeded $6 billion per year, while industry profits have increased by more than $2.5 billion per year, both at 1977 prices. Substantial further gains can be achieved by moving from the current system of airport pricing to the system advocated here, slot auctioning (marketable landing rights) with priority flights. Such a system, combined with general earmarking, has the added advantage of providing better incentives for capacity expansion than the current system.

CONCLUDING COMMENTS

In the first half of this chapter we reviewed the basic model of optimal pricing and capacity of congestable facilities in the absence of distortions. We showed that optimal peak-load (or time-varying) congestion pricing leads to the efficient utilization of capacity. Furthermore, the revenue collected from optimal peak-load tolling generates just the right amount of revenue to finance optimal capacity if a doubling of passengers and capacity expenditure results in the same level of congestion. In metropolitan areas, since airspace is crowded, a doubling of capacity expenditure and traffic is likely to increase the level of congestion, in which case toll revenues would more than cover the costs of optimal capacity; in lightly populated areas, because of the fixed costs involved in constructing a one-runway airport, a doubling of capacity expenditure and traffic is likely to reduce congestion, in which case toll revenues would fall short of optimal capacity expenditure. Although congestion tolling might not result in the air traffic system being fully self-financing, the degree of self-financing would certainly be higher than it is under the current system.

We then showed that the model can be extended straightforwardly to incorporate congestion-related safety factors, by interpreting congestion costs to include not only the time costs of congestion but also the safety costs.

In the second half of the chapter we discussed issues related to the practical implementation of congestion tolling. We argued that the current system of airport finance, which entails a complex system of federal subsidies for capacity expansion, monopoly pricing of airport concessions, weight-based landing fees, and slot allocation is most likely highly inefficient. Not only does it fail to distribute traffic efficiently between peak and off-peak periods, but also, since the congestion externality caused by a plane is only very weakly related to its weight, it fails to price correctly between different classes of aircraft.

We then proposed a congestion-based pricing system which, we think, achieves a sensible balance between practicability of implementation and efficiency. Under this system:

Each flight would have a priority, with higher priority flights being able to move up the landing and departure queues more rapidly.[8]

Airports would allocate a certain proportion of slot time in each period to the various priorities of flight.

Airports would be *required* to sell slot time at the various priorities and for each period by public auction.

The federal government would tax the auction revenue raised at metropolitan airports and use the tax revenue to finance capacity expansion at other airports.

Local airport authorities would be permitted to use their after-tax-or-transfer auction revenue, plus other revenues raised,[9] for operations and capacity, as they saw fit, subject to a break-even requirement and federal safety regulations.

Implementation of this policy would (1) result in a more efficient utilization of aircraft and infrastructure, (2) reduce peak-period congestion and hence congestion-related accidents, (3) help the business traveler by substantially reducing uncertain peak-period delays, especially on high-priority flights, (4) help the casual traveler through a reduction in the cheapest (low-priority, off-peak) fare, (5) make the air traffic system closer to self-financing, (6) result in administrative simplification.

Little in our proposal is new. Economists have been arguing for many years, and with virtual unanimity, for congestion pricing of air travel in one form or another. That congestion pricing has not been implemented suggests that there must be serious problems in the design of institutions vis-à-vis air travel. Thus, research on the redesign of these institutions to give the agents responsible for pricing and capacity policy the incentives to make efficient decisions should be accorded high priority.

We have focused on the role of congestion pricing in improving air travel safety, arguing basically that congestion pricing reduces congestion and reduced congestion improves safety. There are, however, many margins of safety over which congestion pricing is ineffective. For these margins, safety regulation and other pricing policies are appropriate. Since an unfettered market would seriously fail to provide proper safety, the efficient provision of air travel safety requires both congestion pricing and extensive safety regulation.

NOTES

1. To simplify the analysis somewhat, we have assumed "anonymous crowding"—that the level of congestion depends on the total number of users, independent of the composition of users by type. It is straightforward to extend the analysis to treat nonanonymous crowding, but then different types of users should be charged tolls of different sizes. In some contexts, this may not be practical, in which case determination of the optimal toll is a second-best problem. But in other contexts, differential tolling is feasible: For example, the congestion

caused by a plane depends on its performance characteristics, and tolling based on such characteristics is feasible.

2. That is, $\lambda^h Q = f(\lambda x_1, \ldots, \lambda x_m)$ for all λ and all vectors of inputs.

3. Broadly speaking, the primary responsibility for achieving an equitable distribution of income rests with the income tax. Unless it can be persuasively argued that the income tax is inequitable, we consider it generally ill-advised to redistribute through the price system. For a precise statement of this point see Atkinson and Stiglitz (1980).

4. There is a complicated pattern of externalities between flights of different carriers. Suppose, for example, that carrier A has a flight from Washington, D. C. which arrives at New York at 10 a.m. and a flight from New York to Montreal at noon. If carrier B introduces an 11 a.m. flight from New York to Montreal, it will divert traffic from carrier A's noon flight. If, however, carrier A did not have a noon flight, carrier B's 11 a.m. flight would generate additional traffic on carrier A's Washington-New York flight.

Under the hub-and-spoke system, each major carrier concentrates its transfers at a particular airport, and different major carriers tend to use different airports. One probable reason this system has developed is that it internalizes the externalities discussed above. Another is that it reduce intercarrier transfers, which passengers dislike because of bad connections resulting from the difficulty each carrier has in coordinating its flights with other carriers.

5. This is the implicit conceptual exercise performed in most discussions of the efficient pricing of air travel.

6. Differences in airplane speeds and weights give rise to nonconvexities. As a result, pricing by itself cannot in general achieve full optimality. It may be necessary to supplement pricing with some segregation of planes, according to their performance characteristics, by time of day, runway, or airport.

7. The airports would act as competing clubs, and, as is familiar from the literature on clubs and the Tiebout hypothesis (e.g., Scotchmer and Wooders, 1987), the resulting allocation would be efficient.

8. Furthermore, in the event of inclement weather, the lowest priority flights would be the first to be canceled, and so on.

9. Concessions, weight-based landing fees to cover marginal runway damage, and so on.

13

Economic Deregulation's Unintended but Inevitable Impact on Airline Safety

JOHN J. NANCE

In 1986, the seventh year after the airline industry of the United States was deregulated by an act of Congress, the Federal Aviation Administration (FAA) suddenly announced it was seeking a $9.2 million fine against one of the major air carriers in the nation, Eastern Air Lines, for violating FAA regulations a staggering total of more than 78,000 times. While the news was a shock to the industry and the flying public, it was merely an extreme example of what was rapidly becoming a common announcement, major fines for significant violations against a lengthy list of America's finest carriers.

In the previous year, for example, a long procession of major and intermediate-sized airlines had been caught falling below FAA minimum standards in various areas and fined amounts ranging from several hundred thousand dollars (Alaska Airlines) to the staggering fine that Eastern finally paid in full. Violations and fines had been levied before against major airlines, but the frequency with which inspections were turning up a great number of deficiencies, and the seriousness of the deficiencies themselves, were unprecedented. The conclusion was becoming unavoidable that something was amiss in the industry, and the only rational common denominator was the ruinous competition born of deregulation.

Two years before, *Newsweek* (January 30, 1984), in an eight-page article entitled "Can we keep the skies safe?," had exposed to the public the worries of thousands of airline employees, safety experts, researchers, congress persons, and FAA inspectors in the first major crack in the dam of official denial that deregulation and safety were even remotely related issues. Perhaps as a result of the criticism, Department of Transportation (DOT) Secretary Elizabeth Dole two months later launched the air carrier inspectors of the FAA on a massive "white glove" inspection of all certificated airlines. The National Air Transportation Inspection (NATI) was to determine in part whether deregulation was causing any decrease in the level of safety. The year after the various phases of the inspection were complete, an internal FAA report documented the fact that the NATI inspection had uncovered widespread failings on the part of both the FAA and the industry, failings unequivocally resulting from the immense pressures and systemwide competitive chaos created by the Airline Deregulation Act of 1978.

And then came 1985, a year so strewn with aircraft wreckage and deceased passengers worldwide that even though only a few of the airline accidents of that year were related in any way to U. S. deregulation, growing internal worries about the state of the airline safety system began to get the public's attention. Crashes such as that of a Delta Air Lines Lockheed 1011 in Dallas and the slaughter of 249 U. S. servicemen in a military chartered Arrow Air DC-8 in Gander, Newfoundland, both had valid deregulatory involvement, as did several commuter crashes. In Delta's case, the policies, equipment, and capabilities of both the FAA and Delta's training program were called into question by the National Transportation Safety Board's (NTSB) 1986 accident report (National Transportation Safety Board, 1986b), and between the lines of each finding the negative influence of an aviation system strained to the limit by deregulatory pressures could be clearly seen.

In the Gander crash, the airline involved would not have been in existence without deregulation. In fact, the sad reality was that Arrow Air fit the predictable mold of the inevitable deregulatory opportunist: an "airline" that had every intention of operating at the lowest permissible level of maintenance and training cost for the maximum amount of profit, and that apparently had no intention of helping in any way to contribute to the building of the national air transportation system.

By the spring of 1986, many media were beginning to ask worried questions about deregulation's impact on air safety and air traffic control, and several major newspapers such as the *Denver Post,* the *Dallas Times Herald,* and the *Miami Herald* did month-long independent, exhaustive studies of their own; each firmly validating the fact that deregulation has had a widespread, complex, and worrisome impact on airline safety and on air safety in general. Suddenly such investigative reports, national television coverage, leaked governmental reports, and congressional testimony began to remove the air of legitimacy and authority from assurances that deregulation was blameless in matters of airline safety. In addition, the widespread reports of safety level deterioration were confirmed in private by hundreds of pilots, controllers, mechanics, and FAA inspectors—involved operationally sophisticated people who had been reporting such things in vain since 1980.

By mid-1987, the subjective evidence that airline safety had been impacted by economic deregulation had mounted enough to override the Administration's denials. While most technically knowledgeable airline people agreed that the airline industry's operations had not become unsafe, it was also becoming apparent that the safety margin had in fact been reduced. By definition, that meant that the airline industry in general had become less safe than it once was, and less safe than it could be, in ways that were very difficult if not impossible to measure by traditional objective statistical methods.

Two realizations have emerged from the painful process of dawning awareness of trouble: (1) Though economic deregulation had inadvertently, but predictably, created sufficient operational chaos and cost cutting since 1978 to significantly reduce the safety margin of the airline system of the United States, this decline has nevertheless failed to show up as a measurable increase in accidents and/or casualties. (2) Any significant reductions in the safety margin of the airline system must be viewed as potentially increasing the public's exposure to accidents, and any such increases in risk should be viewed as seriously as actual accidents and deaths resulting from airline operations. (3) Therefore, since the traditional statistics based on accident data cannot in practice measure actual reductions in the safety margin unless such margin no longer exists and accidents result, effective national monitoring of airline safety on a real-time basis will be impossible unless and until new standard methods of measuring and reporting fluctuations in the safety margin are developed and used by everyone involved.

THE UNFORTUNATE EFFECTS OF COMPLETE DEREGULATION

But wait a minute—none of this was supposed to happen. Economic deregulation of the scheduled airlines was not supposed to affect safety anywhere, as the man who is widely credited with being the father of airline deregulation, Cornell economics professor Alfred Kahn, has been quick to point out.

In fact, by the very text of committee members' statements in the numerous House and Senate hearings on airline deregulation in 1976 through 1978, it is clear that safety was never supposed to be an issue in the first place. According to the participants themselves, deregulation was designed to lower prices and cure what many considered inefficiencies that had built up under the benevolent dictatorship of the Civil Aeronautics Board (CAB), including overly fat, overly protected, and overly paid airline corporate leaders and their union employees. Airline safety, which was excellent but not perfect in 1978, was not supposed to deteriorate below the level then enjoyed by the flying public. The Congress made that promise in plain and simple language and repeated it often, to quell the extreme worries of the airline industry, which mostly opposed the bill, and of the public, especially the mayors and airport managers of hundreds of smaller U. S. communities, who correctly feared the loss of decent, safe airline service as a result of the measure.

But having convinced itself, Congress brushed off all objections and rushed an industry that many regarded as a public utility into the fires of unfettered, uncontrolled, free-market competition. It was little more than an experiment, a chance taken with an imperfect but incredibly safe, stable, and convenient national transportation system. It was an experiment that had the potential to endanger the lives and welfare of hundreds of thousands of people, employees and passengers alike. And it was an unfortunate fact that any early warnings of slipping safety margins would be sounded by those who had little or no credibility with financially pressured managements or a free-market-dedicated Administration—the technical and operational professionals who are usually members of labor unions.

Unfortunately, that congressional promise to the American public that economic deregulation would not affect safety, however honest in intent, was empty and inept from the first because of an amazing misunderstanding of the role of the FAA and the human nature of the airline business. Congress thought the FAA, through its air carrier inspectors, could simply police the industry and force it to comply with the rules. If airlines comply with the rules, the reasoning went, they will be safe, and if they are safe, they will be as safe as they were when we deregulated.

The rules, however, were the minimums, and the nation's airlines had always maintained a level of safety—in maintenance, training, and operational areas—far above those federally mandated minimums. In fact the established airlines had traditionally set the standards, while the FAA set the basic minimums. More often than not, those two levels were far apart. For example, an airline might provide 290 hours of instruction for each new pilot in ground training as a flight engineer, whereas the federal minimum required only 235 hours. Airlines indulged in excessive training because it was good for safety and reliability, and because they could afford it in a regulated environment. But the airline could choose to reduce those training hours at will as long as they didn't drop below the legal minimum. The resulting course at 235 hours might be quite inferior to the quality of the course presented previously, but it would be legal and less expensive, and the FAA would be powerless to stop the decline.

The point that Congress missed was the fact that the level of airline safety the public had come to expect was built by established airlines that had enough money to be able to indulge themselves in the highest level of airline safety attainable, which translated to levels of compliance that were far above the minimums. In fact, many of the major carriers, pioneer airlines such as United Air Lines, Delta, Braniff Airways, American Airlines, Eastern, Pan American World Airways, Western Air Lines, and Northwest Airlines, to name a few, had spent hundreds of millions of dollars over many decades to build gigantic maintenance, training, and operational centers staffed by hundreds of highly skilled technical people who considered their jobs as careers, and who acted in most cases accordingly. In addition, expensive exchanges of information and equipment, manpower, and technical advances in safety over the years were responsible for considerable progress in the quest for zero crashes and zero passenger casualties.

Why are all these expenditures significant? Because none of them was required by federal rules. The very facilities and programs that were the underpinnings of airline safety in the United States had been built by airlines operating far above and beyond any minimum government standards, and with dollars that came from the stability of a regulated industry.

Back in the summer of 1978 during the hearings into airline deregulation, senior airline executives joined union leaders in trying to educate lawmakers in the realities of the airline business, and in what would happen to the long-run stability of the safety system under a totally deregulated environment. The veteran airline people painted a picture of the type of air carrier deregulation would create, a pseudo-airline that would have little dedication to building a stable airline system and no motivation to spend millions for maintenance and training facilities not required by the FAA—that would be in the business for profit alone. Such a carrier would resemble neither the existing intrastate airlines that had been used as a model by the proponents of deregulation, such as Southwest Airlines of Texas or Pacific Southwest Airlines of California, nor the established airlines. This new type of deregulatory airline would lease older airliners, hire inexperienced or furloughed crew members at cut-rate wages, and even lease their maintenance and training facilities. Once such a company had established a cost base far lower than any established carrier could achieve, it would ignore the marginal or nonprofitable routes in the major airlines' systems and come after the high-density, high-profit routes. The profit from those high-density routes, which had been protected from ruinous competition under regulation, generated enough income for established carriers to support their marginal routes to smaller communities. More important, however, the extra dollars of profit from those routes fueled the safety system itself, providing the money for building and maintaining established training and maintenance facilities, the maintenance and training personnel, and the programs themselves; developing and buying new or advanced electronic equipment; having properly equipped hangars in more than one location for a stable maintenance flow; and a thousand other items necessary to a stable, professional, and safe operation.

The dilution of income from intense competition on their most profitable routes, they warned, would force the established carriers to find places to rapidly cut their costs. That would mean that every aspect of an airline's operational expense would have to come under instant scrutiny for reasons of financial survival, including maintenance and training budgets.

The statements were prophetic and accurate, as a long list of new-entry, deregulatory airlines have demonstrated over the ensuing decade. New airlines such as Hawaii Express (bankrupt), Pacific East Air (bankrupt), Pacific Express (bankrupt), People Express (effectively bankrupt/sold), and Arrow Air (bankrupt); along with reformed or massively expanded carriers such as Capitol Airlines (bankrupt), Frontier Airlines (bankrupt), Air

Florida (bankrupt), Northeastern Airlines (bankrupt), Air Illinois (bankrupt), Provincetown-Boston Airlines (bankrupt/sold to People Express); and many others proved to the consternation of their FAA inspectors that an airline can slash maintenance stocks, the numbers and experience level of maintenance workers, maintenance programs and facilities, and maintenance schedules far below those of the established carriers without ever violating the federal minimums. They showed graphically that marginally qualified pilots and flight attendants can be given the absolute minimum in training and supervision, paid wages that force a perpetually high turnover, low-experience profile in the company's operational ranks, and still be kept just at the legal minimums to fly the public around as a full-fledged airline. They demonstrated conclusively that the use of older airliners, many of them leased for short terms and maintained by a scattering of loosely controlled contract maintenance organizations, meant the commercial use of aircraft that were seldom repaired, monitored, overhauled, or checked with the same degree of care and precision as the fleet operated by United or Delta or American. Though these carriers eventually failed economically, while they were serving the public as certificated air carriers they were providing a demonstrably lower level of safety than the established airlines whose routes they were raiding and whose financial health they were imperiling. They may have remained legal, but they were not as safe.

In operating that way, these airlines were transporting people who blindly trusted that any airline permitted to operate in this country would be as safe as any other. Clearly there were and are new-entry companies that tried to stay far above the minimums from the very first and for the most part succeeded, but they are the exception, not the rule. Airlines that expanded too fast from commuter status, such as Provincetown-Boston Airlines (PBA), provide an instructive example both of how cut-rate philosophies lead to cynical compromise of the rules, and of how difficult if not impossible it is for the FAA to catch cheaters in time to protect the public. PBA continued its expansion and operations for two years after a dismissed crew member tried to tell the FAA of massive violations and cheating. When the FAA was finally forced to investigate, they discovered a total of 13 pages of violations and assessed the death penalty for the airline. PBA's certificate to operate was revoked in October of 1984.

The effects of the deregulatory process were not to stop with the new entry carriers or newly reformed or expanded ones. No aspect of commercial aviation has escaped unaffected, despite the wishful thinking of deregulatory proponents that, since only the economic aspect of the scheduled airline industry was deregulated, only that small area would be altered. The turbulent changes and air traffic pressures born of deregulation have materially affected air traffic control, FAA surveillance activities, airport planning and utilization patterns, and the usability of American airspace by private and military traffic. In addition, all forms of human factors research and human factors training, the advance of judgment training for pilots, the promulgation of advanced electronic collision avoidance aids, and hundreds of other high-priority items have been delayed for years by financial uncertainty, technically uninformed management, and scrambled priorities caused specifically by deregulation. Deregulation has, in other words, affected all of American aviation, not just the scheduled airlines.

As the veteran airline people warned, the effect on established carriers has been profound. As we enter the 10th year of deregulation, we do so with a greatly altered industry consisting of a handful of megacarriers, all of whom have been fighting a constant battle to find places to cut expenses without digging too deeply into the budgets of their safety-related departments. A few of the majors, notably those whose survival is still in doubt, have actually caused a significant decline in their safety standards, as FAA enforcement actions and independent investigations have shown. Some have cut very little, because their size and profitability, aided by the benefits of mergers and consolidations, have kept them from being financially imperiled. All airlines, however, have engaged in cost cutting in maintenance, training, operations, and wages to some extent to stay viable.

While undoubtedly many such cuts are inconsequential in their real effect on airline safety, they are akin to the wielding of a scalpel in a delicate operation: One significant mistake can be fatal. The demurrer from the airline industry to all these points is usually as short and as impassioned as the recent retort from the chairman of a substantial east coast carrier: "Now just a second here," he responded, "responsible airline people would never knowingly do anything to cut or compromise safety!" This is a true statement; never would such people knowingly compromise safety. The key word, of course, is "knowingly."

The natural propensity of responsible airline managers and owners is to have the highest possible level of operational safety. As guardians of a public trust, there is no other acceptable viewpoint. As business people, no matter how ignorant of airline operational realities, there is no other logical course of action. A lesser safety goal for an airline may lead to a reputation as unsafe or less safe, and that could frighten away passengers and raise insurance rates. However, recent experience suggests that the idea that a crash will cause a major public shunning of an airline is false (see Chapter 5 in this volume).

The sometimes frantic cost cutting that occurs when an established carrier suddenly finds its existence imperiled or its profit margins seriously eroded creates an atmosphere of frantic budget slashing in which people are tempted to make subjective decisions using only objective criteria.

For example, when vice presidents of maintenance, especially in smaller carriers, are called in to explain why they have not met their budget goals, they are likely to be asked to cost justify in excruciating detail the major areas of their budget. FAA-required items are, of course, a given element. So, in most cases, are the basic costs of maintenance equipment already in place. But, for example, when asked why maintenance personnel undertake an expensive inspection every 25 hours when the FAA would probably approve an increase to 35 hours, or to justify extra quality circle/quality control training or other courses for workers not required by the FAA; or when forced to defend a work schedule that provides sufficient line mechanics to handle the peak periods of arrivals and departures when less experienced and less expensive part-time contract mechanics could be substituted; or when challenged to justify the presence of one Boeing 727 electrical generator spare on the shelf in each out-station maintenance facility when, in the case of a breakdown, crews might be cajoled into flying their airplane back to the airline's hub city with an inoperative generator on an approved FAA-deferred maintenance basis, the maintenance vice presidents are going to be hard pressed to convince executives who know little or nothing of the operational realities of airline flying and airline safety that nothing should be cut. Executives are struggling with issues of corporate survival, and cost cutting is their principal tool. The personnel costs, especially with work forces defended to some degree by unions against complete impoverishment, can bear only so much of the burden. As one West Coast carrier discovered in 1985 with $69 fares between San Francisco and Seattle, even paying workers nothing would result in the loss of over $9 per seat. If costs are to be cut, the maintenance section has to take its lumps as well, or, as one vice president of a major carrier recently said, "there'll be no one left in this office building to argue [with you] over how much safety is enough."

If the vice presidents in our example are forced to capitulate on each point, not a single item by itself would create an unsafe operation. Not a single change would guarantee a crash. But taken as a whole, such downsizing of the extras in the safety system of an airline, the costly cushion that leaves a large margin for error and for absorbing multiple failures of personnel and material, translates to a diminution in the overall level of air safety. It has been just this sort of pressure over the past ten years that has led to a record number of requests to FAA Air Carrier Inspectors for liberalized interpretations of the rules, raised inspection hours and replacement times on parts, exemptions for inoperative doors, exemptions and exceptions for deadlines on deferred maintenance. These are the hidden pressures, the pressures that the theoreticians who devised the grand plan for deregulatory freedom never understood, and about which they remain ignorant. This is the type of progressive deterioration of the safety buffer that calls to mind a complaint scribbled in a memo from President Lyndon Johnson regarding the wearing effects of a lengthy appoint schedule: "[It's] like being nibbled to death by ducks!"

An uncontrolled competitive environment does not stop at creating efficient operations in an airline, but goes beyond that and threatens the heart and soul of the safety buffer by forcing a never-ending spiral of cost decreases. Airlines must match the lowest fares in order to retain enough of their market share to provide sufficient income to stay in business. Unfortunately, the lowest common denominator of safety that will result in too many cases is the legal minimum. As stated, that has never been the standard, much less the desired, level of safety in the airline system of the United States.

In addition, the vast majority of the federal rules, the Federal Air Regulations (FARs), leave a substantial amount of interpretive discretion to the FAA Air Carrier Inspector who is Principal Operations Inspector (or Principal Maintenance Inspector) for a particular carrier. Inspectors and their supervisors can agree to lower minimums; many rules can be waived, reinterpreted more liberally in the carrier's favor, or otherwise altered at the request of the airline. This is not necessarily an undesirable situation. There is need for nonbureaucratic flexibility and common sense in the monitoring of a complex airline operation. However, the methodology of FAA supervision was designed to serve a stable regulated airline system populated by careful, established, experienced carriers who did not hurry to change things without adequate study. Deregulation has brought about the exact opposite. The attitude of too many carriers during the past eight years has been similar to that of Air Florida in their heady, expansionistic frenzy of the late 1970s: incredible increases in scheduled flights, more personnel, changed training requirements, different technical demands, newly acquired airplanes, constant deadlines and crises, confusion, and changing paperwork. For a team of resident FAA inspectors even to keep track of such an operation, let alone substantively monitor it, is beyond the realm of realism. To expect inspectors to field a constant barrage of requests for exemptions and exceptions, interpretations and approvals, with anything resembling a consistent and careful analysis of each is to expect the impossible. In such circumstances, accidents such as Air Florida's fatal crash into the Potomac River in 1982 are understandable. When the determination of what is and is not safe and acceptable for the traveling public is made on the run, under pressure, and with an economic imperative as the driving force, recent history confirms conclusively that such decisions will not be of the highest quality and reliability.

Pilots, mechanics, and FAA inspectors could have told the committees on Capitol Hill in the late 1970s that there are no simple determinations of what is safe and what is unsafe in the airline business. In fact many tried to do so, but Congress was too enamored of the free-market idea to listen, especially when the protestor had a vested interest in the regulated stability. Many of the training courses and many of the extras in the maintenance department are hard to justify on a cost-benefit basis, but they contribute to safety nonetheless. The majority of the decisions that airline operational people face are subjective: Do we have sufficient reserve fuel for the weather conditions? Should the flight be delayed for deicing? Should the plane be allowed to fly with a possibly defective instrument that we don't have parts to fix?

There are no lights on the forward panel of a Boeing or McDonnell Douglas jetliner that illuminate and say "Safe" or "Unsafe." Human beings make those decisions on the basis of experience and training and availability of parts and maintenance. When any area begins to be cut back because of economic reasons, the safety buffer, the amount of extra assurance that no malfunction or mistake will compound itself so as to metastasize into an accident, is diminished.

The two major domestic airline tragedies of mid-1987, whatever the ultimate determination of probable cause, raise additional questions about deregulation's wearing and disruptive effects on the human beings who actually operate the system. The Northwest Airlines crash of August 16, 1987, near Detroit, Michigan, raises serious questions as to whether the contentious atmosphere of upset and disagreement between two barely merged pilot forces may have contributed to a breakdown of cockpit procedural discipline. The Continental Air Lines crash at Denver, Colorado, on November 15, 1987, raises questions about the shocking, but legal, lack of pilot experience in that particular type of aircraft and calls into question the effects of rapid upgrading of pilots during periods of tumultuous expansion, contraction, or other disruption brought about by deregulatory forces. Of course the people on the front lines of the airline industry have been alarmed by such effects for years. Only now, however, has the nation finally been listening to some extent.

But again, we must return to the very valid statement that airline leaders would never "knowingly" do anything to compromise safety. One of the most profound problems created by deregulation is the entry into the airline business of thousands of bright executives and hundreds of airline owners and leaders who, despite top executive skills and business or financial training, know nothing of airline operations in a technical or practical sense, have never worked for an airline before, and make the serious mistake of considering the process of operating an airline no more complex or different in nature than, as the chairman of one of the nation's largest airlines is reputed to have said, a truck assembly plant.

The airline business is a subjective, human business, delicately balanced on the edge of human experience, human capabilities, human attitudes, and human performance. The airplanes used are designed, built, sold, and maintained, not to mention flown, by humans. The operations of an airline on a day-to-day basis, the dispatching, the servicing and fueling and cleaning, the planning and routing and supervising, are all human functions. Even the making of objective decisions in the executive suite is a human function. And, as an axiom, we have known at the very least for the past 50 years that if we humans have one characteristic in common, it is our predictable propensity for making mistakes.

While it is true that increasing automation in a factory can almost totally eliminate the need for anyone other than a handful of computer operators, it will be decades before similar systems can fully automate aviation. Even Boeing's latest airliners, which feature the advanced cockpit full of automated flight systems and cathode ray tube displays, must have the presence of two pilots to operate them. Certainly the technology exists to do away with both of these pilots and still have the plane arrive safely, provided nothing goes wrong. But multiple flameouts of engines; the infuriating propensity of the weather to defy precise prediction, as evidenced by our continuing struggle with wind shear; and the thousands of variables in airline operations that require the alert attention of trained, stable, experienced, professional people necessitate the presence of human beings if there is to be any hope of nearing the goal of 100% safety.

As Harvard Fatigue Laboratory pioneer Professor Ross McFarland established in the late 1940s and 1950s, the human factor, the interaction and reliability of human beings with their technological creations, cannot be ignored if safety is a consideration. Pilots get tired; so do mechanics, flight attendants, dispatchers, fuelers, security people, baggage handlers, tug drivers, and a host of other players in the performance of a commercial airline ballet of arrival, departure, and flight. Fatigue in that sense breeds mistakes.

Deterioration in attitude also has to be taken into account. Highly trained and professional employees who believe that they are making too little to support their families, who feel betrayed by their employers, and whose futures are uncertain because of the poor financial state of their airline, tend to be less efficient. They are demoralized, disappointed, disaffected, disgruntled. Managers and executives may argue endlessly about the right and wrong, the justice or injustice, the necessity or unavoidable nature of such attitudes in their work force, as they take what they consider to be necessary steps to keep their corporations alive. Certainly it would be short-sighted and financially dangerous to let the desires of the work force control all actions of the corporate leadership. But regardless of causes, the unavoidable point is that when workers react in negative ways to pressures that have been imposed, worker attitudes can reduce the safety margins of companies. An airline that cynically abrogates its union contracts to gain a financial advantage and survive will, without question, lessen its safety margin for years because of the anger and decimated morale of its employees. Even a carrier that places extreme pressure on its people along with laudable communication and explanation in order to win significant contract and pay concessions for the survival of all, as with Western Air Lines in 1984, will still have to battle a diminution of the safety buffer from damaged human attitudes. While troublesome to executives, the conclusion is unavoidable that angry or upset people do not perform as well as those that are satisfied and motivated.

The advocates of intense price competition knew well that deregulation might cause some spectacular bankruptcies and imperil many of the existing, established airlines that had built the air transportation system of the United States. They also knew, and discussed openly, that bankruptcies would translate into significant destruction of jobs, disruption of families, individual bankruptcies, relocations, and turmoil. Therefore, it is not unreasonable to have expected such advocates to have examined and understood the axiomatic connection between massive disruption of established labor relationships and the impact of such disruption on the human element and safety. It should have been obvious to them, and to Congress, that the deregulatory convulsion they were about to unleash would create, at least for some time, a definite, measurable disturbance in the attitudes, outlooks, and futures of all airline employees, which in turn would lessen their ability to perform without mistake, thereby contributing to a lessening of the safety buffer.

It is a sad commentary on the lack of preparation and research into this major realignment of commercial aviation that the one promise made with the greatest solemnity, that safety would not be affected or deregulated, was impossible to keep from the first, for so many reasons.

DEREGULATORY SUBSTITUTION OF LOWER COMMUTER SAFETY LEVELS FOR MAJOR AIRLINE SAFETY LEVELS

Among the many effects and influences of the deregulatory upheaval on the safety system of the airline industry, perhaps none is more profound, and more disturbing, than the naive reliance of the deregulators on commuter airlines to fill the void left by large airlines rapidly vacating unprofitable routes in the face of the looming deregulatory wars. The theory held that such free-market replacement of departing carriers by smaller airlines more attuned and responsive to the needs of the localities would be a good thing. Whenever safety was brought up, the FAA's enforcement of the FAR minimums was cited as an all-encompassing response. The truth is that while the established, large airlines at least had an excellent safety system and much experience, which had been the main factors contributing to the high level of safety Americans enjoyed in 1978, the safety history of commuter carriers was entirely different.

From the time the small airlines came into existence, their safety record has been many times worse than that of the major and regional carriers using larger airplanes. In the early 1970s, the NTSB recognized that the concept of permitting a poorly monitored air taxi operation to provide scheduled service under the relatively lax provisions of what was known as Part 135 of the FARs was to run the risk of slaughter of innocent passengers. They began recommending that the FAA rewrite Part 135 to make it more stringent, and to recognize that what had begun as air taxis had now become commuter airlines. For various reasons, the FAA dragged its feet for seven years before issuing a new Part 135, and even then the new rules were not enforced with any vigor for the first two years following issuance. The NTSB (1980) circulated a scathing report on the commuter industry and the FAA's failure to act to protect innocent passengers. Consumers erroneously thought that if an airline, however small, had a license from the FAA, it was safe enough to entrust with their lives. Partially as a result of that NTSB report and continuing pressure on the FAA, the commuter industry has made great strides in maturing and operating more like an airline system. The Regional (commuter) Airline Association has worked long and hard to get its membership to upgrade as many aspects of commuter aviation as possible.

However, Part 135 carriers, or former Part 135 carriers that have undergone a partial metamorphosis into Part 121 airlines (the section of the FARs governing large airlines), are simply not the equivalent of full Part 121 carriers, no matter how hard they try. This is a point of great importance. Commuter airlines fly smaller aircraft that are maintained often in less-than-optimal conditions, with inadequate FAA surveillance and oversight, and with pilots who are paid so little on average that the majority of the best and brightest stay in commuter cockpits only long enough to acquire enough flying hours to go on to the major airlines. As a result, the turnover rate is extremely high and the level of experience is usually very low. Furthermore, these young aviators usually fly airplanes without pressurization at the low altitudes most subject to weather disturbances; fly without radar, sometimes as a single-pilot operation; and often fly under demanding schedule conditions that promote fatigue. Unlike the major carriers, there are few commuter aircraft for which flight simulators are available and in which young pilots and captains can train. And too often, in the smallest operations, there is too little time, money, and expertise on the part of management to devote much attention to a decent ground training and flight training curriculum. Meeting the minimums and getting the flights through is the order of the day in commuter aviation, and with thin profit margins and high expenses there is great temptation to push the limits wherever the limits get in the way of profitable operations.

Despite the best intentions of owners and managers, these tendencies affect most commuter carriers, or as they are now known, the regional airlines. Nevertheless, the commuter or regional carrier has now become the backbone of airline service to America's small and even middle-sized communities. Despite a frightening, though slowly improving, accident rate, these carriers have become feeders to the "hub" cities of most of the major airlines in the country. One of the problems that this arrangement entails is that the flying public is being systematically deceived by connecting flight systems and shared computer reservation listings, which imply that a flight from Dallas/Fort Worth to Tyler, Texas, for example, will carry you with the same degree of safety as the Boeing 727 or McDonnell Douglas DC-9 which brought you to the Dallas airport. When the airline providing that commuter connection is using Twin Otters, aging Convairs, Shorts 300s, or other smaller aircraft flown by an inexperienced and overstressed commuter pilot force in constant upheaval, the representation is false. To be sure, the level of safety of such carriers is very, very good in most instances, but it does not approach the level of the major carrier because generically it cannot.

The implication of commonality would be disturbing enough, but when small commuter lines enter into marketing arrangements, with a major airline's logo and name on the side of their diminutive airplanes, the implicit misrepresentation borders on fraud. The average airline passenger is not sufficiently knowledgeable of airline operational details to know that a propellor-driven airplane bearing the major airline's paint scheme and its name, even when the word "commuter" is made part of it, is connected to that major carrier by marketing agreement alone. The conclusion will be drawn in a majority of cases that if the name on the side is that of a major carrier, then the pilots, mechanics, quality, and care of that established airline extend to the smaller one. No one will aggressively explain to the passengers that they are about to enter the world of commuter aviation in which accidents occur with more than three times the frequency of mainstream airlines.

The executives and leaders of the major carriers who enter into such marketing agreements without taking on the maintenance and training responsibilities of the feeder commuter airline know that the public will generally reach the wrong conclusions about the safety of the smaller carrier, and that they will do so on the basis of the presence of the large carrier's name, reservations commonality, and terminal feeder location. In fact, it is no secret that the reason for mixing the names of the commuters and majors in the first place is to overcome a historical public resistance to flying on commuter aircraft. With such blurring of identities, if nothing is done by the major carrier to realistically inform and warn the passenger of the difference, then under our concept of the common law we are dealing with a prima facie case of material fraud, practiced throughout the U.S. airline system every day. People are being deceived into entrusting their lives to airlines that do not and cannot offer the level of safety that the large airline's name on the side of the smaller aircraft cynically implies, and this essentially dishonest approach must be changed.

The wholesale substitution of commuter carriers for major airliners is the substitution of an across-the-board lower level of safety for the flying public. This, too, should have been understood by the Congress that was about to make such substitution inevitable. Unfortunately, to this day there is inadequate realization on Capitol Hill of this disparity.

DAMAGING EFFECTS ON THE FAA

One of the most tragic effects of deregulation has been the resultant plight of the federal agency on which Congress placed the task of maintaining the safety levels of 1978, the FAA. Congress gave the FAA no more money, no more manpower, no new laws, and no new autonomy with which to handle the exponential increase in work load that deregulation was about to unleash. The FAA was doomed to fail in the face of the deregulation bill before the ink was dry on President Jimmy Carter's signature.

As several FAA administrators have correctly stated, the FAA is not supposed to be a police force. The FAA could not prevent cheating and rule violations even if it had an FAA inspector on every flight deck or standing beside every mechanic. Safety has to come from within the airlines—it cannot be forced systemwide.

In fact, the proper role of the FAA is that of a coach and a monitor. As a coach, it is supposed to assist, cajole, discuss, guide, and act as the conscience of the industry in matters of safety. Its actions tend to bring about an average level of safety far above the legal minimums. Deregulation, however, tends to substitute cost controls for safety as the prime concern of airline managers, and this, along with the massive work load of initial certifications of new carriers, which consumed the better part of the first five years of deregulation, has forced the FAA to operate as a police force. The long series of recent enforcement actions, fines, and shutdowns is proof of this assertion. The problem is that even if given all the money and manpower the FAA could ever ask of Congress, it would be unable to be everywhere, or to force the retention of the previous high safety margins in an industry constantly motivated by cost cutting and attempts to survive ruinous competition.

Eastern Air Lines' massive fine is an example. The FAA's action against Eastern demonstrated two important points: (1) If the FAA had been able to do an adequate job of continuous monitoring and surveillance of Eastern during the period these violations were committed, the trend, and the violations, would have been caught and addressed years before they reached the level of 78,000 individual counts, and (2) since Eastern had never before permitted itself to sink to such a low level of compliance, whether the violations were principally paperwork or otherwise, and the only major change internally had been the pressures of deregulatory cost cutting, the conclusion that deregulatory pressures caused or materially contributed to the violations is both logical and correct.

Similarly, the FAA failed to identify the increasing lack of coordination and communication in the cockpits of one of the nation's most stable carriers, Delta Air Lines, before a rash of embarrassing incidents placed the airline in the spotlight of media attention during the summer of 1987. In the wake of a massive trans-Atlantic navigational error and near midair collision, an inadvertent twin-engine shutdown in Los Angeles, an incident of landing at the wrong airport, and several other problems, the FAA began a special inspection of Delta's pilot procedures and cockpit discipline. The conclusion of the inspectors' report published several months later validated Delta's overall safety, but identified significant human-factor weaknesses in training and procedures that had been directly exacerbated by the disruption of merging the pilot forces of Delta and Western Air Lines and by the subsequent upgrading and cross-training of hundreds of pilots. The merger, of course, was a clear result of the carrier's battle to stay stable and solvent in the intense price competition brought on by deregulation. This disruption was clearly deregulatory in nature. Again, such disruptions lower safety margins, as the above incidents and finding graphically demonstrated, without showing up in accident and death statistics. In Delta's case, the FAA performed brilliantly after the fact and produced a report of high quality. Delta acted with laudable attitude and speed in adopting the recommendations; still the case reveals that the FAA has been stretched too thin and left too devoid of human-factors sensitivity to detect problems as they occur, or before they imperil the flying public.

If such problems could occur at Eastern and Delta, why should anyone believe they cannot occur at other airlines, especially new-entry, low-cost, poorly controlled, technically immature operations such as those that thundered in and out of business in a hurricane of confusion during 1980 through 1984?

Fortunately in the past three years there have been significant improvements in the FAAs ability to hire, train, and field a more effective inspector work force. While quite properly asking the DOT for still more resources, recent administrators, such as T. Allan McArtor, exhibit a willingness to admit that problems do exist in the system and in the FAA, problems that will take innovative and energetic actions to address. Such improvements, if carried to their logical conclusion, will help unmask and discipline airlines large and small who cheat in the name of profit, but they are merely a beginning. The FAA's abilities and resources have been sorely strained by the deregulatory expansion of the airline system and of American aviation in general. Major substantive restructuring of the agency is vitally needed to increase its funding and legal authority, to eliminate the limiting and contradictory mandate to "promote" as well as regulate, and to raise the FAA to the desired status of an industry teacher and coach as well as referee. And major restructuring will be necessary to give the FAA what it still does not have, an institutionalized method of understanding and dealing with the last major frontier of air safety, human factors.

Similar restructuring and alteration in funding is essential for the FAA's air traffic control functions, which were wholly inadequate even before deregulation and which are now, as then NTSB chairman Jim Burnett pointed out during 1987, in danger of being perpetually inadequate. Again, only Congress can make these critically needed alterations.

CURRENT STATE OF THE INDUSTRY'S SAFETY SYSTEM

In the case of the airline industry itself, the years since 1986, while still speckled with accidents, near midair collisions, airline management wars with employees, near accidents of other sorts, hub-and-spoke capacity problems of continental proportions, and many other maladies, have brought a significant improvement in safety. This is a direct result of the natural shift from the deregulators' dream of perfect competition into the reality of an oligopoly of megacarriers who are in turn locked into feeder arrangements with a growing number of commuter airlines. These megacarriers are increasingly able to overwhelm their respective hubs and dominate their respective markets.

Why on earth is this good for safety? Because it brings back some semblance of profitability, control of yield, and market stability, through force of marketing and operational muscle, and returns the megacarriers to a position in which the future can be defined in terms of years rather than months. For the maintenance and training departments, this means a significant lowering of financial pressure to cut costs, and that means at the very least an arrest of the constant attempts to lower training and maintenance requirements, maintenance stocks, and the salaries of all concerned personnel. It does not mean an elimination of such pressures, merely a lessening, but that decrease, driven by the increased public concern over service and safety, has arrested what had become a free-fall in the airline safety system nationwide, the same free-fall that airline operations people had been watching in scorned silence for the previous seven years.

Under no circumstances does this mean that deregulation's pressures are over, or that the safety margins have been restored and no further actions are needed. It merely means that the thinning of safety margins has been halted for the most part, and that rebuilding has begun, a massive effort that must now, for the first time, incorporate a systemwide appreciation of the role of human factors and the treatment of human beings as the cornerstone of the safety system.

There is, regrettably, one significant exception in the industry to this generally improving trend, Texas Air Corporation of Houston. Texas Air persists in attempts to create fare wars for the purpose of grabbing market share, maintains the work forces of both Continental and Eastern Air Lines in a state of perpetual upheaval in morale and professional status, and tenaciously clings to the philosophy that the airline business is merely a profit-making business that should be driven only by the bottom line. Indeed, Texas Air's actions in effectively declaring war on the employees of Eastern Air Lines has both strained the safety of that once-proud carrier and placed great pressure on the FAA to monitor growing concerns over safety compliance attitudes in all operational areas. It should be noted that not a single violation of FARs needs to take place for such constant upheaval to be contrary to the best interests of the U. S. airline system. It is the corporate philosophy of Texas Air that is at variance with our national interest in maximum safety levels, and it is our own past invitation to such incorrect philosophies formed by the Airline Deregulation Act of 1978 that is at fault. While airlines must make a profit to survive in a capitalist society, their first responsibility should be, as it was prior to deregulation, to build a stable, safe, and efficient public aerial transportation system. Ripping off the maximum profit for the smallest cost with zero consideration for the humanity involved may be good business in a vacuum, but such corporate attitudes are incompatible with high levels of stability and safety required in a national airline system.

SOLUTIONS

What, then, is to be done?

1. We must create a new method of gauging the level of airline safety by taking a realistic, real-time measure of the safety margins. Accident and death statistics are worthless as a gauge of such considerations, since only the complete absence of a safety buffer will show up as a clear trend in the number of passengers killed and the number of aircraft destroyed. Instead, we must formulate specific indicators that will fluctuate immediately and obviously when any trend of increased cost-cutting pressure begins to affect safety margins among the nation's air carriers.

At the recent Northwestern University Transportation Deregulation and Safety Conference, NTSB board member John K. Lauber (1988) made the point this way:

An accident is clear evidence that safety is lacking, but, by itself, the absence of an accident does not demonstrate that safety has been achieved. What is needed is a set of "leading safety indicators," along the lines of the "leading economic indicators," and these should be used to examine further the effects of airline deregulation. For example, in the competitive environment resulting from deregulation, the indicators might include the impact on airline training, safety, and medical departures, as well as maintenance expenditures. Another question might be whether there is excessive pressure to reduce turnaround times at hub airports.

In addition, there must be developed a way to read quantitatively the telltale signs of decreasing morale and increasing internal disruption of a work force whose delicately balanced professional tasks simply cannot be performed as well by upset practitioners. Identification of such symptoms may not in and of itself provide a method of reversing the process, but it can alert the FAA to immediately and substantively increase monitoring, while alerting the public and bringing deserved public scrutiny to bear on a carrier intent on keeping its employees in states of agitation. This is, after all, a democracy, and even the concept of caveat emptor cannot work if the buyer cannot obtain adequate information on the safety of a product.

The excessive time that has been wasted over the previous years in arguments over whether a deregulatory-driven decrease in safety margins did, in fact, exist could have been avoided if an adequate system of safety indicators had existed as early as 1980. The clear slide in safety margins, the increasing financial pressure on maintenance and training departments, the entry of marginally qualified airlines and technically unsophisticated managements, would all have "pegged the meter" by 1982. Trusted indicators could have cut through the government's blind trust in the free market and sparked some early adjustments, which could have prevented the deep damage that was, in fact, done by 1984 and 1985, damage to the FAA and to established airline safety systems as well as to the human beings whose lives and careers have been scrambled and destabilized for the past decade. In the absence of those indicators, however, warning voices have been virtually ignored. Even now, in the face of convincing proof of pressure on the safety margins, the absence of objective statistical data enables doubters and demagogues alike to continue pretending that total economic deregulation has no generic or actual effect on airline safety. The filling of this credibility gap must become a national priority.

2. It may be difficult for today's more business-oriented breed of airline leader to appreciate, but the fact remains that the extremely safe U. S. airline system that we still continue to enjoy (albeit with reduced margins) was built by airlines that, as United Air Lines' pioneer chairman Pat Patterson used to say, would "spare no expense on safety." That tradition was possible because of the past regulated stability and profitability of the airline industry, and because the vast majority of the workers, whether overpaid or not, considered their jobs careers.

There is no doubt that regulation became too tight—the airlines became too protected and inefficient and the fares too unresponsive to competitive influences. Deregulation in moderation was needed. But what we have done with the Deregulation Act of 1978 is to place a vital public utility in the free market. Since one of the aspects of a public utility is regulation for the public's safety, the protection of that function is government's responsibility. The fact that safety cannot be maintained at the previous levels under the present system has been graphically demonstrated and proven by the inability to maintain standards. Therefore it would seem unavoidable that government must reinvolve itself to whatever degree is necessary, and in some other form than that of increased FAA surveillance and enforcement, to restore the previous environment of continuous encouragement of safety standards far above the minimums. Since this can most easily be accomplished by the carriers themselves, some minimal governmental control to give the airlines a stable environment of reasonable profitability seems vital.

We know that the central cause of the deregulatory cost-cutting pressures—uncontrolled competitive wars—can break out again at any time in the current unregulated market, caused by a single airline intent on gaining increased market share at the cost of short-term solvency. We also know conclusively that such wars put pressure on the safety margin. The conclusion is inescapable that this propensity must be eliminated. The only method of preventing the ravages of excessive cost cutting is to eliminate the opportunity to price products below cost and the tendency to overcrowd specific routes to the point that all lose money. Certainly these tendencies indicate the need for a mild degree of reregulation. Such renewed governmental control could easily take the form of a regulatory authority that steps in only when needed, not one that controls all routes and rates. Or perhaps the adoption of something akin to the original 1977–1978 proposals made before those House and Senate hearings for a "zone of reasonableness" in fare controls and route entry, supervised by a separate governmental entity, would be acceptable. However it is done, though, the key to stability and a return to safety levels far above the minimums—the key to major airlines spending hundreds of millions of dollars on nonrequired maintenance, training, and operational structures of high quality—lies in providing airlines the environment in which they can make enough money to indulge their natural tendency to have the greatest level of safety possible. The free market is obviously incapable of providing such stability.

3. Commuter airlines are a crucial link in the modern aviation system. Their operations must be upgraded to the standards of the major airlines. When they have an association with a major airline, that association must be formalized and the major airline must impose its standards on maintenance and training activities.

4. The air traffic control function of the FAA should be separated into a government corporation structure with vastly streamlined procurement and managerial capabilities. The present system, though operating heroically and steadily under crushing pressure, is overloaded and decades behind in equipment, procedures, manpower, human relations, and overall ability to stay ahead of the air traffic demands on the nation's air space.

5. The FAA must be removed from the political straitjacket of the DOT, and must be divested of its mandate to "promote" aviation. Its administrator should be elevated to cabinet rank with Senate approval and the term of office should be at least seven years. There are excellent ideas abounding on Capitol Hill for giving such a reformed FAA complete control over the Aviation Trust Fund, and increasing the yield of that fund to provide independent sources of money for the massive increases in the FAA's budgetary requirements that will be needed in the immediate and intermediate future.

CONCLUSION

The first steps in repairing a problem in a human system are to admit that the problem exists and then to find ways of measuring and gauging the depth and breadth of the problem and its effects. Finally, one looks for the most effective, realistic solutions.

We have wasted a decade in fruitless arguments revolving basically around the ideological theory that there can be no flaw in and no criticism of the free market. Several thousand years of human history have had to be disregarded to reach such a conclusion, since we should have learned long ago that pure ideologies reveal their imperfections when applied to human systems, and that such systems cannot operate without flaws. When crossed with the very human business of airline flying, the free market is certainly no exception. Deregulation has created pressure on the safety margin and reduced it. That is fact. We must now find a new way to measure fluctuation in the safety margin as we encourage the repairs that have already begun and set about the task of changing the economic and regulatory structure of the system to prevent future damage.

14

Summary of Other Aviation Issues

LEON N. MOSES and IAN SAVAGE

At the Northwestern University Conference, nearly 15 papers in addition to those included in this book were presented. Substantive points were made in these papers, which support and enlarge on issues addressed in preceding chapters. In this chapter we summarize the debate on five issues: the air traffic control system; the consequences of increasing demand for airline pilots; the oversight activities of the Federal Aviation Administration (FAA); the pilots' perspective on safety; and finally, whether lower air fares have resulted in less highway fatalities. These are important issues in the current debate, but they fall outside research described thus far.

The authors on which we have drawn most are John K. Lauber, a board member of the National Transportation Safety Board (NTSB); John E. O'Brien, Director of Engineering and Air Safety of the Air Line Pilots Association (ALPA); Herbert R. McLure, an Associate Director of the General Accounting Office (GAO); and John S. Kern, Deputy Associate Administrator for Aviation Standards of the FAA.

AIR TRAFFIC CONTROL SYSTEM

Increasing Demands on the System

Air travel increased considerably during the years of deregulation. Passenger miles have increased by 62% in the period 1978–1986, and the number of scheduled aircraft departures rose by 35% over the period 1978–1987. Of course, as discussed in Chapter 11, the increase in travel has increased aircraft activity in and around the traditionally heavily used areas such as the Washington, D. C.–New York–Boston corridor, and on the West Coast. But the increase in flights has not been confined solely to short-distance trips. By 1985 there had been a 17% increase in long-distance flights handled by "enroute" (dealing with traffic outside the main traffic areas) rather than terminal control towers (Table 14.1).

Table 14.1 Commercial Aircraft Handled
at Enroute Traffic Control Centers,
Fiscal Years 1977–1985

Year	Total	Index (1978 = 100)
1977	25,973,299	93
1978	28,055,382	100
1979	29,909,712	107
1980	30,061,372	107
1981	29,531,111	105
1982	27,854,842	99
1983	29,361,418	105
1984	31,615,489	113
1985	32,708,709	117

SOURCE: Federal Aviation Administration Air Traffic Activity Reports Fiscal Year 1981 and Fiscal Year 1985.

The air traffic control (ATC) system of today emerged in the years when most traffic was of the long-haul, nonstop variety. Deregulation brought the hub-and-spoke system to existing centers, but it also was responsible for the emergence of new hubs such as Salt Lake City, Phoenix, and Albuquerque (Winston, 1988). Of course these airports have control towers, which essentially oversee operations at the airport, but they do not have control facilities that deal with traffic in a 30-mile area surrounding the airport. This is a serious matter, because hub status can result in a significant increase in the amount of traffic and congestion at an airport. Figure 14.1 shows the effects of moving to hub-and-spoke operations. The Charlotte, North Carolina, and Louisville, Kentucky airports had similar levels of activity before deregulation, but diverged sharply when the Charlotte airport became a hub. The existing system now suffers from the greatly increased demands that have been placed on it and the fact that it was not designed to cope with the very significant changes in the structure of the demands it faces. The National Airspace System Plan, launched in 1981, aims to align the ATC system with the network and travel volume that has developed under deregulation.

Controller Work Force Strength

On August 3, 1981, the Professional Air Traffic Controllers Organization (PATCO) declared a strike against the FAA. The strike resulted in the walkout of approximately 13,000 controllers. Although some of the strikers returned, about 11,400 did not and were subsequently dismissed. The action by PATCO and the FAA response left the air traffic control system with significantly reduced capacity. According to the FAA, the poststrike ATC system was initially managed by about 4,669 nonstriking controllers,

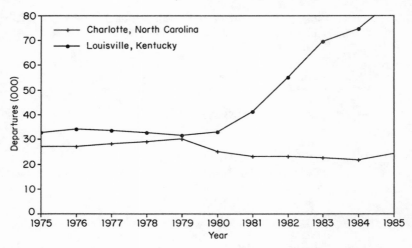

Figure 14.1 Example of Change in Airport Activity After Deregulation. The Charlotte, North Carolina, Airport Became a Hub Airport. SOURCE: Federal Aviation Administration.

3,291 supervisors and management staff, 800 military controllers on detail, and about 1,000 newly hired personnel, totaling 9,760, compared with a 17,275 prestrike work force (National Transportation Safety Board, 1981).

Immediately after the strike, the FAA implemented the National Air Traffic Control Contingency Plan. This plan limited ATC system capacity to 75% of that existing prior to the strike and reduced schedules at 23 major airports by 50%. The restrictions were subsequently lifted, and traffic levels are now much higher than before the strike. However, by March 1987, the total controller work force had grown only to 14,900, of which 11,200, were operational controllers. This is less than the 1981 prestrike work force level of 17,275, which included 13,300 Full-Performance-Level (FPL) controllers.

There is an additional factor to consider when attempting to analyze the ability of the present ATC system to handle demand compared with its prestrike ability. A prestrike FPL controller had much broader qualifications than does today's operational controller. This situation produces a much less flexible work force and ATC facilities that are much less capable of responding to the expanded demands brought about by deregulation.

The FAA planned to have a total work force of approximately 17,200 by the end of fiscal year 1988. It is still in the process of hiring air traffic controllers. These numbers continue to be below the 17,275 that existed prior to the strike (O'Brien, 1988).

McLure of the General Accounting Office supported O'Brien's comments, by reference to work his office had undertaken at congressional request. It found FAA-budgeted staffing levels substantially below those called for by the FAA's own standards in all 16 en ute centers that were checked, and actual full-time staffing below FAA-budgeted levels in 11. GAO also found full-time staffing levels below FAA-budgeted levels in 10 of the 12 terminal control towers. Finally, staffing was so inadequate at two centers and one tower that supervisors had to spend at least half of their time acting as ordinary controllers.

Physical Capacity—The National Airspace System Plan

Even before deregulation, the FAA knew that air traffic would continue to rise through the turn of the century and would place unprecedented demands on the ATC system. It also knew meeting this demand safely and efficiently would require additional facilities and equipment, improved work force productivity, and the orderly replacement of aging equipment. So in December 1981, FAA published a comprehensive National Airspace System (NAS) Plan to modernize, automate, and consolidate the existing system.

In terms of dollars required, the NAS Plan is one of the largest single civil procurement programs ever proposed. It is expected to cost more than $16 billion by the year 2000. FAA believes the plan represents a practical way to achieve a significantly more efficient system while reducing the risks of midair collisions, landing and weather-related accidents, and collisions on the ground.

One NAS Plan goal is to increase the productivity of air traffic controllers. Their productivity is to be increased by an Advanced Automation System and the consolidation of existing work stations. Outdated vacuum-tube electronic equipment will be replaced by solid-state devices in ground-to-air communications, navigation, approach and landing, and surveillance systems. However, according to the GAO, each of these systems has already experienced schedule delays ranging from a year for airport surveillance radar to eight years for the Advanced Automation System, resulting in corresponding delays in anticipated productivity gains.

Operational Errors and Near Midair Collisions

According to critics, the result of more aircraft being handled by fewer, less experienced controllers and increasingly archaic equipment has been an increase in the number of operational errors both on the ground (termed "runway incursions") and in the air (termed "near midair collisions," or NMACs). FAA data on these incidents, reported in Table 14.2, show that incidents per aircraft departure have risen rapidly in recent years, and in the case of NMACs is 40% above the 1980 level. The decline in the number of incidents in 1981 through 1984 requires some interpretation. It is the period, following the PATCO strike, in which restrictions were placed on the number of flights in congested areas. It should also be borne in mind that recent changes in reporting methods may have increased the number of incidents actually reported.

Table 14.2   Near Midair Collisions and Runway Incursions per Departure

Year	Near Midair Collisons/Departure (1980 = 100)	Runway Incursions/Departure (1983 = 100)
1980	100	N/A
1981	70	N/A
1982	56	N/A
1983	81	100
1984	91	79
1985	115	104
1986	119	486
1987	141	N/A

N/A, data not available.

SOURCE: Federal Aviation Administration.

Many of the NMACs are between small private aircraft and commercial jet aircraft in the congested areas surrounding major airports. While private pilot inexperience is certainly one factor in the near collisions, another is the inability of the ATC system to track small aircraft effectively. The reason is that most small aircraft are not fitted with "mode C" transponders that report altitude. Without this device, controllers can track only latitude and longitude, insufficient information to detect aircraft that are dangerously close. All commercial aircraft are fitted with these devices. Their cost is under $2,000. The Department of Transportation is currently proposing that fitting of transponders be compulsory in congested areas (Crandall, 1988).

INCREASED DEMAND FOR AIRLINE PILOTS

The high demand for pilots caused by the growth of airlines shows no sign of abating, according to O'Brien. The Air Line Pilots Association expects commercial pilot hiring to add between 4,000 and 6,000 pilots per year to work force for the rest of the decade.

Fewer experienced military-trained pilots are available for hire by the airlines. In 1980, nearly 83% of all airline pilots had military flying experience, according to a study by the Congressional Research Service. Today only 40% are trained by the military. The overall decrease in the number of military pilots and the increasing demand for commercial pilots have hurt smaller jet and commuter airlines. These airlines have a 30% to 50% turnover per year, with 90% of departing pilots going to work for the major airlines.

The competition between airlines for pilots raises concerns about maintaining the qualification standards for new airline pilots. On average, newly hired pilots still have 3,000 hours of flight time. However, the amount of jet-flying time required by airlines has gone from 2,300 hours in 1983 to

1,600 in 1984 and to 800 in 1985, according to O'Brien. Lederer and Enders (1988) of the Flight Safety Foundation, an independent agency concerned with aviation safety, reported that in 1983 only 8% of commuter airline pilots had fewer than 2,000 hours of experience in the cockpit, but by 1985 the figure had risen to 23%.

Lauber considers the effect this may have had on accident rates. He draws on NTSB accident reports to provide information on hours of flight experience of pilots involved in accidents. Effectively, Lauber argues that the probability of a pilot's being involved in an accident on any given flight is negatively related to the pilot's total hours of experience. The median experience level of the pilots involved in accidents has fallen from 16,500 to 13,000 hours (Figure 14.2). Since the number of less experienced pilots has been increasing, the weighted probability of an accident from this source has risen.

The increasing demand for airline pilots and the high turnover rates at commuter airlines continue to be great concerns to the NTSB, reports Lauber. During a six-month period in 1985 and 1986, the Board investigated three commuter airline accidents: Bar Harbor Airlines in Lewiston, Maine; Henson Airlines near Grottoes, Virginia; and Simmons Airlines in Alpena, Michigan. Flight crew qualifications and training were issues in all three investigations. For example, the captains of both the Bar Harbor and Henson aircraft had relatively few flight hours as pilot-in-command. Both were paired with first officers who also had little experience in their respective positions. In the Simmons case, both pilots had relatively high flight hour totals overall, but the captain had met the pilot-in-command requirements only three weeks before the accident, and the first officer had flown only about 20 hours with the airline.

Figure 14.2 Total Experience of Accident-involved Pilots: Large Jet Commercial Aircraft. SOURCE: National Transportation Safety Board Accident Reports.

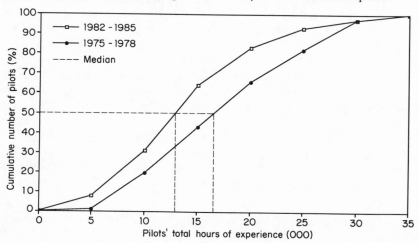

FAA OVERSIGHT ACTIVITIES

Kern of the FAA notes the impact of deregulation on the work load of his agency. Prior to deregulation, strict entry controls had produced an industry that was fairly predictable and stable. The work load of safety inspectors fell because technological improvements in aircraft and in the air traffic control system had produced a downward trend in accident rates. As a result, the resources that the FAA devoted to certification, surveillance, investigation, and enforcement began a gradual decline in the early 1970s. That decline continued unchecked through January of 1984. FAA's Flight Standards' field inspector staff diminished from 1,672 to 1,331, a decrease of more than 20%.

Deregulation changed the nature of the industry. The number of air carriers operating large aircraft increased from 60 in 1978 to 148 in 1985, a rise of 150%. FAA resources were needed to provide initial certification of these carriers. Approximately 4,000 existing operators of small aircraft had to be recertified as commuters or air taxis as a result of revisions in safety regulations adopted in 1978. Moreover, since there was a high turnover rate of these types of firms, the resources they required for certification were even higher than the 4,000 number would suggest. Certification of operators entailed large expenditures of inspector time to ensure that they had the ability, fiscal resources, and organizational structure to train crew members adequately and program the maintenance of aircraft effectively.

The problem was that the FAA had become a shrinking agency in a rapidly growing industry. Fewer inspectors were spread over more airlines. The number of inspectors per operator fell from 4 in 1978 to about 1.5 in 1985, according to O'Brien (Figure 14.3). Kern concludes that, "The industry had changed significantly while the FAA safety regulating programs had not." More important, the entry of new airlines in the late 1970s and early 1980s meant that the FAA focused its limited inspector resources on certification. Kern says that as a result, "routine operations and maintenance compliance (i.e., inspection and surveillance) were mostly left undone."

A report to the FAA by the Allen Corporation of America (1985) highlighted the problem of inspectors' inabilities to complete work under unrealistically short deadlines. The NTSB in its reports on the three commuter airline accidents mentioned in the section above found deficiencies in the FAA's "pilot certification" program and the agency's surveillance of that program.

The problem was summed up by the FAA Administrator of the period, Donald D. Engen, before the House Committee on Public Works and Transportation, Subcommittee on Aviation, on May 22, 1986. Administrator Engen said,

There has been change in the air transport segment of aviation brought on by deregulation. That change was not accommodated, perhaps not recognized, in the

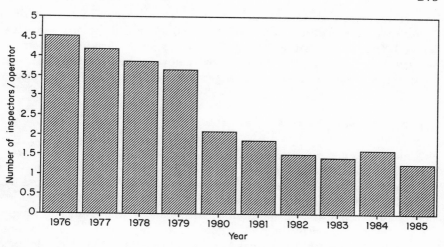

Figure 14.3 Number of Inspectors per Operator (Authorized Staff Levels, Not Staff on Books). SOURCE: Federal Aviation Administration Office of Program and Regulations Management.

1979 to 1983 time frame. Inspectors were becoming burdened, and being good hardworking people they tried to adapt to this change by working harder. The rapid increase in the number of air carriers brought a real burden to the inspection system. New operators did not know how to start an airline and asked the FAA to help them start up. Many lacked the depth of background to build effective maintenance organizations. The mind set of some operators was cavalier—others had labored intensively to be professional. This changing industry came into the headlines in 1983 with a regional airlines accident that showed poor compliance with regulations, and less than adequate oversight and inspection by the FAA.

In 1984 this trend seemed to reverse. The FAA implemented a number of major safety audits of the aviation industry. These audits resulted in the large fines against household-name airlines reported in Chapter 13. The FAA also conducted an internal Safety Activity Functional Evaluation, project SAFE, which made recommendations on increased staffing, an effective evaluation procedure for safety standards, an update of Federal Air Regulations, a unification of inspection procedures, increased data bases for inspectors, and revisions in its internal organizations. As a result of the study, the FAA planned to increase the number of inspector positions in fiscal years 1986 and 1987 by 365, and requested an additional 178 inspector positions in fiscal year 1988.

O'Brien of ALPA believes the underlying cause of the decline in the inspection and enforcement capabilities of the FAA, of the failure to hire sufficient new air traffic controllers, and of the slow progress in modernization of the air system infrastructure has been insufficient funding by the Administration and Congress. The Balanced Budget and Emergency Deficit Control Act of 1985 (the Gramm-Rudman-Hollings Act), which is directed toward reducing the present federal budget deficit by requiring across-the-board savings in government agencies, has compounded the problem. Bailey and Kirstein in Chapter 11 report that the Air Transport Association has led a move to cure the problem by removing the FAA from the Department of Transportation and having its activities supported by user fees.

THE PILOTS' PERSPECTIVE

In June 1986, the Air Line Pilots Association sent a 110-question survey dealing with matters of air safety to its 28,000 members. Over a third of the questionnaires were returned, and a random sample was analyzed by a Washington, D.C. opinion research firm (Fingerhut, 1986). The results help to throw light on the anecdotal claims made by John Nance in Chapter 13. Interpretation of the results needs to be made with care. Working conditions of ALPA members have declined in quality since deregulation, salaries have increased at a lower rate, and new hires are often on a reduced pay scale. Nonunion carriers have entered the industry, and some existing carriers, such as Continental Air Lines, are no longer staffed by ALPA pilots. It would not be surprising if ALPA pilots took a somewhat hostile view of deregulation.

Overall almost half of the respondents felt that deregulation had greatly affected safety, with almost all respondents acknowledging that there had been some impact. Nearly 70% of respondents felt that financial pressure on airlines was partly to blame, 40% felt that inexperienced managers were partially the cause, and 60% said the government and the FAA had a hand in the decline in safety.

A concern voiced loudly in the survey is that increases in airport congestion have increased the risk of collision. Over 40% of the pilots responding listed the capacity deficiencies of the air traffic control system as their *single* greatest aviation fear. This was over 10 times the frequency with which they mentioned concerns about the weather (such as wind shear) or hijackings. Increases in FAA staffing and the rebuilding of the ATC system were listed as higher priorities for the expenditures of federal funds than expenditures on human factors research or wind shear detection equipment.

Over 80% of respondents felt that the "Minimum Equipment List" (MEL) of aircraft, which specifies which combinations of on-board equipment can be *in*operative without the plane being grounded, was too permissive. About 10% of pilots said they were frequently pressured into flying aircraft in contravention of the MEL, with another 40% saying that it occurred sometimes. About half the pilots felt that the aircraft they flew had excessive deferred items of maintenance, with the same proportion believing that the airworthiness of their aircraft had declined since deregulation.

An interesting additional point commented on by the pilots concerned cabin safety. In the years since deregulation, the density of aircraft seating has increased. Lederer and Enders (1988) feel that this has lowered the chance of survival in an accident because the "brace" position cannot now be effectively adopted by larger passengers. In addition, the higher loads mean that emergency egress rates are slowed. Not only are there more passengers in each aircraft, but passengers are increasingly breaking existing guidelines on the amount of carry-on baggage they bring into the cabin, primarily because they believe the probability of losing checked baggage has increased. Approximately 80% of pilots in the ALPA survey felt that excessive carry-on baggage can comprise a safety hazard in turbulent air or emergency landings.

IMPACT ON HIGHWAY FATALITIES

There is no doubt that air travel has increased substantially since deregulation. Some part of this increase may be due to a shift from auto to air travel brought about by the decline in the price of air relative to highway transport. If there is a diversion from the automobile, a form of travel that is many times more dangerous than air travel, then these passengers have experienced a net safety gain as a result of deregulation. Oster and Zorn (Chapter 10 in this volume) have speculated that this may be the case for the additional passengers attracted to short-haul commuter flights.

McKenzie has undertaken research in this area. In initial work with Shughart (1986), regressions were carried out on the level of automobile travel at the aggregate level with a dummy variable to represent airline deregulation. In this work a weak relationship was found, but it was not large and not statistically significant. In later work (McKenzie and Warner, 1987), a more sophisticated model of automobile demand was fitted to the same national data. The new model used data on prices, incomes, population, and average highway speed and a dummy variable structure for the postderegulation years. Their analysis suggested that without deregulation, automobile usage per year would have been greater by 4.6 million miles, or by about 4%. On the basis of this reduction in miles per year, McKenzie and Warner estimated that about 1,700 lives had been saved annually.

There is undoubtedly some logic to McKenzie and Warner's basic proposition. It is not unreasonable to assume that some of the passengers attracted to air travel reduced the amount of their automobile use. However, the magnitude of their fatality savings is surely questionable. Unprecedented oil price shocks in the mid- and late 1970s are argued by traffic engineers to have had major structural effects on patterns of automobile usage and driver behavior. How much of McKenzie and Warner's 4% decline in auto use is due to this factor is open to debate. One way around this problem is to investigate data on previous mode choice of "new" airline passengers. Without data of this sort, the magnitude of the logically appealing "cheaper skies mean safer streets" proposition remains untested.

THE MOTOR CARRIER INDUSTRY

As in the aviation section, the four areas of potential concern identified in Chapter 1 will be addressed here: the effects of reduced profitability, of new entrants, of modal shift changes, and of congestion.

Chow in Chapter 15 analyzes safety-related expenditures by firms near bankruptcy, relative to expenditures by those that are financially solvent. Corsi and Fanara in Chapter 16 examine the accident record of newly certified carriers. Both chapters point out causes for concern with safety performance.

In Chapter 17 Boyer points out that freight transportation is characterized by a degree of intermodal competition. Road transportation is inherently less safe than railroad transportation. If deregulation of the motor carrier industry has led truckers to be successful in capturing traffic from the railroads, then deregulation can have overall safety effects even if the safety characteristics of individual modes remain unchanged. Boyer investigates the evidence for this kind of effect by looking at the movement of manufactured goods, which traditionally has been characterized by intermodal competition.

The issue of congestion is less applicable to motor carrier deregulation. While there are certainly many congested roads in the nation, and increasing truck traffic may increase congestion in certain areas, overall, trucks comprise only a small percentage of vehicles operating on the nation's highways. While the comments of Arnott and Stiglitz in Chapter 12 regarding congestion fees are equally applicable to highways, at this point we cannot determine whether such fees would reduce truck-related deaths even on highly congested roads.

Studies that examine the effects of deregulation on truck safety are hampered in a number of ways: (1) poor data, especially for nonfatal and property damage accidents; (2) significant changes in motor vehicle technology; and (3) important changes in safety legislation in the 1980s. In Chapter 18 Jovanis, who cochaired the trucking sessions at the Northwestern University conference, considers these problems and comments on the substantive evidence described in the nearly 15 other papers presented at the conference by leaders in industry, labor, and government and by consumers.

15

Deregulation, Financial Distress, and Safety in the General Freight Trucking Industry

GARLAND CHOW

The interstate trucking industry was substantially deregulated by administrative rulings beginning in 1977 and later by the Motor Carrier Act of 1980. Reductions in barriers to entry and the loosening of controls on rate making increased the intensity and severity of competition. Numerous studies suggest changes in the structure, conduct, and performance of this industry over the last decade.

At the same time it is asserted that economic deregulation also led to less safety on the highways "by creating economic circumstances that make trucking far more dangerous." (Labich, 1987). It is generally accepted that safety regulation and enforcement was woefully inadequate to police the multiple players in the industry before or directly after deregulation. Recent legislation has helped rectify the deficiencies in enforcing federal safety regulations. But at the center of this argument is the relationship between profitability and the incentive of carriers to improve or encourage safety in their operations. Deregulation has reduced the profitability of many motor carriers, and many of these hard-pressed carriers may be practicing unsafe driving practices, deferring needed maintenance, or not replacing unsafe equipment.

The validity of the assertion that trucking deregulation has negatively affected the level of safety on the highways is very controversial and worth resolving, not to bring back regulation but to understand what can be done to improve highway safety in the future.

LINKAGES BETWEEN REGULATION AND SAFETY

Economic Regulation and Safety

Academic and industry literature suggest that the most important rationale linking economic regulation and safety is the profitability-safety relationship. Decreased profitability of the industry and of individual carriers results in poor financial condition, which provides an incentive for the carrier to reduce expenditures on maintenance and in other areas, such as training, that affect safety, or to engage in operating practices that are not conducive to safety. Baker (1985) described the logic as follows:

The only way for a marginal carrier to stay in business . . . is to cut costs and there are few places to cut costs in motor freight operations. Fuel prices are market set and fuel usage is, in large part, a function of new technology and maintenance, either of which costs money. The only alternative is to run older equipment, pay less in wages, work drivers longer, and/or skimp on maintenance.

Such behavior belies the maxim that safety pays. It is claimed that carriers look at safety activities like any other expenditure or investment and invest in safety when it yields a good return. The return might be reduced insurance and operating cost, and improved reputation and more business.

Research in other transport industries suggests that the relationship is complex. Golbe (1983) observed that profitable railroads had a lower average accident rate than unprofitable railroads. She also found that, for profitable railroads, if there was any relationship at all between accidents and profitability it was a weakly positive one. For unprofitable railroads, accident rates rose as losses rose. It should be noted that most of the econometric tests in Golbe's study indicated that the coefficient of profitability, in the equations explaining accidents, was not significantly different from zero. Profitability was measured by net income per gross ton-mile in this study.

Research was conducted on the profitability-safety relationship in trucking by Corsi, Fanara and Roberts (1984). On the basis of the accident experience of regulated and private trucking in 1981 as reported to the federal Office of Motor Carriers (OMC), they found no significant relationship between net operating income and the number of accidents in which the carrier was involved. Their report did not indicate whether the carriers in the sample represented both profitable and unprofitable carriers.

The definition of safety performance of carriers in all of these studies was some measure of gross accident involvement. In the case of rail, access to the railway is for the most part controlled, so that gross accident involvement fairly reflects the performance of the carriers. The obvious exception for railroads is accidents at highway crossings where the actions of highway drivers may cause the accident. In the case of trucking, the carrier's vehicles share the road with private automobiles as well as with other trucks, pedestrians, and animals. In 1984, 65.9% of all accidents reported to the OMC were collisions with other vehicles, persons, or animals, and trucks are certainly not the cause of all of these accidents. Gross accident involvement is not indicative of the performance of the carrier with respect to safety because the carrier's driver or vehicle may not have been at fault. An indication of the magnitude of the difference between gross truck accident involvement and truck-at-fault accidents is derived from California Highway Patrol statistics. They indicate that trucks were at fault in 50.8% and 50.9% of the accidents in which they were involved in 1985 and 1986, respectively. The insignificant change over the two years indicates the robustness of the estimate. This is somewhat confirmed by experience in Ontario, where 47.2% of truck accidents were caused by the commercial driver and 19.5% by mechanical or cargo-related factors; 33.3% of the accidents were not caused by the driver or equipment. The OMC data base used in the Corsi et al. study did not contain information on which accidents were truck-at-fault and which were not, which may explain the low explanatory power of the Corsi et al. estimates. For example, their final multiple regression model explaining carrier accident rates had an R square of only .0647.

A basic criticism of all of these studies is the use of profitability as an explanatory factor. Profitability is but one dimension of the financial condition of a firm. A firm that is earning less than it spends or is not earning the required rate of return is certainly a firm with financial problems. Yet the prospect of being profitable in the future, whether highly probable or not, may justify continued operation if not continued investment. Short-term profitability, such as earnings in one year, are more variable than cumulative earnings or returns over several years, which reflect the longer run strengths of the carrier. More important, other dimensions of the financial condition of the firm could impact the behavior of the carrier in safety matters. The cash flow of a carrier certainly affects the ability of the carrier to maintain its equipment properly. A carrier with negative earnings, for example, would be able to finance equipment purchases, maintain its equipment, and continue safety programs as long as the funds that had been accumulated in the past were available. Such funds may also be available from external sources such as lenders. The capital structure or debt/interest coverage would affect the willingness of lenders to loan the carrier capital to carry it through unprofitable periods. All of this suggests that current profitability alone does not adequately reflect the financial dimensions that may influence carrier safety performance. A better measure would reflect long-term profitability as well as other dimensions of financial condition, such as cash flow, capital structure, and productivity.

In summary, the relationship between regulation and safety in trucking is worth further investigation. The ideal research design would evaluate safety performance with and without regulation. It would recognize how regulation affects the financial condition of carriers and how safety performance is related to various states of financial condition. This is an imposing task, because there is controversy over the impact of deregulation on profits and financial condition, as will be discussed below.

Profitability of Trucking since Deregulation

The number of new entrants into different geographic or commodity markets has accelerated since deregulation. At the same time, so have the num-

Table 15.1   Trucking Failure Trends, 1978–1985

Year	Trucking Failures[a]	Line-Haul Trucking Failure Rate[b]
1978	162	27
1979	186	32
1980	382	48
1981	610	78
1982	960	104
1983	1,228	165
1984	1,409	185
1985	1,533	236

[a]Trucking figures include the following: SIC 4212 (Standard Industrial Code)—local trucking without storage; SIC 4213—trucking, except local; SIC 4214—local trucking with storage.

[b]Failure per thousand companies: includes SIC 4213 only.

SOURCE: American Trucking Associations (various years).

ber of exits. Carriers exit the industry for numerous reasons and in many ways. To the economist, the primary rationale for exit from an industry are levels of profitability and return that are below the opportunities available elsewhere. It is asserted that there has been a serious decline in profitability of motor carriers due to deregulation, which has led to numerous bankruptcies. There are clearly more bankruptcies and exits in the trucking industry since 1980, as indicated in Table 15.1.

The profitability of the trucking industry has also clearly declined. Financial statistics for the total trucking industry as reported by the American Trucking Associations are shown in Table 15.2. Each measure of financial performance has declined since 1978. The bottom of the decline appears to be 1982—the trucking industry has increased its financial performance since then. However, industry performance continues to be be-

Table 15.2   Financial Performance of Trucking 1978–1985

Year	Debt-to-Equity	Net Profit Margin (%)	Operating Ratio (%)	Return on Equity (%)
1978	0.49	4.72	94.67	14.51
1979	0.57	3.00	96.25	10.70
1980	0.57	2.67	96.32	9.91
1981	0.44	2.55	96.94	7.51
1982	0.43	1.01	98.23	2.21
1983	0.41	3.55	95.90	11.24
1984	0.45	3.13	96.19	10.10
1985	N/A	N/A	96.49	N/A

N/A, data not available.

SOURCE: American Trucking Associations (various years).

low historic highs. Of course the pattern will differ from segment to segment of the industry.

There is substantial evidence that trucking industry earnings have decreased and exits have increased, but there is much controversy over the role of deregulation in these trends. Many experts take inherently contradictory positions. Proponents of the competitive marketplace indicate that deregulation has decreased prices by increasing the efficiency of the marketplace. This includes eliminating inefficient carriers. The market works by having continuous entry and exit to guarantee long-run efficiency. At the same time, the effects of the recession that affected the nation beginning in 1979 reduced traffic volumes or expected volumes. This would have had an impact on industry profitability regardless of the regulatory environment. Thus, it is difficult to separate the influence of regulatory change from the influence of the economy; it is beyond our scope here to make a final determination. We conclude that deregulation certainly reduced the profitability of carriers because they were less protected from the rigors of the marketplace, but to what degree is not known. The latter is not of much concern since the regulatory environment has been changed and the changes are likely to endure into the foreseeable future. Knowing how deregulation affected the financial well-being of the trucking industry is less important than knowing how "financial well-being" is related to safety. As carriers exited, did they perform poorly or well with respect to safety? Did the carriers that survived perform better or worse with respect to safety as their profits declined, notwithstanding the impact of deregulation versus the economy? Deregulation has come and is here to stay. The issue is whether the remaining carriers perform better or worse with respect to safety under the new regime of the marketplace. These more pragmatic questions become the basis for this research.

Safety Performance and Safety-related Practices

Factors that directly cause accidents and affect the severity of truck accidents can be classified into three groups: human, vehicle and equipment, and environment.

Driver factors

Driver error is cited in numerous studies as the cause of an overwhelming majority of truck-at-fault truck accidents. There is substantial evidence that such errors increase when the driver is speeding, driving under the influence of drugs or alcohol, driving past normal maximums on hours of service, or is fatigued (Chaumel et al., 1986; Harris and Mackie, 1972; McDonald, 1984). Over 90% of the truck-at-fault accidents in California were classified as driver related (California Highway Patrol, 1987). Approximately 80% of truck-caused accidents in Ontario were observed to be primarily the fault of the driver (Chaumel et al., 1986).

Many carrier decisions can produce or create situations conducive to driver error. Gander (1986) showed that when income is dependent on speed, drivers will speed. This is particularly true for owner-operators whose annual income and productivity are a function of the utilization of their equipment. Most drivers of regular-route operations are paid on a mileage basis. But such drivers, as well as truckload drivers, may be encouraged to speed or to drive excessive hours by the schedules given to them by the carrier. Beilock (1985) found that over one third of drivers surveyed had to violate either speed or hours of service regulations to meet their schedules.

Motor carriers engage in safety programs to improve the quality of the driver. A comprehensive program would include selection of driver personnel, driver training, award or incentive plans, general safety information, inspections, and road supervision of drivers. There is ample evidence that such practices improve safety. Corsi et al. (1984) found that carriers with high accident rates have "a disproportionate number of accidents in which the driver is under 30 years old, has less than 2 years experience, and does not use his seat belt." They also found that several aspects of compliance with federal safety regulations were related to accident rates for the smaller class III carriers. Specifically they found that noncompliance with hours of service, accident reporting, driver qualifications, and vehicle inspection and maintenance requirements were related to higher accident rates. However, they did not find a similar relationship for the larger class I and II carriers.

Vehicle and equipment factors

Vehicle and equipment failures are obviously less important than driver-related causes of truck accidents. Nevertheless, they contribute to both the frequency and the severity of accidents. The probability of such failure is determined by the age of the equipment and the maintenance level. Preventive maintenance can have two influences on equipment failure. First, it can increase the life of vehicle components, reducing the incidence of failures. Second, the inspection aspect of maintenance detects failures and therefore minimizes the time during which such failures present a hazard. McDole and O'Day (1975) concluded after extensive study that defect-related accidents are reduced by inspection and repair. They also observed that poorer maintenance practices are associated with the smaller and private companies. This is consistent with Corsi et al.'s (1984) findings on the relationship between accidents and noncompliance with regulations on vehicle inspection and maintenance for the class III carriers.

Vehicle age affects the probability of accidents in several ways. First, newer vehicles may incorporate new technology not available on older vehicles because of the cost of retrofitting the older model. Second, newer vehicles have newer components, which are less susceptible to failure. McDole and O'Day (1975) reported that in one state, vehicle defects were the causative factor in 10% of accidents involving trucks 10 years old or older. In contrast, trucks less than three years old averaged 4% for the same statistic. Thus, carriers can influence the safety of their operations through maintenance and fleet replacement decisions.

Driver employment policy

Carriers can influence their accident rates through their utilization of owner-operators. Owner-operators can also directly enter the industry as small carriers competing for traffic. In both cases the decision process of the owner-operator with respect to safe driving and preventive maintenance is relevant. Most owner-operators are paid on the basis of a fixed percentage of revenue. Thus, the owner-operator bears most of the impact of price discounting, since the percentage of revenue paid to the owner-operator is typically between 70% and 80%. The basic economics of the full-load carrier and the owner-operator are similar. Baker (1985) estimated from his data base that, given representative revenues and costs of operating line haul vehicles, the full-load driver must either drive far beyond the federally allowable hours or speed in order to make a decent living. He also suggested that drivers who try to make a living under such conditions may not be the quality of driver that is desirable on the highway. In addition to owner-operators having economic incentives to practice unsafe driving practices, their violations of safety laws are the hardest to discover and control.

There is evidence that owner-operators do practice unsafe driving practices to a greater degree than company drivers. Wyckoff (1979) reported on several driving practices from a survey of 10,500 driver interviews. His results, which appear in Table 15.3, show that owner-operators consistently violate various safety regulations or are involved in accidents more than company drivers. Only the speed of driving is comparable between these two groups of drivers. Part of the reason is that company drivers can be controlled more easily, but the other factor is that owner-operators are paid less than company drivers for comparable work. A recent survey by Capelle and Beilock (1988) found that owner-operators are perceived to speed more than drivers of carrier-owned fleets, violate hours of service regulations, and are viewed as more likely to be drug and alcohol abusers. Corsi et al. (1984) found that a 1% increase in a carrier's owned-vehicle miles as a share of total vehicle miles will reduce its accident rate by almost 6%.

Other factors

Expenditures on insurance and safety, maintenance, and newer equipment all represent controllable factors on the part of the carrier, given the particular market. But the discussion shows that many decisions affecting safety cannot be deduced from simple direct expenditures. Methods of rewarding drivers, the work environment, and the scheduling of trips are decisions that have no recordable cost or activity measure. In addition, decisions with regard to the design of equipment (visibility, quality of components), and utilization of equipment (overloading) will also influence the rate and

Table 15.3 Driving Practices by Types of Driver

	Common Carrier		Contract Carrier	
Variable	Company Driver	Owner-Operator	Company Driver	Owner-Operator
Cruising speed (mph)	58.85	60.26	59.39	60.95
Moving violations/100,000 vehicle miles	0.41	0.77	0.76	0.74
Reportable accidents/100,000 vehicle miles	0.19	0.31	0.26	0.33
Percent of drivers using multiple logbooks	1.87	11.89	4.90	10.22
Percent of drivers who "regularly mispresent logs"	4.27	33.91	20.26	33.56
Percent of drivers who "regularly drive beyond the 10-hour limitation"	2.48	11.30	13.05	27.23

SOURCE: Wyckoff, 1979.

severity of accidents. We must recognize that any direct expenditure measure will not capture the true effort and attitude of a carrier toward safety.

Most environmental factors that cause accidents are, in a sense, also influenced by the carriers' choice of markets to compete in, but can be considered external to the issues of regulation, profitability, and safety. The carrier that competes in truckload (TL) markets with few fixed terminals requires a different type of line-haul operation than regular-route carriers competing in less than truckload (LTL) markets. The TL carrier does not have the regular turnarounds of the LTL carrier. Therefore it tends to expose its drivers to potentially longer and more irregular hours. This points to the need to recognize different types of trucking operations in making firm-to-firm comparisons of safety performance.

Literature Summary

There are several intermediate links between economic regulation and the safety performance of the trucking industry. One paradigm suggests that deregulation has increased the number of carriers in financial distress in the trucking industry. The literature reviewed indicates that there has been a serious decline in the health of the trucking industry in the immediate years following deregulation, but the causal links have not been verified. The next link is the proposition that carriers in financial trouble tend to reduce expenditures in safety-related activities or to encourage unsafe operations. This is the focus of the empirical analyses below. Unsafe operating practices in turn result in less safety on the highways. We have cited a small part of a substantial body of literature supporting this latter link.

The discussion also suggests that a study linking the financial condition of carriers directly with safety as measured by total accidents must be interpreted carefully. Whereas studies linking driver and vehicle programs with accidents typically used truck-at-fault accidents, the few studies linking financial performance with accidents did not. This deficiency is crucial and may lead to misleading conclusions. The discussion also suggests that single year measures of profitability are inadequate measures of the financial dimensions of the carrier, which influence the carrier's efforts related to safety.

This study investigates the hypothesis that there is a relationship between financial condition and safety. The study could be accomplished at two levels. First, the relationship between the carrier's efforts to be safe (safety conduct) and financial condition could be explored. Second, the relationship between financial condition and actual safety performance measured by truck-at-fault accidents could be explored. We focus on the first relationship because of the unavailability of data on truck-at-fault accidents in the OMC database.

METHODOLOGY

Carrier Population

This study's empirical focus is on class I and II general freight carriers. The operations of such carriers are significantly different from those of nongeneral freight carriers, who typically compete in TL markets. Further, as noted above, the behavior of smaller carriers such as those classified as class III is different, and is not addressed here.

Measurement of Financial Condition

Financial analysts have long searched for more accurate methods of gauging financial health and forecasting insolvency. Single financial ratios used individually are not adequate measures of the financial condition of a firm. A carrier with negative earnings may have enough liquidity to stay in business long enough to return to profitability. A carrier with positive earnings may not have the cash flow to meet current obligations. Most lenders and credit analysts will look at a variety of financial indicators to make a balanced assessment of credit worthiness. These indicators are often grouped into separate categories, each of which measures a particular aspect of financial strength. The following is one such breakdown:

Profitability Ratios. Ratios in this group gauge the rate of profit relative to some base. Common ratios used in the trucking industry are operating ratio and return-on-transportation investment.

Turnover Ratios. These statistics indicate the efficiency of assets, capital, and labor. The widely used ratio in this group is the capital turnover ratio.

Leverage Ratios. This category of ratios indicates the extent to which a firm used debt to finance its asset base. The excessive use of debt can create significant profitability and liquidity problems should revenues not reach break-even traffic volumes. Interest charges are fixed charges that are unavoidable. Excessive debt is thus a direct cause of bankruptcy risk.

Liquidity Ratios. These ratios measure the ability of the firm to pay its current debts (current liabilities) as they become due. The less liquid the firm, the greater the danger that creditors will force the firm into involuntary cessation of operations. Liquidity ratios include the current ratio, debt coverage, and cash flow.

All of these types of ratios can be derived from basic financial statements of motor carriers. The challenge is to determine what importance should be placed on each dimension of financial performance. The use of predictive models that combine financial statement analysis (ratio analysis) with statistical techniques is one approach to overcoming this difficulty, as is done in the Altman Z family of models (Altman, 1983).

A Model of Financial Distress for Trucking

Altman Z models have been applied to a variety of industries. Chow and Gritta (1985) demonstrated that the Z″ version provides reasonable estimates of the financial condition of selected segments of the trucking industry. The Z″ version was found to be superior to the Z or Z′ version since it does not rely on the market value of equity or include an asset turnover ratio. Most trucking companies are not publicly held firms and have fewer assets than manufacturing firms. Thus the Z″ version better reflects the economic and ownership characteristics of the trucking industry.

The most current version of the Z model is claimed to be applicable to a wide range of industries, since it was developed from a cross section of bankruptcies of publicly held firms. In 1982 it was used to suggest that Carolina Freight Carriers was a candidate for bankruptcy. Hindsight verifies what most industry observers believed at the time: Carolina Freight was in fact one of the outstanding carriers in the trucking industry in terms of financial strength and future profitability. A model that is estimated from the data of carriers in the trucking industry may have made a better prediction, for the same reasons that the Z″ model was superior to the Z or Z′ models. Furthermore, when the economic and regulatory environments change, the factors that best explain financial distress may change. Chow, Gritta, Adrangi and Ebelt (1986) have shown that the financial indicators that best predicted financial distress in the U. S. trucking industry before deregulation are different from those after deregulation in 1980.

A model of financial distress that we shall call the "C" model was estimated (Chow, Vasina, Adrangi, & Gritta, 1987). Actually, a set of models was developed in which each model was specific to different segments of the U. S. trucking industry. Motor carriers were classified as financially distressed if they had gone bankrupt or were technically insolvent. The financial distress model was developed using multiple discriminant analysis (MDA) with a dependent variable of 0 to 1 indicating whether the carrier was financially distressed (1) or not financially distressed (0). Independent variables were financial ratios, as discussed previously. MDA provided statistically determined weights for financial ratios which significantly characterized firms in financial distress.

One of the early versions of the C model was calibrated from a sample of U. S. general freight carriers in financial distress between 1982 and 1986. It has the following form:

$$C = a + b_1 X_1 + b_2 X_2 + b_3 X_3 + b_4 X_4 + b_5 X_5 \tag{15.1}$$

where

C = financial distress score
X_1 = retained earnings/total tangible assets
X_2 = total liabilities/total tangible assets
X_3 = shareholder equity/total tangible assets
X_4 = current liabilities/total tangible assets
X_5 = working capital/total tangible assets

The C scores were normalized so that interpretation of the C model is that a positive C score corresponds to a probability of bankruptcy of *less* than 0.50, while a negative C score corresponds to a probability of bankruptcy of *greater* than 0.50. Viewed in this manner, a C score of 0 corresponds to a 0.50 probability of bankruptcy, and the higher the C score the lower the associated probability of bankruptcy. This "probability of bankruptcy" is based on the degree to which a firm "resembles" a firm that is known to have failed or to be in technical insolvency.

Measurement of Implied Safety Conduct

We concluded that the ideal relationship to quantify is the correlation between financial condition and actual safety performance. However, the data bases available for this study did not allow the proper quantification of such a relationship. Past attempts to link profitability and safety were found to be deficient. There is no dearth of a priori and empirical evidence, however that links various safety-related operating and management strategies with safety performance. If we take such relationships as given, we can infer safety performance from the linkages between financial condition and safety by knowing the relationship between financial condition and safety-related operating strategies. It is the latter relationship that our empirical focus is on.

Several aspects of safety operating strategies can be measured from the operating and financial statements of motor carriers. Motor carriers engage in safety programs or risk management programs to improve safety. The usual emphasis is on drivers. A comprehensive program would include selection of driver personnel, driver training, award or incentive plans, general safety information, inspections, and road supervision of drivers. The majority of these expenses are accumulated in the functional cost account, Insurance and Safety, in the uniform system of accounts prescribed for carriers reporting to the Interstate Commerce Commission (ICC). The functional account, Insurance and Safety, includes direct and allocated expenses from numerous natural cost accumulation accounts such as salaries, supplies, communications, and so on. The insurance component does not include any insurance paid for revenue vehicles or for cargo loss and damage. Thus, the use of this functional account does not present any identification problems. The insurance component is primarily for supervisors' automobiles and not for cargo or vehicle liability, although administrative costs of self-insurance may be hidden in this account. Some safety-related expenditures may not be reflected in this account, such as the cost of mechanical devices included in the purchase price of equipment.

Maintenance expenditures should reflect the maintenance required to safely operate the carrier's fleet, given the level of vehicle utilization and the age of the equipment. Some forms of maintenance marginally affect the safety potential of the carrier. If a carrier fails to change the oil properly, this may reduce the efficiency of the truck or prevent it from operating at some point, but safety is affected marginally. In contrast, brake checks and adjustments, tire checks and adjustments, and so on are essential to safe driving conditions. Carriers accumulate maintenance expenditures in a functional activity account, and this should reflect the carrier's overall maintenance effort. It is apparent that no accounting system can precisely account for all maintenance costs and, in particular, for safety-related maintenance costs, since many maintenance-related activities are undertaken by the drivers, whose salary is accounted for separately. At best, this measure will be an imperfect representation of maintenance resources utilized to maintain safety.

Required maintenance should be a function of the level of vehicle activity. This factor is accounted for by expressing maintenance expenditures as a cost per intercity vehicle mile. Maintenance should also reflect the age of the vehicle. Examples of conscious strategies that trade off vehicle maintenance and fleet replacement costs are the early equipment replacement cycles used by some fleets. By replacing their vehicles within two to three years, these businesses assure that their vehicles never reach the point of requiring engine overhauls or maintenance other than preventive maintenance. Hence, their maintenance expenditures per vehicle mile always appear small relative to the industry average.

When maintenance costs are reported as a functional activity for aggregating costs in the carriers report to the ICC, these expenditures reflect costs expended in local service and terminal structures. An alternate measure of maintenance expenditures that relates more closely to line-haul activity is the summation of the following natural expense categories for line haul only: vehicle repair and service wages, miscellaneous paid time off for vehicle repair and service wages, and supplies.

The age of equipment can reflect a myriad of decisions of the firm. A carrier with equipment all the same age could show an aging fleet simply because its planned equipment replacement date has not been reached. This is most likely to be the case for small carriers, where the size of the fleet does not permit as many opportunities to diversify the fleet age. Observation of industry practices indicates that most large fleets consist of a portfolio of vehicle vintages such that certain portions of the fleet require replacement at yearly or biyearly intervals. This allows the average age of the fleet to be relatively constant and would suggest that a trend toward an older- or younger-age fleet reflects other factors.

New technology that makes existing equipment technologically obsolete or tax law changes may encourage early replacement of equipment. For example, the ability to use twin trailers has caused many carriers to replace perfectly usable conventional trailers with these multiple units in order to achieve operational efficiencies. On the other hand, financial difficulties may prevent carriers from making equipment replacement decisions that would improve both safety and operational efficiency. Carriers in financial distress are the least likely to be able to obtain financing at reasonable cost or without substantial increase in financial risk. Equipment suppliers are, of course, reluctant to grant long-term financing to businesses that are not likely to be able to pay their interest burdens much less the principal on the debts incurred in the purchase of new equipment. Therefore, it is difficult for carriers that are in financial difficulty to obtain the capital required to purchase new equipment. Finally, equipment replacement may be deferred simply because the forecast of business volume may not justify the purchase. In the recessionary period following deregulation this may have been the case. In summary, the age of the equipment fleet reflects many factors, some of which relate to the financial condition of the firm.

The age of the fleet cannot be measured directly from the operating statistics in the financial statements of the carrier. However, the statistics do report the depreciated and book value of operating assets of the carriers. A surrogate for the age of the fleet is the ratio of net operating property to total operating property. This assumes that the bulk of the assets is road equipment. As the ratio rises, the fleet age is lower.

A shift from company-owned equipment and employee drivers to owner-operators can reflect a shift in the burden of safety prevention. We noted above that owner-operators are frequently identified as vehicle operators with a greater propensity to violate federal safety regulations, and it is inferred that their incentives to practice deferral of maintenance and other safety-related activities are high, given the economics of their business. Thus a shift from company-owned equipment to owner-operators may result in more accidents. This can be measured by the ratio of vehicle miles rented with driver to total vehicle miles.

ANALYSES AND RESULTS

Aggregate Trends in Safety Conduct and Financial Condition

General freight carriers are divided into four groups: Instruction 27 (I-27)/ class I; I-27/class II; Instruction 28A/class I; and I-28A/class II. I-27 carriers are general freight carriers that earn 75% or more of their revenues from intercity movement of general commodities. I-28A carriers are those carriers that do not meet this criterion but earn more than 50% of their revenues from general freight. Aggregate statistics for various measures of safety conduct are shown for these groups from 1976 through 1985 in Table 15.4. Insurance and safety costs per vehicle mile and maintenance costs per vehicle mile were deflated by the producer price index.

Insurance and safety costs for all general freight carriers decreased (in deflated dollars) between 1976 and 1985, from 1.61¢ to 1.35¢ per vehicle mile. Since 1980, the rate has fluctuated above and below 1.4¢ per vehicle mile. There appear to be slight differences in these costs between size class and type of general freight carrier. Maintenance costs also show an oscillating pattern, with the highest maintenance expenditure since deregulation (1977) being in 1983 and the lowest in 1984. Type of carrier differences are very significant.

The ratio of net carrier operating property to total carrier operating property is our measure of the newness of the assets being used. The higher the ratio, the lower the age of the assets used. An interesting pattern emerges over every five-year period as the ratio moves from a low of .50 in 1976 upward to .55 by 1979, than declines back to .50 in 1982 and than begins to rise back to .54 by 1985. This suggests a natural aging and replacement cycle of the total fleet over a five-year cycle. That deregulation occurred with the aging portion of this cycle is coincidental, and this may provide an alternative explanation for the aging of the fleet. The class I, I-27 carriers consistently own the newest fleets.

Finally it can be observed that the utilization of owner-operators increased significantly from 1976 to 1985. This is in spite of the general consensus that the supply of owner-operators, especially good owner-operators, has declined due to low earnings spurred by rate competition.

These figures suggest that there are size (class I vs. II) and type (I-27 vs. I-28A) differences in safety expenditures. We do not know how financial performance affected these expenses. Therefore, ordinary least squares estimates were made of the relationship between these safety conduct variables and the C score and dummy variables representing size and type class. Each of the four groups is a mutually exclusive segment of general freight yielding 40 observations. The class grouping represents an economies-of-scale measure. The type grouping accounts for differences in operations of carriers that affect these expenditure levels. The C score represents the aggregate financial condition of each group.

The results are summarized in Table 15.5. The coefficients for the C score variable are positive and statistically significant in every relationship

Table 15.4 Trends in Safety Conduct by Size Class and Type of General Freight Carrier

		Carrier Group				
		I-27		I-28A		All General
Measure of Safety	Year	Class I	Class II	Class I	Class II	Freight
$ Insurance and safety	1976	.015	.016	.026	.014	.016
expense/vehicle mile[a]	1977	.015	.016	.023	.014	.016
	1978	.014	.017	.013	.014	.014
	1979	.014	.016	.013	.014	.014
	1980	.015	.015	.014	.017	.014
	1981	.015	.010	.015	.013	.014
	1982	.015	.014	.013	.015	.013
	1983	.014	.012	.014	.015	.014
	1984	.013	.011	.015	.012	.012
	1985	.015	.010	.015	.019	.013
$ Maintenance expense/	1976	.026	.031	.065	.050	.029
vehicle mile[a]	1977	.025	.032	.064	.044	.028
	1978	.024	.029	.030	.048	.025
	1979	.025	.027	.038	.040	.026
	1980	.027	.027	.029	.054	.026
	1981	.026	.026	.041	.039	.026
	1982	.027	.028	.029	.040	.026
	1983	.027	.032	.029	.047	.029
	1984	.026	.025	.035	.040	.025
	1985	.028	.022	.029	.043	.025
Net/book carrier operating	1976	.520	.410	.479	.443	.507
property (fleet age)	1977	.535	.418	.463	.466	.518
	1978	.552	.464	.495	.499	.546
	1979	.559	.483	.512	.548	.553
	1980	.539	.460	.523	.481	.534
	1981	.526	.496	.498	.471	.521
	1982	.503	.421	.503	.506	.500
	1983	.483	.348	.545	.441	.485
	1984	.521	.382	.587	.466	.523
	1985	.546	.408	.546	.463	.539
Miles rented with driver/	1976	.074	.132	.266	.297	.095
total vehicle miles	1977	.079	.130	.242	.324	.098
	1978	.092	.139	.475	.383	.118
	1979	.095	.180	.486	.491	.136
	1980	.109	.186	.631	.372	.160
	1981	.095	.142	.563	.412	.159
	1982	.141	.207	.559	.342	.211
	1983	.159	.189	.576	.393	.251
	1984	.160	.270	.517	.364	.245
	1985	.153	.287	.506	.403	.248

[a] Figures deflated by producer price index.

examined. As each general freight segment improved in financial health, so did per-vehicle-mile expenditures for insurance and safety and maintenance. The healthier the industry segment, the newer the fleet. However, the stronger the financial condition, the greater the use of vehicles rented with drivers, which includes owner-operators mileage. The fleet age vari-

Table 15.5   Determinants of Safety Conduct of General Freight Carriers[a]

	Type of Safety-related Conduct			
	Insurance and Safety Expense/ Vehicle-Mile	Maintenance Expense/ Vehicle-Mile	Net Carrier Operating Property/ Carrier Operating Property	Vehicle-Miles Rented with Driver to Total Miles
R^2	0.92	0.96	0.89	0.89
C score	0.02	0.02	0.68	0.16
	(11.624)	(3.739)	(9.598)	(3.353)
Fleet age	N/A	0.02	N/A	N/A
		(3.449)		
Class II	0.0007	0.006	0.02	0.03
dummy variable	(0.564)	(2.714)	(4.73)	(0.854)
Type I-28A	0.007	0.02	0.23	0.35
dummy variable	(6.004)	(7.28)	(5.175)	(11.493)

[a]t statistics in parenthesis.

N/A, not in equation.

able was included in the maintenance regression and its coefficient was positive and significant, which indicates that more maintenance is expended on younger fleets. The only class size effect is found in maintenance, where such expenditures are generally higher for class II carriers. Finally, the I-28A carriers appear to be spending more on both insurance and safety, and maintenance, use more owner-operators, and have older fleets.

In summary, the evidence suggests that increased general freight carrier expenditures on safety-related activities and more frequent vehicle replacement are positively correlated with better financial condition. There also appears to be a positive correlation between the dependence on rented vehicle miles and financial condition. Greater utilization of owner-operators results in increased highway hazards. This industry has, however, gone through rationalization, as carriers have exited and entered this segment of trucking. The data represent the surviving carriers in each year. The next step in the analysis is to distinguish between carriers in financial distress and those that are not.

Financially Distressed versus Nonfinancially Distressed Performance

As noted above, a carrier is defined as financially distressed when it is known to have declared bankruptcy or to be in a prolonged state of technical insolvency. Between 1982 and June 1986, 108 general freight carriers reporting to the ICC were classified as financially distressed. Through 1986 344 general freight carriers were confirmed to be nonbankrupt. The C-score value of the nonfinancially distressed population is observed to be significantly higher than the C score of the bankrupt population. A simple comparison of the performance of the two populations is shown in Table

15.6. The table also shows that the arithmetic mean performance of the nonfinancially distressed carriers is uniformly superior with respect to activities related to accident prevention. The average financially distressed carrier spends less on insurance and safety and maintenance despite having older equipment. Furthermore, there is some tendency to use more owner-operators on the part of the bankrupt carrier group. However, only the fleet age and use of owner-operator variables are statistically significant.

Safety Conduct within Nonbankrupt Firms

The previous analysis compared safety performance between the bankrupt and nonbankrupt segments of the general freight motor carrier sector. Next we focus on the behavior within the nonbankrupt carrier segment. Cross-tabulations were used to identify patterns in safety conduct with respect to financial condition. The nonbankrupt carrier population was broken down into quintiles according to their C score for one set of comparisons, by net profit margin for another set of comparisons, and by leverage for the final set of comparisons. Leverage is defined as total debt to total shareholders equity. Each category of safety conduct was also broken down into quintiles so that exactly one fifth of the population made the highest level of expenditures on, for example, safety and insurance per vehicle mile, one fifth made a lower level of expenditure, and so on. The categorizations are not shown here and are available on request.

Each grouping of safety conduct variables was cross-tabulated with each grouping of the financial performance variables. The chi-square statistic

Table 15.6 Safety Conduct of Financially Distressed versus Nonfinancially Distressed General Freight Carriers[a]

| Variable | Population (sample size) | | |
	Nonfinancially Distressed	Financially Distressed	T Value
C score	0.463 (344)	−1.17 (108)	10.34
$ Insurance and safety expense/ intercity vehicle mile	0.128 (295)	0.067 (86)	0.48
$ Maintenance expense/intercity vehicle mile	0.294 (295)	0.121 (86)	0.48
$ Line-haul maintenance expense/ intercity vehicle mile	0.216 (295)	0.128 (86)	0.31
Net carrier operating property/ total carrier operating property	0.402 (342)	0.350 (108)	2.57
Number of trucks rented with drivers/total trucks employed	0.108 (336)	0.153 (103)	−1.47

[a]1985 financial statistics used for nonfinancially distressed carriers. Financial data refer to year prior to bankruptcy for carrier exiting industry for financial reasons. If such data are not available, financial data two years prior to bankruptcy are used. For technically insolvent carriers, 1985 financial statistics were used.

SOURCES: Tabulated from American Trucking Associations (various years).

was used to identify those cross-tabulations where the probability of the observed safety conduct frequency relative to financial performance had a very small chance of occurrence. Significant chi-square statistics indicate that C scores and leverage seem to be related to fleet age and use of owner-operators, whereas net profit margin seems to be related to expenditures on maintenance and to fleet age. No financial measure appears to have any relationship with safety and insurance expenditures. A significant chi square simply indicates that the observed frequency by which carriers appear in each safety conduct group, given the financial performance, is not likely to be by chance. The actual pattern must be viewed to determine the economic implications. A strong monotonic pattern was not evident in any of the cross-tabulations. For example, a larger percentage of the carriers with the lowest profit margins were in the top quintile of maintenance expenditure, but when the two highest levels were aggregated, the percentage in these two groups did not differ between carriers with highest and lowest profit margins. Carriers with the lowest C scores clearly owned the largest percentage of old equipment, but the carriers with the highest C scores did not have the newest equipment. Similarly, the carriers with both the highest and lowest C scores made minimum use of owner-operators. The biggest users were in the third and fourth quintiles: average to below-average financial condition. We conclude that within the nonbankrupt carrier segment, there is no strong relationship between financial condition and safety conduct.

Safety Conduct of Bankrupt Firms

Multiple regression analysis was utilized to estimate the relationship between financial condition and safety-related expenditures of carriers that were bankrupt or were classified as technically insolvent. Carriers that went bankrupt in the period 1983 to June 1986 were considered in this analysis. For each bankrupt carrier, the safety conduct and financial performance of the carrier was obtained for the three years prior to bankruptcy. There were 81 bankrupt carriers but only 229 observations, since data were not available for every year for each carrier.

The regressions used to explain safety-related expenditures had the following general form:

$$Y = b_0 + bX_1 + b_2X_2 + b_3X_3 + b_4X_4 + b_5X_5 \tag{15.2}$$

where

$Y =$ a variety of variables described in the text below

$X_1 =$ the measure of financial performance, either C score, net profit margin, or leverage

$X_2 =$ log of vehicle miles included to account for economies of scale in this activity

X_3 = average weight/shipment (average load is substituted for average weight for I-28A carriers, which do not report this figure) to account for differences in operations

X_4 = average length of haul to account for differences in operations

X_5 = vector of six dummy variables for each year after 1979 through 1985 to account for price level changes and other time-related effects

Three models were estimated, one for each measure of financial condition using insurance and safety expenditure per vehicle mile as the dependent variable. The results are summarized in the first line of Table 15.7 where only the coefficients for X_1 are shown. There appears to be little explanatory power in the model, much less a relationship between financial condition and expenditures on safety.

The regressions utilized to explain maintenance expenditures were more revealing. Both measures of maintenance are somewhat negatively related to net profit margin, and total maintenance is positively related to higher leverage carriers. Thus, the more profitable of the bankruptcy candidates will tend to spend less on maintenance, and more highly leveraged carriers tend to spend more on maintenance. No relationship is observed between maintenance and the overall measure of financial health, the C score.

The greatest amount of explanatory power was found in the estimation of fleet age, measured by the ratio of net carrier operating property to carrier operating property. Thus, the higher the ratio, the newer the assets being utilized by the carrier. The coefficients for all three financial performance variables are highly significant. Carriers with the best C scores and net profit margins tend to have newer fleets, whereas carriers with large amounts of debt tend to have older fleets.

The final behavior to explain was the usage of owner-operators, defined as the ratio of owner-operator trucks to owned trucks. The results are not easily explained. There is less dependence on owner-operators when the carrier is in better overall financial condition, as indicated by the positive coefficient for the C score, but there appears to be more dependence on owner-operators when the net profit margin is higher. In addition, the leverage coefficient suggests that the more highly leveraged carriers use more owner-operators.

The first and the last observations are consistent. A more highly leveraged, indeed overleveraged carrier, does not have the capacity to finance its own equipment and must depend on vehicles supplied by owner-operators. On the other hand, more profitable carriers tend to have higher C scores in the long run. But each observation in the population under study represents a carrier that will go out of business in one, two, or three years. Short-run profitability has nothing to do with the ability to finance and acquire equipment. This confirms the need to recognize more than simple profitability in trying to identify carriers likely to be less safe.

In summary, these regression estimates of the safety behavior of carriers nearing bankruptcy show that as they approach their eventual demise, as

Table 15.7 Summary of Ordinary Least Squares Estimates of Safety Conduct Relationship to Financial Performance for Bankrupt Carrier Group

Dependent Variable	C Score		Net Profit Margin		Leverage	
	R^2	Coefficient	R^2	Coefficient	R^2	Coefficient
Insurance and safety expense/vehicle mile	.02	-2.24×10^{-4}	.02	0.093	.02	-6.07×10^{-4}
Maintenance expenditure/ vehicle mile	.22	-0.004	.23	-0.122	.23	0.024^c
Line-haul maintenance expenditure/vehicle mile	.15	-7.76×10^{-4}	.18	-0.184^b	.15	0.002
Net carrier operating property/total carrier operating property	.38	0.024^a	.37	0.484^a	.35	-0.047^b
Number of trucks rented with drivers/total trucks employed	.07	-0.035^a	.03	0.336^c	.03	0.052^c

[a] Significant at the 1% level.

[b] Significant at the 5% level.

[c] Significant at the 10% level.

measured by an overall index of financial health, the C score, they tend to let their fleet age and to depend more on owner-operators. No relationship could be observed between the C score and maintenance or between the C score and insurance and safety expenditures. In contrast, it appears that the more profitable carriers in this group spend less on maintenance. We must remind ourselves though that these are carriers in financial difficulty and that in the short run, being profitable and deferring maintenance expenditures is not inconsistent. One is reminded that a large amount of variance in safety behavior is left unexplained in the models estimated.

Statistical Results

The empirical investigations of this chapter focused on the segment of the motor carrier industry known as general freight carriers. An index of financial health or fitness of these carriers was developed from a C score model. We observed that group expenditures for insurance and safety and for maintenance parallel financial condition. The industry segment spent more per vehicle mile in these areas as financial health improved. The age of the industry's assets moved in reverse direction to financial health. As financial health improved, the ratio of net carrier operating property to carrier operating property increased. A positive correlation between the relative use of subcontracted line haul and financial health was observed. A significant "size of carrier" and "type of general freight carrier" effect was also found in some areas of safety conduct.

The next step was to analyze the behavior of general freight carriers at the carrier level. We were able to distinguish between carriers in financial distress (eventually bankrupt or technically insolvent) and carriers not in financial distress. Nonfinancially distressed carriers typically spend more on insurance and safety and on maintenance, have newer fleets, and use fewer owner-operators than financially distressed carriers. Of course, there are always above-average and below-average performers in each group. When the nonfinancially distressed carrier group was investigated at the firm level, no identifiable pattern between financial condition and safety-related activities could be observed. In contrast, further analysis of the financially distressed group at the firm level indicated that fleet age and use of owner-operators were significantly related to the firm's financial fitness. As financial fitness improved, equipment was replaced and less use was made of owner-operators. In the financially distressed group, we also observed that the carriers with the highest net profit margins tended to spend less on maintenance.

CONCLUSIONS

This chapter examined the safety performance of the general freight segment of for-hire trucking and focused specifically on the relationship between the financial condition of these carriers and their safety performance. The ideal measure of safety performance is accidents for which the carrier is at fault, but such data were unavailable. Therefore, efforts of carriers to prevent accidents were used to infer safety performance. These included expenditures on safety, expenditures on maintenance, the age of the fleet, and the use of owner-operators. Measures of the aspects of safety conduct were derived from annual report information submitted to the ICC.

Profitability is only one dimension of the financial condition of a carrier. A measurement of the financial condition of the carrier was represented by the C score, a weighted combination of financial ratios. The weights were statistically determined so that low values of the C score are associated with carriers that have gone bankrupt.

The aggregate conduct of the general freight segment of trucking is generally consistent with the hypothesis that safety performance deteriorates with the declining financial health of the industry. Less is spent on safety and on maintenance and the fleet ages. However, it also appears that more owner-operators have been used as the industry's financial condition improves. When carriers in financial distress are distinguished from those that are not, a more consistent picture emerges. The average carrier that eventually goes bankrupt spends less on safety and maintenance, has older equipment, and depends more on owner-operators. Finally, it appears that as financially distressed carriers approach their eventual demise, as implied by a worsening C score, they spend even less on new equipment and more on subcontracted line haul. This is not unexpected. We did not find similar relationships in the nonfinancially distressed group of carriers.

Federal and state safety enforcement agencies have limited resources to use in their efforts to improve highway safety. They must focus their preventive efforts on individual carriers with the greatest likelihood of causing accidents. One such characteristic is the carrier's financial condition. This study has shown that carriers that have gone bankrupt exhibit conduct that is least conducive to highway safety. This may be obvious to many enforcement officials, who need to identify such carriers well before bankruptcy. The financial models used to develop the C scores can be used to forecast probable bankruptcies several years in advance and could provide an early warning device for enforcement agencies.

Policymakers involved in safety prevention should use predictions of the financial condition of the trucking industry or of specific segments of trucking in their strategic planning. If the current rate of carrier bankruptcy and turnover is expected to continue, government resources allocated to safety will have to be increased.

16

Effects of New Entrants
on Motor Carrier Safety

THOMAS M. CORSI and PHILIP FANARA, JR.

Since the passage of the Motor Carrier Act of 1980 (MCA), the number of motor carrier firms has more than doubled from about 18,000 in 1980 to approximately 37,000 by the end of 1986, a net gain of 19,000 carriers. An important policy debate has developed concerning the impact of the MCA on motor carrier safety.

One particular aspect of the policy debate is whether new motor carrier entrants, those who received operating certificates from the Interstate Commerce Commission only after passage of the MCA, have accident rates significantly higher than those of the established carriers. Some assert that the inferior safety performance of the new entrants, many of which have small-volume operations measured in terms of vehicle-miles, results from their substandard maintenance and training programs combined with operating practices that result in driver violations of federal hours-of-service regulations. Others deny the assertion that the new entrants have safety records significantly different from those of the established carriers.

A recent article in *Fortune* set forth the dimensions of the newly heightened concerns about truck safety (Labich, 1987):

The number of U. S. highway accidents involving heavy trucks has rocketed in recent years. . . . The growing safety problem is a lesson in the perils of deregulation. . . . Deregulation has compounded the problems by creating economic circumstances that made trucking far more dangerous. Price competition forced hundreds of large and medium-sized companies out of business. The smaller outfits and independent owner-operators that took their place are nimbler, but these new entrants have a hard time making money. . . . Result: Many hard pressed truckers have plenty of incentive to spend excessive hours at the wheel and overlook maintenance requirements.

Thomas Donohue, President and Chief Executive Officer of the American Trucking Associations, has supported the argument that a direct link exists between the MCA and the deteriorating safety performance of motor carriers, especially resulting from the inferior safety records of the new entrants. In testimony before the House Public Works and Transportation Committee, Donohue stated (Donohue, 1985):

When Congress passed the MCA, it retained the standard that an applicant for operating authority be fit, willing, and able to provide service. Some predicted that the relaxation of entry requirements and of rate making controls would result in safety problems. They were concerned that increased competition, reduced controls and rate cutting would induce motor carriers to defer maintenance and repair of vehicles. . . . A study prepared for the ATA Foundation for Traffic Safety indicates that smaller profit margins will have a deleterious effect on safety. Specifically, the study stated that "changes resulting from deregulation are threatening the safe operation of motor carrier equipment on the highways and endangering the lives of motorists and truckers alike." Of even more importance is the fact that of all Class III motor carriers (revenues below one million dollars annually) with an unsatisfactory safety rating assigned by the Department of Transportation (DOT), almost 70 percent received their initial operating authority since 1978, when entry was liberalized. . . . One can only conclude that this higher proportion of new carriers with unsatisfactory safety ratings indicates a relationship between deregulation and safety.

Another, contrasting, opinion stated by Philip Haseltine, then Deputy Assistant Secretary for Policy and International Affairs of the DOT, is that there exists no evidence linking passage of the MCA and the resultant safety problems with motor carriers. Also in testimony before the House Committee on Public Works and Transportation, he strongly argued (Haseltine, 1985):

Questions have been raised about the effect of relaxed motor carrier entry on highway safety. We have carefully monitored the trucking industry's safety record since implementation of the Motor Carrier Act of 1980, and, as we expected prior to deregulation, have found no valid statistical evidence linking the presence or absence of economic regulation with the safety performance of motor carrier operations. Truck accident rates are about one accident per million vehicle miles for unregulated (exempt) carriers as well as for common and contract carriers.

Despite the wide divergence of opinion on this critical issue, there exists a paucity of reliable research comparing the safety records of new entrants with those of the established carriers. Furthermore, no effort has been made to account for any of the factors that may explain any of the observed differences in the accident rates between new entrants and the established carriers. Existing studies, such as the one mentioned by Donohue in his testimony, are largely conjectural. They cite the higher number of overall motor carrier accidents and then conclude that new carriers operating on low profit margins are directly responsible for the adverse results.

Yet there is no statistical evidence comparing accident rates of new carriers with those of established carriers controlling for the effects of other influencing factors. While we have evaluated the impact of carrier compliance with federal regulations on carrier accident rates controlling for the effects of other influencing factors such as carrier financial and operating performance (Corsi et al., 1984), we did not directly addressed the issue of the new entrants and their safety records.

STUDY OBJECTIVES

This study represents a statistical evaluation of the safety performance, measured in terms of accidents per 100,000 vehicle-miles, of new motor carrier entrants in comparison to the performance of carriers established prior to the passage of the MCA, with control of other influencing factors on accident rates that might bias the results. It specifically addresses the following research questions: Is there a statistically significant difference between the accident rates of new entrants and those of established carriers? To what extent does a carrier's size, measured in terms of vehicle miles operated; its record of compliance with federal safety regulations governing operating practices such as driver hours of service and vehicle inspections and repair; and type of commodity hauled influence its safety record? Are there any differences between the new entrants and established carriers in the manner in which these variables influence accident rates? Even controlling for the influence of these variables, are the accident rates of the new entrants still significantly higher than those of the established carriers?

UNIVARIATE ANALYSIS

Discussion of Variables

This study focuses on the relationship between motor carrier accidents and the following three variable sets: (1) date on which the carrier established its operation in relation to date of the passage of the MCA, (2) compliance by the carrier with a wide range of federal safety regulations, and (3) carrier operating characteristics, including type of commodity hauled and total vehicle miles of operation.

Date established
Carriers were grouped according to the date on which they applied for their initial Interstate Commerce Commission (ICC) certificate in relation to date of the passage of the MCA. An important study objective is to determine whether new entrants, due to competitive pressures and low or negative profit margins, have maintenance and operating practices that adversely impact their safety records. As a corollary, the study will investigate whether, despite the new competitive environment, the pre-MCA carriers have continued established maintenance and operating practices and, as a result, have significantly safer operations than those of the new entrants.

While we recognize that prior to applying for an ICC certificate new entrants could have been conducting operations that would not have required such a certificate, for example, exempt hauling or leasing to regulated carriers, we contend that the receipt of an ICC certificate by new entrants constitutes a major and significant expansion of their previous activities. The expansion involves the handling of different commodities over new routes. It is the purpose of this investigation to assess whether the expansion of activities associated with obtaining an ICC certificate has safety implications for the new entrants.

Compliance with federal safety regulations

An important study goal is to assess the influence carrier compliance with a wide range of federal safety regulations has on its accident rate. An argument to be investigated is that new entrants have operating and maintenance practices that deviate from industry norms and are not in compliance with the federal regulations. We anticipate that this noncompliance will be associated with a higher level of accidents among the new entrants.

Operating characteristics

The expected statistically significant inverse relationship between carrier size and accident rate will be examined in the study. Smaller carriers, in general, have less sophisticated safety training programs and maintenance operations than do the larger carriers. There is an anticipated inverse relationship between accident rates and expenditures in these areas. Furthermore, smaller carriers are likely to have longer equipment replacement cycles, resulting in the operation of older, less safe equipment.

Finally, the study will investigate the expected positive relationship between certain types of carrier activities and higher accident rates. It is anticipated that the hauling of certain types of commodities involves distinct operating patterns that increase accident probability. For example, the long-haul, cross-country operations associated with produce hauling and return loads of regulated freight are most likely associated with long hours and tired drivers, which in combination are expected to lead to greater accident propensity. Therefore, firms specializing in such types of operations are more likely to have higher accident rates than are firms with distinctly different operating patterns.

Source of Data

In this study we rely on data collected by auditors from the Bureau of Motor Carrier Safety (now the Office of Motor Carriers), Federal Highway Administration, and reported on form MCS-32. We used a computer file of these MCS-32 forms compiled by the Bureau during calendar years 1985 and 1986. During the two-year period, the Bureau audited 837 new entrants and 1,082 established carriers. Only the most recent audit was used in the analysis for any carrier with multiple audits during the 1985–1986 period.

Motor carriers were identified as being new entrants or established carriers by examining their ICC certification number. If a carrier's number exceeds the one given by the ICC on July 1, 1980, the carrier applied for its certificate after passage of the MCA. The certificate numbers are handed out by the ICC sequentially on the basis of date of application.

The MCS-32 data base is an extensive, thorough evaluation of motor carrier safety. The Bureau of Motor Carrier Safety (BMCS) personnel conducted their audits on a random basis over the two-year period. In addition, auditors revisited carriers if their initial visit uncovered safety problems. Any biases introduced by the Bureau's audit selection process would presumably have the same impact on both the new entrants and the established carriers, since carrier membership in either group had nothing to do with the selection process employed.

The MCS-32 form includes a significant amount of valuable information that enabled us to perform our statistical analysis and investigate each of the issues presented above. The form provides information on a carrier's accident rate, measured in terms of number of accidents per 100,000 vehicle-miles based on the operating experience during the 12 months prior to the audit date. This serves as the dependent variable in the analysis.

In addition, the form provides data on all the variables used in the models to explain observed differences in carrier accident rates. These include a carrier's annual vehicle miles; its principal cargo hauled; and its compliance with various federal regulations such as hours of service, vehicle inspection and maintenance, and so on. Auditors ranked a carrier as having acceptable (a rank of 1), marginal (2), or unacceptable (3) compliance with each regulation.

For the purposes of this study, the particular federal regulations examined were (1) driver qualification regulations (Code of Federal Regulations Part 391-R391); (2) driving of motor vehicles regulations (R392); (3) accident reporting requirements (R394); (4) hours-of-service regulations (R395); (5) vehicle inspection and repair regulations (R396). Part 393, entitled "Parts and Accessories Needed for Safe Operations," was left out of the analysis since BMCS auditors rated less than 2% of the carriers with an unacceptable level of compliance with this regulation.

Mean and Correlation of Variables: New versus Established Carriers

Safety performance and explanatory variable differences
Table 16.1 substantiates the differences in accident rates between new entrants and established carriers. There is a significant (at the 1% level) difference in the number of accidents per 100,000 miles between new entrants and established carriers. On the basis of data from 1,082 established carriers and 837 new entrants, established carriers had an average accident rate per 100,000 vehicle-miles of 0.13, while the comparable average among the new entrants is 0.165, 27% higher.

It should be noted that, without data on entrants from other time pe-

Table 16.1 Differences in Safety Performance and Compliance: New Entrants versus Established Carriers

Measure	Established Carrier Mean (S.D.)	New Entrant Mean (S.D.)	Difference (T-Statistic)
Accident rate (accidents/100,000 miles)	0.130 (1.91)	0.165 (3.26)	0.035 (2.93)[a]
Compliance with driver qual. regs. (R391)	1.61 (0.767)	1.96 (0.836)	0.35 (9.24)[a]
Compliance with driving of motor vehicle regs. (R392)	1.13 (0.386)	1.15 (0.420)	0.02 (0.69)
Compliance with accident reporting requirements (R394)	1.56 (0.822)	1.67 (0.894)	0.11 (2.58)[a]
Compliance with hours-of-service regs. (R395)	1.87 (0.831)	2.05 (0.837)	0.18 (4.57)[a]
Compliance with vehicle inspec. and repair reqs. (R396)	1.33 (0.625)	1.46 (0.714)	0.13 (4.14)[a]
Vehicle-miles (million miles)	8.65	2.47	6.18 (4.16)[a]

SD, standard deviation.

[a] T statistic significant at 1% level.

riods, it cannot be concluded that the new entrants emerging post-MCA are worse, or better, than were entrants at other times in terms of their overall safety record. However, the MCA has, as noted, intensified any problem by facilitating an unprecedented increase in the number of new entrants. Managers of new entrants face significant challenges as they initiate expansion into different commodities over new routes. The expanded operations involve hiring additional drivers and adapting safety, fleet maintenance, risk management, and driver training programs to a larger system. During the learning period, the accident levels for the new entrants are likely to be higher. As the skills of new-entrant management increase, their accident rates will tend to resemble those of the existing carriers.

Indeed, if the new entrants are grouped on the basis of the year in which they applied for their certificate, it is clear that the accident rate problem is particularly intense among the "newest" of the new entrants. Table 16.2 provides an analysis of variance, comparing the within-group variances with the between-group variances for the established and new carriers divided into seven groups. The established carriers are divided into two groups: (1) those with certificate applications prior to January 1, 1979, and (2) those with certificate applications between January 1, 1979, and June 30, 1980. The new entrants are divided into five groups: (1) those with applications between July 1, 1980, and December 31, 1981, (2) those with applications in 1982, (3) those with applications in 1983, (4) those with applications in 1984, and (5) those with applications in 1985.

The analysis of variance results indicate that the mean differences among the groups are significant at the 2% level of confidence. Particularly interesting is the significantly higher accident rate for the new entrants established in 1985—0.246 accidents per 100,000 miles—compared with that of established carriers with pre-1979 certificates—0.128. Among the other groups of new entrants there is little variation in rate, the groups averaging about 0.158 accidents per 100,000 miles. These results certainly focus attention on the specific activities of the "newest" of the new entrants with less than two years experience.

There are also differences between new entrants and established carriers in the values of the variables selected to explain observed differences in accident rates. New entrants have significantly higher noncompliance with federal safety regulations than do the established carriers. Recalling that carriers received a ranking of 1 for regulatory compliance, of 2 for marginal compliance, and 3 for unacceptable compliance, it is readily apparent from data in Table 16.1 that the new entrants have worse compliance, as suggested by their higher average compliance scores, than do the established carriers.

Difference-of-means tests indicate a statistically significant compliance difference between the new entrants and the established carriers with respect to the following regulations: compliance with the driver qualification regulation (R391), compliance with accident reporting requirements (R394), compliance with hours-of-service regulations (R395), and compliance with vehicle inspection and repair regulations (R396). Only with respect to compliance with the driving of motor vehicle regulations (R392) are the differences between the new entrants and the established carriers not significant.

It is particularly interesting to note that the average compliance score for the new entrants on the driver qualification regulations and the hours-of-service regulation stands at 2.0—a level indicating overall only marginal compliance with the regulations.

Table 16.2 Differences in Safety Performance among Carriers on the Basis of Year of Certificate Application

Year of Application	Mean Accident Rate[a] in 1985 or 1986	No. of Carriers
Pre-1979	0.128	950
1/1/1979–6/30/1980	0.147	132
7/1/1980–12/31/1981	0.167	313
1982	0.156	152
1983	0.150	159
1984	0.158	159
1985	0.246	54
Overall mean	0.145	1,919

Analysis of variance F ratio = 2.55; significant at 2% level of confidence.

[a] Accidents per 100,000 vehicle-miles.

Table 16.1 also shows that the new entrants and the established carriers differ significantly with respect to their average vehicle miles. As noted, new entrants have mean total vehicle-miles of 2.47 million, while the comparable figure for the established carriers is 8.65 million. However, it is important to note that some new entrants have high total vehicle miles, while among the established carriers are some small operators. Among the new entrants, 4% have total vehicle-miles that exceed the overall mean for the established carriers; among the established carriers, 10% have total vehicle-miles below the mean vehicle-miles for the new entrants. The two carrier groups are not mutually exclusive with respect to size of operations.

Correlation analysis

For the purposes of further analysis, certain data transformations were employed. Due to the distribution of carrier accident rates (approximately a log-normal distribution), a logarithmic transformation was performed on the dependent variable. Since some carriers in the data base had no reportable accidents, a value of 1 was added to each carrier's accident rate to facilitate the logarithmic transformation. Although the transformation will bias the multiple regression coefficients, adjustments in the coefficients are made, in the multivariate analysis, to correct for some of this bias. Moreover, given the large sample size used, the property of consistency is virtually assured.

The categorical independent variables (carrier compliance with regulations and the type of commodity hauled) are transformed into dummy variables. Thus, carriers receive a value of 0 if they comply with a specific regulation and a value of 1 if the auditor assesses their compliance with that regulation as being either marginal or unacceptable. Several distinct commodity operations—general freight; tank operations; building materials; and produce, fruit, and seafood hauling—are singled out for special consideration in developing the independent variable set. Thus, if a carrier hauled general freight, it received a value of 1 on this variable and a value of 0 on all the other commodity dummy variables. If a carrier handled a commodity other than those listed, it received a value of 0 on all the commodity dummy variables. The continuous independent variable, carrier vehicle miles, was expressed in logarithmic form.

A correlation analysis showed that for new entrants noncompliance with each of the safety regulations is significantly positively associated with accident rates. For established carriers, noncompliance with driver qualification and accident reporting is significantly associated with higher accident rates. This suggests that greater carrier noncompliance leads to higher accident rates.

The correlation coefficients also indicate that for new entrants there is an inverse relationship between vehicle miles and accident rates, suggesting that increases in vehicle miles are associated with lower accident rates. However, the correlation coefficient between vehicle miles and accident rates for the established carriers is statistically insignificantly different from zero.

The coefficients between certain types of carrier operations and accident rates are both positive and negative, although most are not significantly different from zero. For the established carriers, none of the types of operations have a statistically significant correlation with accident rates. Among the new entrants, general freight hauling is significantly positively related to accident rate, and tank operations are significantly negatively related.

Obviously, the key to a better understanding of the relationship between the independent variables and carrier accident rates for both the new and established carriers rests on a multivariate analysis, where the effects of variables can be assessed while the effects of other variables are simultaneously controlled. However, the multivariate analysis depends on the evaluation of potential multicollinearity problems between the independent variables. The correlation analysis indicates it is unlikely that the multivariate models will have any problems with multicollinearity since the highest correlation coefficient among independent variables is 0.35.

MULTIVARIATE ANALYSIS: EXPLANATORY MODELS FOR ACCIDENT RATES

Models Described

This section presents the results of multivariate statistical models linking carrier accident rates to a set of explanatory variables on carrier size, compliance with federal regulations, and type of commodity hauled, for new entrants and for established carriers. Separate regression models for each of the two groups, are compared and subjected to statistical testing to determine whether there is a structural difference between the two equations, in other words, to determine if the accident rate of new entrants cannot be predicted on the basis of coefficients in the model for established carriers, or vice versa.

Additionally, individual coefficients in each model are compared and tested to determine whether the observed differences in coefficients are statistically significant. Finally, controlling for the set of independent variables, a statistical test is performed to determine whether the accident rates of the new entrants are significantly different from those of the established carriers.

The interpretation of the regression coefficients depends on whether the independent variables are continuous or categorical. For the continuous variable, a positive regression coefficient denotes that increases in the value of the variable are associated with a higher carrier accident rate, and vice versa. Since both the dependent and independent continuous variables have had logarithmic transformations, the regression coefficients may be interpreted directly as elasticities. For example, a regression coefficient of positive 0.6 would indicate that a 1% increase in the independent variable would be associated with a 0.6% increase in the dependent variable.

For the categorical variables, the coefficients show the difference in carrier accident rates conditioned on the presence or nonpresence of a situation. For example, a positive statistically significant coefficient for a categorical variable indicating carrier noncompliance with a particular regulation establishes a significant link between noncompliance and higher accident rates. To interpret the coefficients of the categorical variables in our regression models more readily, mathematical transformations were performed to convert the coefficients for the categorical variables to show the percent change in the carrier accident rate conditioned on the presence of that particular situation. These transformations also provide adjustments for biases in the coefficients resulting from the semilog form of the equations.[1]

Established Carriers: Model Results

Table 16.3 presents the results of the linear multiple regression linking carrier accident rates to carrier size, compliance with federal regulations, and type of commodity hauled for the established carriers.[2] These results

Table 16.3 Explanatory Models for Accident Rates: Established Carriers and New Entrants

Variable	Established Carriers (974 Firms)			New Entrants (742 Firms)		
	Coeff.	% Change	T Value	Coeff.	% Change	T Value
Compliance with driver quals. (R391)	−.001	−0.11	0.11	−.027	−2.65	1.90
Compliance with driving of motor vehicles (R392)	−.014	−1.41	1.07	.029	2.53	1.49
Compliance with accident reporting (R394)	.094	10.36	10.17[a]	.155	18.18	11.39[a]
Compliance with hours-of-service (R395)	.033	3.44	3.61[a]	.005	0.47	0.33
Compliance with vehicle inspect. & repairs (R396)	−.017	−1.69	1.70	.011	1.10	1.22
Type of commodity						
Tank oper.	.005	0.54	0.39	−0.23	−2.26	1.12
Building materials	.009	0.90	0.51	.032	3.26	1.22
Product, fruit, & seafood	−.015	−1.51	0.89	.043	4.40	1.82
General freight	.007	0.69	0.74	.063	6.66	4.36[a]
Vehicle miles	−.005		1.75	−.027		5.66[a]
Constant	.135			.423		
Adjusted R square		.125			.186	
F statistic		14.894[a]			17.891[a]	

[a]Significant at 1% level of confidence.

show a statistically significant linkage between carrier accident rate and compliance with selected safety regulations for the established carriers. They do not, however, provide support for the linkage between carrier accident rate and either the carrier size or the type of commodity hauled by the established carriers. That is, for established carriers, the type of commodity hauled or the number of vehicle miles operated has no significant influence on accident rate.

Among the established carriers, the entire set of independent variables explains 12.5% of the variation in carrier accident rate around its mean. Indeed, the F statistic, assessing the overall explanatory power of the entire set of independent variables taken together, equalled 14.89, a statistically significant value at the 1% level of confidence.

A review of individual coefficients suggests that the overall significant (at the 1% level of confidence) results are based on the explanatory power of the following two variables: compliance with carrier accident reporting requirements and compliance with hours-of-service regulations.

The positive coefficient for carrier compliance with accident reporting requirements indicates that carriers with a propensity for noncompliance in reporting requirements manifest significantly higher accident rates even on the basis of incomplete counts. In fact, the coefficient value denotes an accident rate about 10% higher for carriers who do not comply versus the rate for those who do comply.

Another significant regulatory compliance variable linked to the accident rate of the established carriers is compliance with hours-of-service regulations. This coefficient indicates that carriers with marginal or unacceptable compliance with hours-of-service regulations have accident rates that are about 3.4% higher than are the accident rates of the compliers. The importance of this finding should not be understated. The significant association between accident rates and hours-of-service violations points out the importance of improving the enforcement of this regulation as a direct, positive step toward reducing truck accidents.

New Entrants: Model Results

Table 16.3 also shows the results of the multiple regression linking accident rates to carrier size, compliance with federal regulations, and type of commodity hauled for the new entrants. The results indicate a significant association between carrier accident rate and the three independent variable sets.

Among the new entrants, the entire set of independent variables explains 18.6% of the variation in carrier accident rate around their mean. Indeed, the F statistic, assessing the overall explanatory power of the entire set of independent variables as a whole, is 17.89, a statistically significant value at the 1% level of confidence.

A review of individual coefficients denotes that the overall significant results are based on the explanatory power of the following three variables: compliance with carrier accident reporting requirements, hauling of general freight, and carrier vehicle miles.

The positive coefficient for carrier compliance with accident reporting requirements shows that carriers who fail to comply with reporting requirements have significantly higher accident rates than do carriers who comply with this regulation. The coefficient for this variable shows that among new entrants, carriers not complying with accident reporting requirements have accident rates about 18% higher than do the carriers who comply.

A second variable statistically significant at the 1% level denotes that new entrants hauling general freight have accident rates about 7% higher than do new entrants who haul other commodities. The significant coefficient for the general freight variable and the insignificant ones for the other commodity types indicate that the hauling of certain specialized types of commodities has no positive impact on accident rates when other influencing factors are controlled.

The third significant variable, carrier size, has a inverse impact on carrier accident rate. Among the new entrants, as carrier size increases so too does the likelihood of a lower carrier accident rate. The coefficient denotes that a 10% increase in carrier vehicle miles is associated with about a 0.3% decrease in carrier accident rate.[3]

Comparative Model Results

The fact that the regression equations for the established carriers and the new entrants have substantial differences in the significant explanatory variables and the manner in which the variables impact accident rates leads us to infer that there is a structural difference between the model for the new entrants and that for the established carriers.

The accident rate of the new entrants cannot be predicted on the basis of the coefficients in the model for established carriers, or vice versa. To confirm this statement statistically, we performed a Chow test of the equality between sets of coefficients in two linear regressions.[4] This test produced a statistically significant (at the 1% level) F value of 4.89. This establishes a structural difference between the equation for the new entrants and the one for the established carriers.

Having established that there is a structural difference between the two equations, we performed statistical tests to ascertain which of the individual coefficients had a statistically significantly different impact on accident rates in the two regression equations. The results, reported in Table 16.4, indicate that the following set of four variables has a different impact in the two equations: compliance with regulations dealing with accident reporting; hauling of general freight; hauling of produce, fruit, and seafood; and vehicle miles operated. In this analysis, positive coefficients indicate that the variable measuring the difference in impact has a more positive impact on the accident rate of new entrants than it does on the accident rate for established carriers. Negative coefficients indicate the reverse.

Table 16.4   Difference in Coefficients: New Entrants versus Established Carriers

Variable	Coefficient Difference
Compliance with driver qualification regulations (R391)	−.026
Compliance with driving of motor vehicles (R392)	.043
Compliance with accident reporting requirements (R394)	.061[a]
Compliance with hours-of-service regulations (R395)	−.029
Compliance with vehicle inspection and repairs regulations (R396)	.027
Commodities hauled	
Tank operations	−.028
Building materials	.023
Produce, fruit, & seafood	.058[b]
General freight	.056[a]
Vehicle miles	−.022[a]
New entrant	.287[a]

[a]Significant at the 1% level of confidence.
[b]Significant at the 5% level of confidence.

As shown in Table 16.4, new entrants have a significantly stronger linkage between their regulatory noncompliance with accident reporting requirements and accident rates than do the established carriers. The positive-difference coefficients for general freight haulers denotes a stronger positive association between accident rates and the hauling of general freight or produce, fruit, and seafood by the new entrants in comparison with the experience of the established carriers. Finally, the negative coefficient for vehicle miles shows that the carrier size variable has a significantly stronger inverse association with accident rate for the new entrants than it does for the established carriers.

One final comparative test involves the question of whether (controlling for carrier size, carrier compliance, and type of commodity hauled) there is a significant difference between the accident rates for the new entrants versus that for the established carriers. The positive coefficient for this variable, as shown in Table 16.4, is significant at the 1% level, thus showing that the new entrants have significantly higher accident rates than do the established carriers. Indeed, new entrants have an accident rate that is 32.9% higher (based on conversion of the coefficient to percent change, as indicated above) than it is for the established carriers, when the effects of the other explanatory variables are controlled.

CONCLUSIONS

The analysis presented here advances the study of the impact of the MCA on industry safety through an evaluation of differences in the safety of new entrants versus established carriers.

The study results show that, on the basis of approximately 2,000 safety audits during 1985 and 1986, new entrants have significantly higher accident rates than do established carriers. On average, the accident rate for new entrants is between 27% and 33% higher than is the accident rate for the established carriers.

Among the new entrants, the "newest" carriers have the worst accident rates. In fact, the accident rate for carriers established in 1985 is 0.246 per 100,000 vehicle-miles, or 70% higher than is the accident rate for all carriers combined.

The new entrants also have significantly worse compliance records than the established carriers. The new entrants complied significantly less than did the established carriers with regulations covering driver qualifications, accident reporting requirements, hours-of-service regulations, and vehicle inspection and repair regulations.

The multiple regression analysis employed served two basic functions: (1) It investigated the linkage between a carrier's accident rate and its compliance record, its size, and the type of commodity it hauled; and (2) it assessed differences in these relationships for the established carriers and the new entrants.

An analysis of the multiple regression equations for the new entrants and the established carriers showed that carrier accident rate was significantly linked to carrier size, carrier regulatory compliance, and type of commodity hauled for the new entrants. It also demonstrated significant structural differences in the equations for the new entrants versus the one for the established carriers. The implications of these differences have very important dimensions.

The regression model for the new entrants shows a higher explanatory power than does the model for the established carriers. All three types of variables—carrier size, compliance with regulations, and type of commodity—contribute to the significantly higher explanatory power of the model for the new entrants, whereas the model for the established carriers substantiated only a linkage between carrier accident rates and regulatory compliance.

Specifically, the new entrants' regression model indicates that vehicle-miles is a significant variable linked to accident rates, as opposed to the influence of that variable in the established carrier model. The negative sign on the difference coefficient suggests that new entrants with low vehicle-miles warrant, as a result of their high accident rates, a particularly strong focus of attention by safety monitors. It is important to emphasize this finding. Among the established carriers, size is not significantly linked to accident rate. Among the established carriers, then, and controlling for the influence of other variables, size of carrier does not have a statistically significant link to accident rate. The opposite is true for the new entrants.

The difference in the coefficients between the new-entrant and established carrier models also directs monitoring effort to two particular industry segments among the new entrants. The difference coefficients indicate a significantly more positive relationship between hauling of general freight or of produce, fruit, and seafood and accident rates for the new entrants than exists for the established carriers. Thus, the results suggest that the new entrants should command careful monitoring attention from safety auditors. In particular, new entrants who are small with low total vehicle miles, as well as those in the general freight or produce, fruit, and seafood hauling sectors, should command special safety monitoring.

Regarding the compliance variables, the models indicate that carriers with the highest accident rates have the greatest propensity for noncompliance with federal accident-reporting requirements. This serves to indicate the importance of improving accident reporting, since the ability to thoroughly investigate the accident rate issue is dependent on the collection and analysis of accurate data. The very carriers who are the source of this problem constitute a direct obstacle to the gathering of accident data. The federal government needs to take direct corrective measures to improve the accident data collection process. If necessary, there should be stiff fines for noncompliance, to make the cost of not reporting high enough to ensure complete compliance.

The model for the established carriers showed a significant inverse association between hours of service compliance and accident rates. Although the new-entrants model did not provide such a finding, when the effects of the other variables were controlled for the new entrants had significantly worse compliance with the hours-of-service regulations than did the established carriers. The problem of hours-of-service violations cannot be understated. These results substantiate a direct linkage between compliance with this variable and accident rates.They should serve as a basis for supporting federal measures to ensure greater compliance with this safety regulation.

POLICY RECOMMENDATIONS

Data presented in Tables 16.1 and 16.2 present convincing evidence that there is a safety problem associated with the vast influx of new entrants into the trucking industry since passage of the MCA. While taken together the new entrants have significantly higher accident rates than do the established carriers, the accident rate problem is particularly severe among the "newest" of the new entrants, whose accident rate is about 70% higher than the overall average accident rate for all carriers. New entrants established between July 1, 1980, and the end of 1984 have accident rate levels that are only moderately above the level for the established carriers. Thus, on the basis of the evidence presented in Tables 16.1 and 16.2, it takes carriers several years for their managements to bring into practice the type of safety, driver training, fleet maintenance, and risk management programs needed to bring accident rates down to the average levels for established carriers.

These facts indicate that government safety resources should be directed toward enhancing the safety practices of the "newest" of the new entrants. Specifically, we recommend that the ICC certification process include as a certificate prerequisite the implementation of a comprehensive risk management program by the prospective entrants.

The components of a comprehensive risk management program must include stiff driver qualifications and extensive driver training, vehicle maintenance and inspection programs, and rigid compliance with existing federal safety regulations. The federal government should also substantially increase the level of fines for noncompliance with these regulations. It should pay particular attention to the hours-of-service regulation. Our results show a linkage between this regulation and higher accident rates. We also find a higher level of noncompliance with most safety regulations among the new entrants in comparison with the compliance record of the established carriers.

The documented safety problem among new entrants suggests the need for further investigation of the operating practices and expenditure policies of new entrants to determine the factors most closely associated with higher accident rates. Such a careful examination would serve as a basis for corrective recommendations. Obviously, such an investigation is beyond our scope here. It would be a serious mistake to interpret our study's results as a plea for reregulation of the motor carrier industry. To the contrary, they should serve as a basis for prioritizing safety monitoring efforts among certain industry groups and along specific lines: enhanced accident reporting accuracy and a requirement that comprehensive risk management programs be instituted prior to certification of a new trucking company.

NOTES

1. The computation that converts the coefficients to percent change is as follows: $D = \exp [B - 0.5V\,(B)] - 1$, where D = percent effect; B is the regression coefficient of the dummy variable; and V is the variance of B. For the proof of this transformation see Kennedy (1981) and Halvorsen and Palmquist (1980).

2. The differences in the sample size between Table 16.3 and Table 16.2 are accounted for by incomplete data in the audit report. Only carriers with complete data on all independent variables were included in the multiple regression analysis.

3. Additional regressions were run to test whether the coefficient on vehicle miles was autocorrelated with the year the new entrant was established. No such autocorrelation was found to exist.

4. The Chow (1960) test measures the equality between sets of coefficients in two linear regressions. In our case we seek to establish that there is a significant difference between the coefficients in the regression model for the new entrants and those in the model for the established carriers. The test statistic is defined as follows:

$$F = \frac{[\text{sse}_n - (\text{sse}_1 + \text{sse}_2)]/k}{(\text{sse}_1 + \text{sse}_2)/(n - 2k)}$$

where k = number of parameters in the model being estimated
   sse_n = sum of squared residuals of entire sample with the set of independent variables
   sse_1 = sum of squared residuals for regression model of new entrants
   sse_2 = sum of squared residuals for regression model of established carriers

For our case the following numbers are applicable:

$$F = \frac{[36.65 - (20.08 + 15.44)]/11}{(20.08 + 15.44)/(1716 - 22)} = 4.89$$

This F value exceeds the critical value of F at the 1% significance level.

17

The Safety Effects of Mode Shifting Following Deregulation

KENNETH D. BOYER

Freight transportation is characterized by a degree of intermodal competition. Where the modes have unequal accident rates, deregulation of one or more modes of transportation can have overall safety effects without changing the safety characteristics of individual modes. This chapter evaluates the effects on freight transportation safety of intermodal traffic shifts that followed the simultaneous deregulation of motor carriers and railroads in 1980.

The provisions of the Motor Carrier Act of 1980 are well known. This law provided almost unlimited freedom of entry into the trucking industry along with greater freedom to act outside the authority of the rate bureaus (Motor Carrier Ratemaking Study Commission, 1983). Most for-hire motor freight now moves under individually tailored rates that are discounted from the official posted price (ENO Foundation, 1985; Pustay, 1983). The Interstate Commerce Commission (ICC) has also diverted resources from enforcement of remaining economic regulations, making the force of these regulations questionable. Although there is some residual control on less-than-truckload carriers, and although much of the structure and form of the old economic regulation are still in place, for the purpose of analyzing intermodal relationships the motor carrier industry can now be considered effectively free of price and entry controls.

The provisions of the Staggers Act of 1980 are more complex than those of the Motor Carrier Act and accomplished somewhat less sweeping deregulation of the railroad industry (Keeler, 1983). As with the trucking industry, the deregulatory statutes did not dismantle the structure of regulation, but changed the interpretation of conditions that trigger regulatory intervention. In both industries, the legislation primarily ratified administrative reinterpretations of rules that had been made during the three or four years prior to 1980.

Railroads now have almost unlimited rights to lower rates and to raise them where the carriers are not "market dominant." The ICC has been reluctant to exercise its authority to impose rates, preferring to encourage the negotiation of traffic contracts between shippers and carriers as an alternative to rate regulation (Tye, 1983). The ICC retains some residual effective control on rail rates, but in practice, rate-setting powers are exercised only over the shipment of certain bulk commodities that are not subject to truck competition. Rail shipments for which there is truck competition are completely deregulated. In fact, the existence of truck competition in a particular market is taken as evidence that railroads are not dominant in that market and thus cannot be regulated.

For purposes of intermodal analysis, then, both railroads and motor carriers are now deregulated and were deregulated at approximately the same time. It is unclear, however, whether that time should be considered 1980, when the deregulation statutes were passed, or sometime in the late 1970s, when the ICC changed the character of the administration of the laws.

The parallelism of the timing and form of deregulation of the two modes of transportation is not reflected in the financial consequences for the industries. Railroads have thrived financially under deregulation. Profitability has jumped from a 1% return on equity in 1975 to 10% in 1984 (Association of American Railroads, 1985) (the former figure is computed under betterment accounting, which tends to depress reported profitability [Boburg, 1985]). Railroads are reported to be amassing cash in excess of investment opportunities in the industry (Standard and Poors, 1986). By contrast, the profitability of the motor carrier industry has dropped, and several well-known carriers have gone bankrupt (Rose, 1985).

The unequal financial response of the two surface transport modes to deregulation is reflected in somewhat different paths of accident rates. As discussed later in this chapter, motor carrier accident rates have shown no trend over the last decade, whereas railroads have become much safer following deregulation.

The mode-shifting consequences of deregulation on safety are also evaluated in this chapter. A sample of commodities carried by both modes is analyzed to get an estimate of the amount of freight that was shifted by deregulation. With use of the estimates of the overall safety of the two modes, an estimate of the safety consequences of intermodal shifting is also given.

SAFETY STATISTICS OF HIGHWAY FREIGHT

The basic types of accidents in which motor carriers are involved are fundamentally different from those of railroads. Table 17.1 lists the types of accidents incurred by motor carriers during 1984, the most recent year for which statistics are available. The great majority of truck accidents involve collisions with other motor vehicles. More than half of the deaths and

Table 17.1 Accidents Involving Private and For-Hire Motor Carriers, 1984[a]

Description	Deaths	Injuries	Accidents
Collision accidents: Collided with			
Commercial truck	208	2,708	4,104
Fixed object	106	1,037	3,069
Automobile	1,314	12,204	12,299
Pedestrian	113	106	208
Bus	19	336	107
Train	12	87	135
Bicyclist	16	44	55
Animal	5	85	222
Motorcycle	64	134	167
Other	159	1,128	1,203
Noncollision accidents			
Ran off road	170	1,664	2,289
Jack-knifed	23	981	2,196
Overturned	100	1,910	2,572
Separation of units	3	31	162
Fire	0	25	212
Cargo loss or spillage	2	47	139
Cargo shift	0	48	104
Other	4	24	63
Total	2,318	22,594	29,302

[a]The statistics listed are for over-the-road trucks only and exclude data for local pickup and delivery. Figures for deaths and injuries include truck drivers and those outside the truck. All figures are for those accidents required to be reported under the Department of Transportation Act. "Reportable accidents . . . are those involving a motor vehicle operated by a motor carrier subject to Federal Motor Carrier Safety Regulations resulting in (1) the death of a human being; (2) bodily injury to a person . . . , and/or (3) total damage to all property aggregating $2,000 or more . . ." (p. i).

SOURCE: Federal Highway Administration (various years).

injuries from truck accidents involve collisions with an automobile. Only 18% of truck fatalities involved noncollisions or collisions with fixed objects, in which the driver is the only one at risk.

The number of accidents involving trucks has been remarkably stable since 1973, as can be seen in Table 17.2. As shown by the totals for all carriers, deaths from truck accidents have been relatively constant at approximately 2,500 to 3,000 per year between 1973 and 1984. The number of accidents per year has been somewhat more volatile, varying from a low of 24,274 in 1975 to a high of 36,854 in 1984. A similar picture emerges when the truck accident figures are deflated by the amount of truck traffic, as shown in Table 17.3. Except for the anomalous year 1973, the motor carrier accident statistics are remarkably constant around an average of 0.057 deaths, 0.667 injuries, and 0.660 accidents per million truck miles. Alternatively deflated, the averages from 1974 through 1984 are 4.88 deaths, 52.44 injuries, and 56.27 accidents per billion ton-miles.[1]

Table 17.2   Motor Carrier Accidents, 1973–1984

Carrier Type	Year	Deaths	Injuries	Accidents
Private	1973	709	5,837	5,233
	1974	601	4,794	4,434
	1975	620	5,320	4,898
	1976	634	5,123	5,017
	1977	730	5,891	5,781
	1978	798	6,375	6,493
	1979	792	6,227	6,872
	1980	616	5,409	6,323
	1981	711	5,725	6,330
	1982	618	5,200	6,341
	1983	541	5,117	5,781
	1984	594	4,976	6,152
ICC authorized[a]	1973	1,938	26,254	22,825
	1974	1,709	21,396	20,233
	1975	1,496	20,416	18,791
	1976	1,773	21,125	20,073
	1977	2,162	25,365	23,726
	1978	2,103	25,853	26,955
	1979	2,212	25,529	28,206
	1980	1,856	21,501	24,724
	1981	2,028	22,495	25,588
	1982	1,712	19,869	24,493
	1983	1,851	20,754	24,849
	1984	2,009	23,273	29,549
Exempt	1973	99	652	618
	1974	88	442	408
	1975	101	476	430
	1976	99	443	467
	1977	76	398	385
	1978	84	464	497
	1979	60	295	404
	1980	47	183	301
	1981	66	272	353
	1982	127	616	821
	1983	111	692	851
	1984	79	728	932
Total, All Carriers	1973	3,058	35,245	30,911
	1974	2,429	26,911	25,358
	1975	2,232	26,374	24,274
	1976	2,520	26,794	25,666
	1977	2,983	31,698	29,936
	1978	2,988	32,757	33,998
	1979	3,072	32,126	35,541
	1980	2,528	27,149	31,389
	1981	2,810	28,533	32,306

Table 17.2 Motor Carrier Accidents, 1973–1984 (continued).

Carrier Type	Year	Deaths	Injuries	Accidents
Total, All Carriers	1982	2,479	25,779	31,759
	1983	2,528	26,692	31,628
	1984	2,721	29,149	36,854

ᵃICC authorized includes all carriers with an ICC certificate. It is not the same as "regulated" carriers, a term usually restricted to companies holding certain kinds of ICC certificates. The classic regulated sector consists of those firms with regular-route common carrier status.

SOURCE: Federal Highway Administration (various years).

Table 17.3 Motor Carrier Accident Statistics Deflated by Traffic Levels

	Year	Deaths	Injuries	Accidents
Per million truck miles	1973	0.090	1.039	0.911
	1974	0.053	0.586	0.552
	1975	0.049	0.578	0.532
	1976	0.053	0.560	0.536
	1977	0.058	0.616	0.582
	1978	0.055	0.600	0.623
	1979	0.058	0.604	0.668
	1980	0.056	0.601	0.695
	1981	0.069	0.705	0.798
	1982	0.062	0.643	0.792
	1983	0.056	0.589	0.698
	1984	0.058	0.620	0.784
Per billion ton-miles	1973	6.06	69.79	61.21
	1974	4.91	54.37	51.23
	1975	4.92	58.09	53.47
	1976	4.94	52.54	50.33
	1977	5.37	57.11	53.94
	1978	4.99	54.69	56.76
	1979	5.05	52.84	58.46
	1980	4.55	48.92	56.56
	1981	5.33	54.14	61.30
	1982	4.77	49.58	61.08
	1983	4.40	46.42	55.01
	1984	4.50	48.18	60.92

SOURCES: Total deaths, injuries, and accidents from Federal Highway Administration, *Accidents of Motor Carriers of Property* (various years). Vehicle miles are from Federal Highway Administration, *Highway Statistics* (various years). Truck vehicle miles are assumed to be the same as rural vehicle miles of combination trucks, as reported in Table VM-1 of *Highway Statistics*. The same assumption is made in Transportation Policy Associates, *Transportation in America* (various years). Motor carrier ton-mile statistics are drawn from page 6 of this publication.

The latter figures will be used in later in this chapter as the marginal safety consequence of shifting a ton-mile of freight from or toward highways.

SAFETY STATISTICS OF RAILROADS

Unlike trucking accidents, many railroad mishaps more closely resemble industrial accidents in which only the worker is at risk from the activity. This leads to a confusing terminology in which accident reports classify railroad accidents into "train accidents," "train incidents," and "nontrain incidents" (Federal Railroad Administration, various years). Nontrain incidents are the most numerous type of railroad accident. However, they do not involve the operation of equipment and so are not directly comparable to motor carrier accidents. Most nontrain incidents do not involve fatalities. The largest source of such nontrain incidents is servicing of equipment or maintenance of way and structure.

While all casualties and damage in the railroad industry are the result of transportation activities, nontrain accidents and incidents will be ignored in the remainder of this chapter because our ultimate interest here is the marginal safety consequences of a transfer of a ton-mile of freight between rail and road. Most nontrain mishaps are related to the level of fixed facilities and thus should not vary directly with the level of traffic. In addition, casualties associated with the maintenance of trucks and public highways are not reported in the motor carrier accident data. To include them in the railroad accident figures would overestimate the relative danger of railroads.

More similar to motor carrier accidents are "train accidents" and "train incidents." The difference between the two categories, both of which cover the operation of railroad on-track equipment, is whether damage to railroad equipment exceeds a reporting threshold, which varies according to the price level and stood at $4,900 in 1985. As seen in Tables 17.4 and 17.5, while derailment is the most frequent type of operations mishap, most fatalities occur from trespassers being struck by trains or from motor vehicles being struck at grade crossings. These sources of railroad accidents are generally not under the direct control of the railroad operator. Thus, even if deregulation provides incentives to railroad companies to put more or less effort into railroad safety, it is not clear that it should have a substantial effect on railroad safety records. In fact, in view of the importance of grade-crossing accidents, by changing the number of motor vehicles operating on highways, trucking deregulation may have had a more important impact on railroad safety statistics than railroad deregulation had on the same series.

Train accident data underestimate the marginal safety consequence of shifting a ton-mile of freight between railroads and motor carriers, because accidents from a number of the ancillary activities that do vary with the

Table 17.4   Description of Railroad Accidents, 1985

Type	Fatal	Injury	Total[a]
Collision	2	45	366
Derailment	2	61	2,495
Rail-highway crossing	33	71	155
Miscellaneous	0	24	414

[a]Previous two columns plus accidents involving property damage only.

SOURCE: Federal Railroad Administration (various years).

traffic level are excluded from the data series. On the other hand, including all train incidents will overestimate the danger of a marginal ton-mile of railroad freight transport by including a series of accident types that are excluded from the corresponding motor carrier statistics. In the discussion here, statistics for both train accidents and combined train accidents and incidents are reported as high and low estimates of the safety of railroad transportation.

Figure 17.1 shows the relative movements of death rates per billion ton-miles from motor carrier accidents and from train accidents and incidents

Table 17.5   Description of Railroad Incidents and Nontrain Incidents, 1985

Type	Train Incidents			Nontrain Incidents, Total
	Fatal	Nonfatal	Total	
Coupling cars	0	316	316	691
Coupling hoses	1	36	37	300
Operating locomotive	0	720	720	554
Operating rail motor cars	1	32	33	66
Operating handbrakes	1	123	124	457
Operating switch	0	47	47	1,325
Hit while on car	1	105	106	44
Getting on/off car	10	1,047	1,057	1,172
Collision/derailment	10	99	109	0
Rail-highway crossing	428	1,765	2,193	54
Struck by locomotive (not at hwy)	355	392	747	26
Service equipment	1	109	110	6,055
Maintaining way & structure	3	120	123	9,019
Freight/baggage/mail	0	19	19	110
Window/doors of equipment	0	114	114	412
Passenger car doors	0	7	7	38
Stumbling, slipping	5	229	234	2,527
Flying objects, etc.	0	423	423	616
Operating on-track work equipment	2	107	109	101
Assault	1	85	86	199
Other	7	697	704	2,380

SOURCE: Federal Railroad Administration (various years).

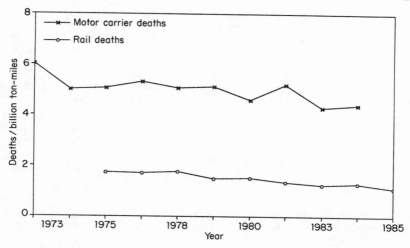

Figure 17.1   Rail and Motor Carrier Deaths. SOURCE: Motor carrier line drawn from data in Table 17.3; Rail statistics from Table 17.8.

over the previous two decades. Unlike trucking accidents, which generally have been constant over the years for which we have data, railroad safety appears to have improved very rapidly. More detail on railroad accident rates is given in Table 17.6. This table shows that deaths, injuries, and accidents all fell by one half between 1975 and 1985. The worst years for railroad safety were apparently the late 1970s at the end of the period of regulation; every year since deregulation shows safer operations than the year before.

It is natural to ask what the reasons for this improved safety are. A clue is provided in Table 17.7, which deflates train accident rates by various measures of railroad inputs and outputs, and in Table 17.8, in which a similar deflation is made on combined train accidents and incidents. All series show a declining accident rate per man hour, train mile, and ton-mile. The series that show the least movement are those deflated by train mile or man-hour. Deaths from train accidents and incidents per train mile and per man-hour have shown virtually no movement over the decade covered by the data. A reasonable hypothesis is that the declining railroad accident rates have been caused by the dramatic drop in railroad employment and train miles that have accompanied railroad deregulation. By contrast, deregulation has been associated with much heavier freight trains, which has kept ton-miles approximately the same before and after deregulation (Association of American Railroads, 1985). Thus, deregulation appears to have made railroads safer primarily by encouraging a reduction in employment and by causing fewer but heavier freight trains to be run. This is not the whole explanation, however. Table 17.7 shows that train accidents, which exclude most grade-crossing accidents, industrial-type ac-

Table 17.6 Trends in Train Accidents and Incidents

Year	Train Accidents Only			Train Accidents and Incidents		
	Deaths	Injuries	Accidents	Deaths	Injuries	Accidents
1975	82	1,220	8,041	1,323	12,999	19,818
1976	152	1,279	10,248	1,411	13,927	22,616
1977	108	985	10,362	1,448	15,861	24,973
1978	139	1,911	11,277	1,566	16,925	26,195
1979	100	1,275	9,740	1,345	16,689	24,899
1980	97	860	8,451	1,339	13,662	21,137
1981	63	562	5,781	1,220	11,067	16,266
1982	49	472	4,589	1,053	8,498	12,736
1983	56	502	3,906	1,022	8,144	11,601
1984	63	893	3,900	1,181	9,138	12,246
1985	52	647	3,430	982	8,031	10,848

SOURCE: Federal Railroad Administration (various years).

cidents, and casualties of trespassers, have noticeably declined over the decade, even when deflated by man-hours or train miles.

THE SAFETY EFFECTS OF DEREGULATION-INDUCED MODE SHIFTING

By any reasonable measure, railroad freight traffic is much safer than high-way freight traffic. While there were approximately 2 deaths from train accidents and incidents per million train miles in 1984 and only about 0.05 deaths from truck accidents per million truck miles, the average freight train is 70 cars long and carries 2,543 tons. This compares with an average of between 12.3 and 13.8 tons for ICC authorized motor carriers (Association of American Railroads, 1985, Table III-E-2; American Trucking Association, 1985, p. 52). In terms of ton-miles, motor carrier transportation is between 4 and 75 times more likely to produce a fatality compared with the same ton carried on a railroad. These figures are derived by dividing the 1984 motor carrier fatality rate per ton mile of 4.5 from Table 17.3 by the corresponding figures of 0.06 deaths from train accidents (Table 17.7) and 1.12 deaths from train accidents and incidents (Table 17.8) in the most recent reported year. As explained above, train accident statistics probably underestimate the marginal safety consequences of railroad freight traffic, while statistics for combined traffic incidents and accidents overestimate the marginal hazards of rail traffic. There are similar large differences in the safety of the two modes measured in injuries per ton-mile or total accidents per ton-mile.

Has mode shifting due to deregulation improve or worsen safety of the surface transportation activity? To the extent that deregulation encouraged the growth of rail traffic at the expense of motor carrier traffic, overall transportation safety should have been enhanced by removing traffic from the relatively unsafe mode. Many economists believed that deregulation would have this effect. A large literature developed around the time of transport deregulation that attempted to measure the cost that economic restrictions on railroads and motor carriers imposed on the economy. This literature assumed that (1) ICC regulation kept railroad rate levels uneconomically high, and (2) the modal split for freight transport was sufficiently price sensitive that lowering railroad rate levels would produce a substantial increase in the rail share of surface freight transport. The cost of regulation was computed as the decreased total costs of freight transportation that would result if railroads were permitted to increase their share of the freight market by reducing rate levels.

There are reasons to doubt the validity of both of the key assumptions

Table 17.7 Trends in Deflated Train Accidents

Accidents	Year	Deaths	Injuries	Accidents
Per million	1976	0.15	1.22	9.80
man-hours	1977	0.10	0.94	9.84
	1978	0.13	1.81	10.68
	1979	0.10	1.26	9.63
	1980	0.10	0.85	8.36
	1981	0.07	0.59	6.09
	1982	0.06	0.59	5.73
	1983	0.08	0.69	5.35
	1984	0.08	1.20	5.25
	1985	0.07	0.93	4.92
Per million	1976	0.20	1.65	13.23
train miles	1977	0.14	1.31	13.82
	1978	0.18	2.54	15.00
	1979	0.14	1.78	13.57
	1980	0.14	1.20	11.78
	1981	0.09	0.83	8.55
	1982	0.09	0.82	8.00
	1983	0.10	0.90	7.00
	1984	0.11	1.51	6.58
	1985	0.09	1.13	6.01
Per billion	1976	0.19	1.61	12.91
ton-miles	1977	0.13	1.19	12.54
	1978	0.16	2.23	13.14
	1979	0.11	1.40	10.66
	1980	0.11	0.94	9.20
	1981	0.07	0.62	6.35
	1982	0.06	0.59	5.75
	1983	0.07	0.61	4.73
	1984	0.07	0.97	4.23
	1985	0.06	0.74	3.92

SOURCE: Federal Railroad Administration (various years).

Table 17.8   Trends in Deflated Train Accidents and Incidents

Accidents and Incidents	Year	Deaths	Injuries	Accidents
Per million man-hours	1976	1.35	13.31	21.62
	1977	1.38	15.06	23.72
	1978	1.48	16.03	24.81
	1979	1.33	16.51	24.63
	1980	1.32	13.51	20.91
	1981	1.29	11.66	17.14
	1982	1.31	10.61	15.90
	1983	1.40	11.16	15.89
	1984	1.59	12.30	16.48
	1985	1.41	11.52	15.56
Per million train miles	1976	1.82	17.97	29.19
	1977	1.93	21.14	33.29
	1978	2.08	22.50	34.83
	1979	1.87	23.25	34.69
	1980	1.86	19.03	29.45
	1981	1.80	16.36	24.05
	1982	1.83	14.82	22.21
	1983	1.83	14.59	20.78
	1984	1.99	15.42	20.66
	1985	1.72	14.06	19.00
Per billion ton-miles	1976	1.78	17.54	28.48
	1977	1.75	19.20	30.22
	1978	1.82	19.72	30.53
	1979	1.47	18.27	27.25
	1980	1.46	14.87	23.01
	1981	1.34	12.16	17.87
	1982	1.32	10.65	15.96
	1983	1.24	9.86	14.04
	1984	1.28	9.92	13.29
	1985	1.12	9.17	12.38

SOURCE: Federal Railroad Administration (various years).

of such regulatory cost calculations. First, why should rail rates have been too high? Before railroad deregulation in 1980, the industry applied for and was granted numerous general rate increases. No mechanism in ICC regulations would have required the industry to accept higher general rate increases than the industry desired (Locklin, 1972). In addition, the legislative history of the railroad deregulation act shows that lawmakers anticipated an industry rate increase, not a decrease. The main policy controversy was how far and how fast railroad rates should be allowed to rise.

The second main assumption, that the market for rail and motor carrier freight services overlapped substantially, has been widely criticized. While it seems logical to blame the ICC for the decline of U. S. railroads, rail freight share losses in the 1950s and 1960s are easily explained by such factors as the increase in the extent of the interstate highway system and the decline in the economic importance of heavy industry (Task Force on Railroad Productivity, 1973; Wilson, 1981).

The inaccuracy of the first assumption is shown in Table 17.9. Real railroad rates have not declined substantially following deregulation. In fact, if the effect of deregulation is inferred as an unexplained change in real rates away from those that would be expected from time trends and the increased length and weight of freight trains that accompanied deregulation, rail rates can be shown to be between 0 and a statistically insignificant 2% above what they would otherwise have been had deregulation not occurred in 1980 (Boyer, 1987). This is generally in agreement with the expectations expressed in the legislative history of the deregulation bill. It contradicts the assumptions of those who expected deregulation to bring about a major change in intermodal relationships.

Movements in real truck rates shown in Table 17.9 should be interpreted with care. Until the period in which administrative deregulation began, the figures show a decline in real rates charged by ICC-certified motor carriers; there has been a slight rise in rates since then. The indi-

Table 17.9 Average Real Rail and For-Hire Truck Rates Before and After Deregulation

Year	Average Rate (1967 Cts per Ton-Mile)	
	Rail	Motor Carrier
Prederegulation years		
1970	1.30	7.73
1971	1.39	8.16
1972	1.38	8.12
1973	1.27	7.66
1974	1.25	7.03
1975	1.25	7.12
1976	1.28	7.02
1977	1.26	6.98
1978	1.20	6.84
1979	1.19	6.97
1980	1.15	7.29
Postderegulation years		
1981	1.18	7.41
1982	1.14	7.39
1983	1.09	7.72
1984	1.06	7.62

SOURCE: Transportation Policy Associates (various years).

Table 17.10   Rail and For-Hire Truck Tonnage Before and After Deregulation

Year	Railroads	Tonnage (millions of tons) ICC-Regulated Motor Carriers	Unregulated Trucks
Prederegulation			
1970	1,572	661	1,167
1971	1,472	707	807
1972	1,531	771	1,163
1973	1,616	830	1,198
1974	1,619	800	1,155
1975	1,471	714	1,030
1976	1,477	808	1,166
1977	1,467	877	1,266
1978	1,481	925	1,335
1979	1,600	917	1,323
1980	1,589	816	1,191
Postderegulation			
1981	1,547	797	1,167
1982	1,351	717	1,074
1983	1,377	757	1,160
1984	1,522	832	1,293
1985	1,439	816	1,323
1986	1,422	852	1,363

SOURCE: Transportation Policy Associates (various years).

cated rise in motor carrier rates following deregulation appears to contradict the reports of widespread discounting among motor carriers. An alternative interpretation of the reasons for the movement in average rate is a loss of low-rated and truckload traffic by the formerly regulated sector of the motor carrier industry. Since the average rates charged on less-than-truckload (LTL) shipments is higher than that on truckload shipments, the increasing specialization in LTL shipments will appear as a rise in average charges. There have been no studies at the national level to determine the effect of deregulation on average motor carrier rates.[2] At the state level, two studies have found that motor carrier deregulation in Florida and Arizona was associated with lower rates to shippers (Beilock & Freeman, 1985; Blair, Kasserman, & McClave, 1986).

Since deregulation has not led to decreased railroad rates, the expectation that it should lead to an increase in rail share of the surface freight transportation market is questionable. Aggregate statistics do not support the expectation of a dramatic increase in rail traffic following deregulation. Table 17.10 shows no obvious breaks in data on tonnage following deregulation. The annual tonnage of each mode of transportation appears to fluctuate with the level of traffic produced by the industries served by the mode. Thus, the very low level of traffic for all modes in the early years of deregulation is the likely result of the recession in heavy industry in the early 1980s. If there have been intermodal shifts in freight traffic due to deregulation, the aggregate data in Table 17.10 suggests that they have been minor.

However, aggregate statistics like those in Table 17.10 are sensitive to the composition of freight traffic. Since railroads and motor carriers specialize in different commodities, any attempt to determine the amount of freight shifted between modes and the safety consequences of such shifts must control for the commodity composition of shipments. If one does not control for traffic composition, it will appear as if there has been a massive shift of traffic from railroads to trucks over the last 50 years when, in fact, the primary cause has been a relative decline in those industries that produce traffic in which rails have a large market share (Task Force on Railroad Productivity, 1973). Unfortunately, controlling for traffic composition is possible only for certain manufactured commodities that are sampled by the *Census of Transportation,* which is taken every five years.[3] For other sectors, data disaggregated by commodity type do not exist. Motor carrier traffic is much more heavily concentrated in this sector than is rail traffic, which tends to be bulk commodities like agricultural and mining products.[4]

To estimate traffic shifts associated with deregulation, an exhaustive sample of three-digit commodities was culled from the *Census of Transportation* between 1963 and 1983 for which there was a complete data series of reported rail figures and in which the definition of commodities was constant between years.[5] In evaluating the importance of deregulation in rail/motor relationships, it is important to calculate changes in modal shares of individual commodities rather than changes in aggregate market share. The reason is that changes in aggregate shares can result from changes in the composition of industrial output, and the latter changes may be independent of regulatory factors. Column 2 of Table 17.11 lists the tons of each of the sampled three-digit commodities shipped in 1983 by rail, private truck, and commercial motor carrier as reported by the *Census of Transportation.* The proportion of that tonnage carried by the two trucking modes in 1983 is listed in the third column.[6] The 854 million tons of freight in column 2 represents 40.3% of the total rail and motor freight traffic of manufactures in 1983.

A naive measure of the effect that the combination of trucking and railroad deregulation had on the trucking share of freight traffic is simply the difference between the most recent prederegulation truck share of each commodity (from 1977) and the postderegulation market share (1983). These figures are listed in the fourth column of Table 17.11. All but eight of the commodities show increases in the trucking industry share of reported freight traffic. Multiplying the percent shares by the total tonnage figures produces the figures in the fifth column. This is a naive approximation of the number of tons of each commodity shifted from rail to road

Table 17.11 Effect of Deregulation on Shipments of Manufactured Commodities

SIC No.	1983 Tons (1,000)[a]	1983 Truck Share	Change in Truck Share 1977–1983	Change in Truck Tons 1977–1983	Trend in Truck Share 1963–1983	Unexplained Change in Truck Share 1977–1983	Unexplained Change in Truck Tons 1977–1983
201	34,374	.968	.016	536	.016	−.029	−983
202	43,022	.988	.007	282	.012	−.003	−131
203	31,625	.826	.107	3,368	.013	.022	681
204	107,500	.767	.144	15,484	.022	.029	3,090
206	14,286	.741	.106	1,510	.019	.004	59
208	71,197	.876	.051	3,649	.006	.005	353
221	991	.983	.037	36	.007	−.007	−7
227	2,439	.903	−.021	−51	.005	−.009	−22
228	2,084	.995	.011	23	.002	.001	3
229	2,057	.911	.092	188	.004	.013	26
242	65,208	.787	.056	3,670	.016	.007	433
249	16,554	.703	−.002	−34	.012	−.030	−500
251	4,287	.943	.104	445	.009	.041	175
261	6,357	.391	.129	821	.018	.031	196
262	27,632	.527	.052	1,449	.009	.007	195
263	22,808	.384	.072	1,639	.005	.027	606
264	16,371	.813	.171	2,796	.017	.050	825
266	1,464	.801	.104	152	.015	.048	70
282	26,652	.603	.056	1,494	.000	.029	786
284	15,782	.908	−.001	−22	.008	−.008	−119
285	5,394	.962	−.019	−100	.004	−.012	−63
287	36,125	.612	.064	2,296	.008	.025	919
289	18,474	.822	.059	1,092	.013	.012	222
301	3,358	.753	.146	490	.009	.047	158
314	216	.977	−.010	−2	.000	−.004	−0
322	12,526	.981	.029	362	.010	−.015	−182
324	61,924	.881	−.020	−1,255	.012	−.038	−2,345
325	13,397	.850	.093	1,246	.010	.012	158
329	28,439	.804	.162	4,609	.016	.062	1,767
331	77,103	.732	.109	8,376	.016	.011	857
332	6,860	.866	.105	718	.005	.030	207
333	8,770	.356	−.193	−1,696	.010	−.098	−858
335	9,068	.846	.025	223	.008	.016	144
336	1,053	.977	.076	80	.004	.059	61
342	2,003	.862	.103	207	−.002	.031	62
346	9,366	.643	.200	1,877	−.008	.132	1,238
351	861	.880	.014	12	.007	−.006	−4
352	2,550	.897	.097	247	.014	.003	7
353	3,596	.895	.134	480	.020	.009	31
356	2,476	.969	.027	66	.007	−.006	−14
364	3,122	.903	.010	30	.004	.008	25
365	944	.792	.093	88	−.001	.034	31
366	498	.942	.077	38	.010	−.003	−1

SIC No.	1983 Tons (1,000)[a]	1983 Truck Share	Change in Truck		Trend in Truck Share 1963–1983	Unexplained Change in Truck	
			Share 1977–1983	Tons 1977–1983		Share 1977–1983	Tons 1977–1983
369	2,856	.954	.016	45	.002	.005	15
371	29,418	.594	.106	3,126	.006	.048	1,412
372	425	.939	−.003	−1	.006	−.016	−6
384	529	.947	.053	27	.002	.034	17
Total:	854,041			60,116			9,594
Change:			.070			.0112	

[a]Tons of all sampled three-digit commodities shipped by rail, private truck, and commercial motor carrier (excludes minor amounts shipped by water, air and unknown modes).

SOURCE: 1983 *Census of Transportation*.

by surface deregulation, because it attributes to deregulation *all* changes in market share between 1977 and 1983. The 7% figure at the bottom of the fourth column is calculated as the sum of these shifted tons (60,116,000) as a proportion of total reported 1983 tons (854,041,000).

But such shifts have been occurring for more than 70 years. While changes in the composition of output between heavy and light industry has led to some increases in the aggregate share of motor carriers (Task Force on Railroad Productivity, 1973), the data in Table 17.11 show that over the 20 years between 1963 and 1983 there was a general shift of most manufacturing commodities from railroads to motor carriers. The rate at which individual commodities have shifted from rail to road is shown in the column "Trend in truck share 1963–1983." These figures report the coefficient on the year variable in regressions of the form

$$\text{Truck share}_t = \alpha + \beta \cdot \text{year}_t + \mu_t \qquad (17.1)$$

where truck share is the combined share of private and commercial trucks as a proportion of all traffic in the listed commodity carried by these two modes and by railroads in the census years 1963, 1967, 1972, 1977, and 1983. Year is a variable whose value is the number of years since 1963.[7] All coefficients are estimated with five observations. The β coefficients listed in the sixth column are thus estimated annual changes in the highway freight share of reported traffic over the 20 years from 1963 to 1983. Except for three commodities, all of the trend coefficients are positive, indicating a trend from rail to highway; the three negative coefficients are quite small in magnitude.

By removing the trend effect on postderegulation changes in market share, a more accurate estimate of the effect of deregulation on truck market share can be made. We calculated forecasting errors for truck market shares in 1983. In mathematical terms, the numbers in the next to last column of Table 17.11, listed as unexplained changes in truck share, were calculated for each commodity with an equation of the following form:

$$Y_{20} - \hat{Y}_{20} = Y_{20} - \hat{\alpha} - 20 \cdot \hat{\beta} \qquad (17.2)$$

where Y_{20} is the truck market share in year 20 (1983), α and β are as given above, and a hat above a variable or coefficient represents least squares estimates from data for that commodity.

Not surprisingly, the figures show considerably smaller effects of deregulation than those in the fourth column, since a large portion of the change in truck share are now attributed to simple trend effects. The last column shows the number of tons of each commodity that were carried by trucks in 1983 minus predicted tonnages that would have been carried in the absence of deregulation, but with the trend in highway freight shares continuing unchanged. The figure of 0.011 at the bottom of the next-to-last column shows that the additional 9,594,000 tons carried by trucks in 1983 above what they were predicted to have carried, given the trend in market shares, is 1.1% of the 1983 total reported tons of 854,041,000. In the discussion that follows, columns 4 and 7 are used as upper and lower bounds on the effect that deregulation has had on the modal splits of rail and highway freight traffic.

In 1984, trucks and railroads carried an estimated combined total of 700 billion ton-miles of manufactures.[8] Extrapolating from our sample of manufactures traffic, an upper bound estimate of the amount of all manufactures traffic shifted from rail to road by deregulation is 49 billion ton-miles. The lower estimate is 7.7 billion ton-miles.[9] This pair of estimates is combined with the upper and lower bounds on the relative safety of motor carriers and railroads discussed earlier in this chapter to produce the effect of deregulation on safety summarized in Table 17.12. The highest estimates in the table blame 236 extra deaths, 2,533 injuries, and 2,565 extra accidents on shifts in freight traffic from railroad to truck following deregulation. The low estimates in Table 17.12 suggests that deregulation caused an extra 29 deaths, 349 injuries, and 337 accidents per year. While both estimates were made for traffic in manufactures alone, it is primarily

Table 17.12 Estimates of Increased Annual Deaths, Injuries, and Accidents Due to Deregulation of Manufactures Traffic

Estimates of Relative Safety Disparities per Billion Ton-Miles	Amount of Mode Shifting Due to Deregulation	
	High Estimate (49.0 billion ton-miles)	Low Estimate (7.7 billion ton-miles)
High estimate		
Deaths (truck: 4.88; rail: 0.06)	236	37
Injuries (truck: 52.44; rail: 0.74)	2,533	398
Accidents (truck: 56.27; rail: 3.92)	2,565	403
Low estimate		
Deaths (truck: 4.88; rail: 1.12)	184	29
Injuries (truck: 52.44; rail: 9.17)	2,218	349
Accidents (truck: 56.27; rail: 12.38)	2,150	337

in this sector that railroads and motor carriers compete for traffic. Although there may be an underestimate of the effects of mode shifting inherent in ignoring other types of traffic, this undermeasurement is not likely to be substantial.

SUMMARY AND CONCLUSION

Deregulation does not seem to have had a major effect on safety statistics of the motor carrier industry. The discussion in this chapter shows that deaths, injuries, and accidents per million truck miles and per billion ton-miles have been virtually constant over the past decade. There is no distinct break in the data series in the late 1970s, when the industry was administratively deregulated, or following legislative deregulation in 1980.

A more subtle effect of surface transport deregulation does, however, appear to have affected overall traffic safety. By any measure, railroads are considerably safer than motor carriers, and increasingly so following deregulation. In large part due to a dramatic reduction in personnel and the number of train miles traveled, railroads doubled their safety rate of the previous decade. One effect of deregulation was to shift some traffic from railroads to motor carriers. Between 1% and 7% of total manufactures traffic carried by rail and motor carriers shifted from rail to road following deregulation. The likely cause of this shift was a lowering of the cost of the services of the trucking industry, whereas rail rates, adjusted for service quality, rose modestly after 1980. The difference in estimates derives from alternative treatment of the long-standing trend from rail to truck traffic. The lower figure removes the trend before calculation of modal shifts, while the higher figure treats all change in modal shares between 1977 and 1983 as the result of deregulation.

Since some freight traffic has been shifted from a safer mode of transport to a less safe mode, overall traffic safety appears to have been worsened by deregulation. If the safety of railroad freight transportation is measured solely by train accidents (ignoring most grade-crossing accidents, injuries to trespassers, and industrial-type accidents), and if the maximum amount of freight traffic is assumed to have been shifted by deregulation, as many as 236 extra freight transportation deaths per year can be blamed on deregulation. A more reasonable estimate of the amount of traffic shifted following 1980, and inclusion of a broader range of rail accidents in calculating train safety, yields an estimate of 29 extra deaths, 349 extra injuries, and 337 extra accidents per year due to shifting traffic from the relatively safe rails to the more dangerous highways.

NOTES

1. While the accident statistics are based on actual accident reports, which were required beginning in 1973, the truck mile and ton-mile statistics are estimates by the Federal High-

way Administration (FHWA). The agency does not describe the means by which it makes estimates of truck miles. Ton-mile figures are constructed by multiplying the FHWA estimate by an estimated average load of a commercial truck. Thus, there are two reasons to doubt the accuracy of the data. There are, however, no alternative estimates available.

2. I attempted to control similarly for trend and an analog of truck weight, LTL percentage, in an attempt to discover the effect that deregulation had on the rate series in Table 17.9. The best guess was that deregulation caused motor carrier rates to rise approximately 4.5%. This figure is statistically insignificant and highly unstable with respect to other included variables. See Boyer (1987).

3. Only partial and unofficial tabulations of the *1983 Census of Transportation, Commodity Transportation Survey* (CTS) have been released. The Bureau of the Census warns that statistics below the level used in this study are likely to be unreliable. In addition, the 1983 CTS collected data on the basis of the Standard Industrial Classification (SIC) of the industries shipping the goods rather than the classification of the goods shipped by those industries. Four commodities appeared to have been especially affected by the change in reporting means and, as reported hereafter, were removed from this study; for the remaining commodities, the tonnages and modal shares appeared to be reasonably consistent between 1983 and previous years.

4. The Bureau of the Census estimates that in 1983, 27% or 312 million tons of the 1,140 million tons carried by railroads were manufactures. For commercial and private trucks, the figures are 54% or 1,802 million of the 3,334 million tons carried.

5. The commodities are indicated by the SIC in Table 17.11. All commodities from the 1983 *Census of Transportation, Commodity Transportation Survey*, Table 2, with three-digit numbers beginning with 2 or 3 and with reported rail tonnages were considered. The following commodities were eliminated for lack of comparable data in other years: 205, 226, 252, 276, 283, 286, 304, 323, 334. In addition, the following commodities were eliminated for apparent change in definition between years: 207, 209, 295, 327.

6. Since safety statistics are not broken out by private and commercial trucking, our research does not deal with intratrucking industry changes in the distribution of traffic associated with deregulation. For a discussion of such changes see Boyer (1987).

7. A simple linear form was chosen over a bounded dependent-variable model because, in all cases, the amount of variation in the dependent variable was sufficiently small as to justify the use of a linear approximation.

8. This figure is estimated by computing the average ton-miles per ton for rails (614 miles), ICC motor carriers (231 miles), and non-ICC trucks (278 miles) from Transportation Policy Associates (various years), and then multiplying the reported *Census* 1983 rail (312 million), motor carrier (761 million), and private truck (1,040 million) tonnages of manufactures by these mileages. The figure reported in the text is the sum of the ton-mileages calculated for each mode.

9. These figures are calculated by multiplying estimates of 0.07 and 0.064 from Table 17.11 by the 1,540 billion ton-miles.

18

A System Perspective on the Effects of Economic Deregulation on Motor Carrier Safety

PAUL P. JOVANIS

The preceding three chapters deal primarily with the effects of deregulation on carrier financial condition, safety performance, and the makeup of the motor carrier industry in terms of the mix of preregulation and post-deregulation carriers. Additionally, Boyer seeks to describe how deregulation has combined with broad economic cost factors to shift freight from rail to truck. He attempts to isolate the effects of the shift in changes in fatalities.

Many additional papers at the Northwestern University Conference examined the connections between deregulation and motor carrier safety. This chapter attempts to synthesize the findings from these papers for the purpose of developing a more complete understanding of the potential deregulation-safety linkages.

A SYSTEM PERSPECTIVE

Economic deregulation itself does not affect truck accidents in a proximate sense. Rather, it works indirectly through a variety of factors to influence, along with other forces, safety performance. These issues were dealt with in detail by Jovanis (1988). What is important at this time is to understand how economic deregulation works its way through the transportation system to affect road accidents.

Figure 18.1 characterizes the ways that deregulation may be linked directly or indirectly to safety performance. Deregulation directly affects three factors that influence motor carrier safety performance: the amount of truck travel, decisions by carriers in providing service, and the influence of ship-

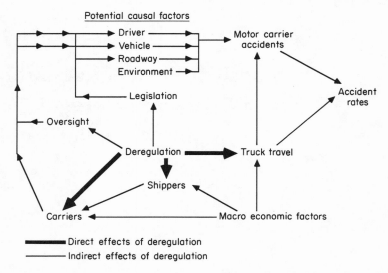

Figure 18.1 Simplified Process Model of Deregulation Effect on Motor Carrier Safety Performance.

pers in negotiating terms for service. In addition to these direct effects, there are a number of indirect linkages.

Deregulation along with broader economic forces, directly affects the amount of truck travel. The amount of truck travel by itself influences the number of motor carrier accidents and the accident rate, usually expressed as annual accidents per million truck miles of travel. In addition to the amount of truck travel, the number of motor carrier accidents is generally perceived as being related to the indicated proximal causal factors: the driver, vehicle, roadway, and environment. What we are concerned with is to understand the linkages between deregulation and these proximal causes. A limited number of these have been discussed in prior chapters; others are introduced here.

Deregulation can influence carriers directly in terms of their financial performance and thus in subsequent decisions that they make in hiring and scheduling drivers and in purchasing and maintaining equipment. This is potentially the strongest and most direct effect of deregulation. These hiring, scheduling, and purchasing decisions take place in the context of government oversight activities and ongoing motor carrier legislation. During the period of deregulation, safety regulations, operating rules, and oversight activities changed significantly. Carrier decisions regarding drivers and vehicles are made in the face of these changes, which confounds any attempt to measure a pure deregulation effect.

Deregulation can also affect shippers by giving them greater influence in negotiations with carriers. Some have charged that shippers force carriers into unsafe practices by negotiating unreasonably low rates, rates that match the rates of marginal, generally new-entrant, carriers. These connections are clearly of interest and are discussed further in the section on carrier management.

Deregulation also has an indirect effect on the oversight activities of government and enforcement agencies. Because there are changes in the number and mix of carriers on the road, oversight activities can, or should, change in response. The level and quality of oversight directly influence the type of drivers and vehicles used by the industry. These issues are also discussed in this chapter.

Last, deregulation may have had an effect on the motor carrier safety legislation that passed into law during the deregulation period. Discussions of these laws are contained in each of the following sections.

STATISTICAL RESULTS: ACCIDENT RATES

National accident statistics show either a declining or level accident rate since the initiation of economic deregulation. Fatality statistics, the truck accident data that are the most reliably recorded, show a generally decreasing rate per million vehicle miles during the 1980s. Using data from the national Fatal Accident Report System (FARS), Schweitzer (1988) showed that fatal accidents involving heavy trucks have hovered between about 3,700 and 4,600 per year from 1978 through 1986 (see Table 18.1). When one computes an accident rate considering the amount of mileage driven by trucks, there is a safety improvement from 0.065 fatal accidents per million vehicle miles in 1978 to 0.054 in 1983–1985. This reduction in accident rates has occurred despite a steady national increase in auto travel mileage during the same time period. More auto traffic would be expected to lead to more truck-auto collisions and thus to a higher accident rate for truck-related fatalities.

Table 18.1 Fatal Heavy Truck Accidents per Million Vehicle Miles

Year	Fatal Accidents Involving Heavy Trucks	Combination Vehicle Miles (Billions)	Fatal Heavy Truck Accidents/Million Miles
1978	4,351	67.3	.0647
1979	4,597	66.5	.0691
1980	4,036	60.0	.0673
1981	4,121	59.0	.0698
1982	3,754	60.3	.0623
1983	3,968	73.6	.0539
1984	4,200	76.9	.0546
1985	4,241	79.4	.0534

SOURCES: Accidents: National Highway Traffic Safety Administration. Miles: Federal Highway Administration. From Schweitzer, 1988.

Table 18.2   Accidents Reported to the Department
of Transportation by Interstate Motor Carriers

Year	Reported Accidents	Reported Accidents Adjusted for Inflation
1978	33,998	32,077
1979	35,541	32,322
1980	31,389	28,220
1981	32,306	27,772
1982	31,756	27,001
1983	31,628	26,032
1984	36,854	29,579
1985	39,156	29,068
1986	24,740[a]	27,740

[a]DOT adjusted its property-damage-only reporting threshold to compensate for inflation effective 1986.

SOURCE: U.S. Department of Transportation. From Schweitzer, 1988.

Data describing motor carrier accidents other than fatalities are likely to be a less reliable description of industry safety performance because of two threats to validity. While national interstate motor carriers are required to report all fatal, injury, and some property damage accidents to the federal Office of Motor Carriers (OMC), intrastate carriers are not required to report accidents to OMC. Therefore the national motor carrier accident data base cannot be considered to be reflective of national motor carrier safety performance. Further, there is colloquial evidence that the self-reporting system of motor carrier accident data collection is being abused. Many large national carriers have well-structured safety programs; a system for reporting accidents is part of their routine operations. This may not be so with all carriers. Because FARS is a census of all fatal accidents (auto, pedestrian, and truck) it is considered to be a more reliable indicator of national safety performance for motor carriers.

The other problem with historical OMC data is that the property damage limit for a reportable accident was fixed at $2,000 from 1973 through 1985. Inflation effects alone would therefore result in more reportable accidents each year even if accidents and mileage remained constant. Recently, the Department of Transportation (DOT) began using a deflator on its reportable limit; Schweitzer showed that the effect is to reduce 1985 OMC reported accidents from 39,156 to 29,068 (see Table 18.2).

While one can try to derive many measures of safety performance from the accident data, one point is clear: We do not know reliably how many accidents occur each year involving heavy trucks. Furthermore, we do not know where they are located or how they happened. If we use fatal accidents as a rough guide, comparison of FARS and OMC fatalities indicate about 40% fewer fatalities in OMC files. While some of this is attributable to legitimate differences in reporting requirements (OMC represents fatalities for interstate carriers only, whereas FARS adds intrastate), it is clear that OMC data substantially underreport heavy truck accidents. Thus, even before one attempts to associate economic deregulation with safety one must consider that data on the primary safety measure, accidents, are generally poorly collected and described for U. S. motor carriers.

In addition to a direct accounting of accident occurrence, there are important shortcomings in the quality of data contained in a truck accident report. Police who complete these reports are generally trained to assign blame to someone, frequently the truck driver, and may ignore potential contributing factors such as roadway design and vehicle equipment malfunction. This is markedly different from the detailed comprehensive examinations by trained investigators that characterize aviation accident data. Any attribution regarding motor carrier accident causality must recognize these limitations.

One is led to conclude that there is no objective evidence to link economic deregulation directly with a change in safety performance in the U.S. motor carrier industry. This position is supported by the apparent decline in fatalities per million truck vehicle miles from 1978 to 1985. A strong caution is advised, however, because of the number of potentially confounding events that occurred during the period of deregulation, which make it virtually impossible to isolate empirically a "deregulation effect." Any empirical assessment of accident or fatality rates is subject to the confounding events that occurred during deregulation: a major recession, a major economic recovery, and changes in safety oversight. Given the further uncertainties regarding the quality of U. S. motor carrier safety data, it is unlikely that any empirical assessment will be able to quantify accurately any actual "cause" of motor carrier accidents. I therefore recommend that the primary policy response be to continue safety regulation and oversight adjustment until a better understanding of deregulation's effects on safety is available. While this is treating the symptom rather than the cause, it seems the only valid response at this time.

There is a strong need for comprehensive, accurate national-level data on truck accidents that are free from the reporting biases and inaccuracies that currently exist. The accident data need to be matched by comparable exposure data, so that high accident rates can be differentiated from high occurrence due to exposure. Both the accident and exposure data should contain the relevant policy data concerning driver, vehicle, and roadway characteristics. Industry, government, and academia can and should work together to develop this plan.

INTERMODAL COMPETITION AND ACCIDENT STATISTICS

Boyer (in Chapter 17) adopts a different perspective concerning accident statistics. He attempts to estimate the additional accidents that occur due to freight shifting from rail to truck because of economic deregulation, after controlling for an underlying sectoral trend of mode shift during the last 20 years (due to the interstate highway system and other factors). The calculations assume that rail and truck are equally competitive over all lengths of haul. Because trucks will be generally more competitive over shorter distances, Boyer's method will overestimate total truck miles and thus total truck accidents due to mode shift. I believe, therefore, that Boyer's estimate is high, if significant at all, and that the mode shift effect is small at best.

CONGESTION

Traffic on the nation's highways continued to increase in the 1980s. Deregulation and a concurrent economic recovery have brought about an expansion of the miles operated by the trucking industry. No studies are extant that have determined if the expansion of truck operations has caused additional congestion at bottlenecks and whether the resulting congestion has led to more accidents. Trucks are still a small percentage of the vehicles operating on our nation's highways, and thus it is likely that at the margin, congestion effects are minimal for trucking. This is in direct contrast to aviation, in which relaxation of entry to markets and an evolution in network structure has led to greatly increased congestion at several major hub airports.

The increase in truck and auto mileage has resulted in a greater sharing of roadway space by trucks and autos throughout the day. This could lead to a greater public awareness of trucks and an increased perception of their operations. Some of the public concern over motor carrier safety in general may be attributed to this increased awareness of truck operations.

CARRIER MANAGEMENT

Chow (Chapter 15) and Corsi and Fanara (Chapter 16) have sought to connect changes in the safety performance of the motor carrier industry with economic measures that are directly or indirectly affected by deregulation. In general, these studies fail to find a *consistent* connection between economic deregulation and safety performance in the U. S. motor carrier industry. There is evidence of inferior safety compliance and higher accident rates for new entrants versus established carriers. There is limited evidence linking financial condition to safety investments but, unfortunately, no direct connection to safety outcomes (i.e., accidents).

Capelle and Beilock (1988) also tried to connect safety with deregulation by interviewing truck drivers and asking them to compare current conditions with those existing prior to 1980 by recall. Drivers reported that there were a variety of economic pressures that caused them to operate less safely, and they also felt that the economic pressures on them had intensified in the postderegulation period. Unfortunately, it is difficult to assess the accuracy of recall over a six- to seven-year period, so the findings are questionable. A 1970 survey (Fellmuth, 1970) reveals economic pressures very similar to those reported currently. These comparisons cast doubt on the use of the driver survey as evidence of a link between economic pressures derived from deregulation and industry safety performance.

Industry sources have identified another possible change in safety input due to deregulation. Rate competition has resulted in wage concessions by unionized labor. At the lower wage rates, carriers can have greater difficulty in finding qualified drivers. The problem has been compounded by the aging of the current driver workforce, for which adequate replacements will need to be found over the next 10 years.

As illustrated in Figure 18.1, deregulation can influence accidents by forcing drivers to exceed current hours-of-service regulations in order to remain competitive. Senator Adams in Chapter 3 of this volume recommends revision of current driving hours regulation or enforcement of the regulation with tachographs or other electronic equipment. Jovanis (1988), using postderegulation data, finds the accident rate to be level from approximately the third through the ninth hour of driving, the highest accident rate occurring during the first hour. One is led to conclude that accidents due to driver fatigue cannot be conclusively linked to deregulation. To the extent that driver fatigue is a problem, it should be dealt with through enhanced discipline of violators, not through economic reregulation.

MOTOR CARRIER SAFETY OVERSIGHT

Oversight activities are the principal means by which government ensures that safety regulations are followed. In the United States these activities include audits of accident records and of financial and operating data by federal officials at terminals and company offices. In addition, the states have a series of roadside inspections to verify compliance with vehicle maintenance and driver scheduling regulations. The direct effect of these activities is the monitoring of equipment and drivers that operate trucks on U. S. highways.

There has been a significant increase in safety oversight since deregulation and a shift from direct federal participation to increased state participation with federal funding support. Three significant pieces of federal legislation have been passed since deregulation (in 1982, 1984, and 1986) that deal substantially though not exclusively with motor carrier safety. While it is not clear that deregulation was the *cause* of this safety regulation, economic deregulation was most likely an issue when the legislation was proposed and passed. These changes in safety oversight have important implications for policy, so they are briefly reviewed here.

A complex set of economic, operational, and regulatory forces culminated in the passage of the Surface Transportation Assistance Act (STAA) in 1982. This Act substantially increased federal user fees paid by motor carriers, including an increase in the diesel fuel tax and an increase in the registration fee. States quickly followed suit in raising their diesel fuel taxes. The objective of these increases was to provide funds for roadway reconstruction and rehabilitation, at least partially as a public works stimulus during an economic recession. In partial compensation for the fee increases, the trucking industry was granted a more uniform system of federal vehicle size, weight, and configuration regulations. Double combination tractor-trailors were allowed to operate on all interstates and designated state highway systems; the weight limits were set at 80,000 pounds gross vehicle weight and vehicle width limits at 102 inches.

The 1982 STAA also initiated the Motor Carrier Safety Assistance Program (MCSAP). The Program provides federal funds to states for motor carrier enforcement-related activities: on-road vehicle inspections, enforcement officer inspector training, and other support activities. A 1986 Act revised the user fee structure and provided additional funds for MCSAP. As a result of the MCSAP program and its related activities, road safety inspections increased from 159,000 in 1984 to 1.2 million in 1987.

The 1984 Motor Carrier Safety Act aimed at establishing uniform national standards for drivers and equipment, and provided for heavier penalties for violations. Previously, safety rules could vary substantially from state to state. This act was enhanced by the Commercial Motor Vehicle Safety Act of 1986, which established a framework for the national licensing and enforcement of commercial drivers. While precise guidelines are to be established during the late 1980s, the Act will require a single license for all motor vehicle drivers. Currently, driving licenses are issued by individual states and unscrupulous drivers can obtain licenses from several states to mitigate against license revocation in one state. The legislation severely restricts or bans the driving of intercity buses and trucks by individuals under the influence of drugs or alcohol and the holding of multiple licenses. The restrictions will take the form of mandatory license suspensions, and, in the case of multiple convictions, permanent license revocation. The laws recommended under the 1986 Act are not fully operational but, if implemented as they appear to have been intended, will have a substantial effect on road driver behavior and the composition of the driver work force.

There is contemporary concern about a significant gap in the current safety regulatory structure that allows Commercial Zone operations, motor carriage that occurs within defined areas that surround major metropolitan areas, to be exempt from federal vehicle maintenance, driver licensing, and work hours regulations. These Zones may serve as a dumping ground for bad drivers and inferior equipment. Repeal of the Commercial Zone exemption is a positive step toward improving the safety performance of the industry and is currently being undertaken administratively by the DOT.

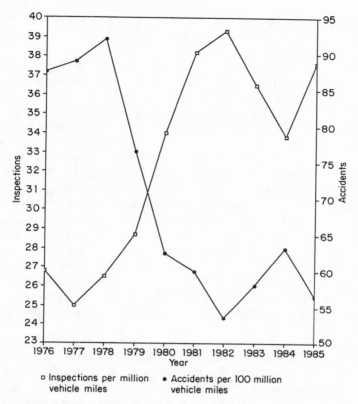

Figure 18.2. California Highway Patrol Truck Inspection and Truck-at-Fault Accident Rates, 1976 through 1985. SOURCE: California Highway Patrol.

An association of U. S. state and Canadian provincial enforcement agencies, the Commercial Vehicle Safety Alliance (CVSA), has been formed and is playing an important coordinating role in safety regulation (Daust & Cobb, 1988). CVSA has programs to uniformly train inspection officers; coordinate inspections to reduce the incidence of multiple inspections of a safe vehicle; and maintain a data base, SAFETYNET, on violations and accidents. This coordination at the state level represents an important difference from aviation, which has inherently interstate (i.e., federal) oversight. There appears to be a consensus that the new arrangement of oversight responsibilities facilitated by MCSAP, including the activities of CVSA, has resulted in more inspections and generally safer highways. Despite this positive evaluation, it is not clear if more efficient use could be made of MCSAP funds as there has yet to be a thorough independent evaluation of the MCSAP program.

Some of the current evidence linking accidents and road inspections seems almost too convincing. Figure 18.2 is taken from Schweitzer (1988) and shows trends in the truck-at-fault accident rate superimposed on the vehicle inspection rate as reported by the California Highway Patrol (CHP). There is clear evidence throughout the time series of an inverse relationship between accident rate and inspection activities: in years with many inspections, the accident rate was low; in years with fewer inspections, the rate was high. It is the scale of the fluctuations that is surprising: A 30% increase in inspections yields a nearly identical and almost immediate *reduction* in the accident rate. Given the complex set of interrelationships illustrated in Figure 18.1, it seems unlikely that enforcement alone would have this strong an effect. For there to be a change in the on-road accident rate, carriers would have to respond strongly and immediately to the oversight by making changes in driver work schedules, hiring, and vehicle equipment maintenance. Any change in the accident rate would have to reflect the marginal change in these three factors. It strains credibility that a change in oversight could alter California's truck accident rates so dramatically and quickly. It is not the association that is questionable but the scale of the effect and the rapidity and effectiveness of industry response. One wonders if it is more likely that increases in road inspections by the CHP implied fewer officers to report accidents; the figure, therefore, may reflect changes in reporting levels as well as changes in on-road safety performance.

Albu and Glover (Chapter 19 in this volume) provide an interesting counterpoint in their description of the United Kingdom experience with deregulation. Deregulation was initiated in 1970 along with more stringent controls on hours-of-service. Additional safety regulations regarding driver licensing and vehicle inspection performance were passed in 1972. Still further revisions on driver licensing and vehicle occurred in subsequent years as part of the move to join the European Economic Community.

There appears to be, in both Britain and the United States, a substantial increase in safety regulation following economic deregulation. This may be due to the broad safety performance of the motor carrier industry at the time. There is no evidence that it is a direct response to economic deregulation, but it certainly is a consistent coincidence. Particularly in the United States, safety oversight activities changed tremendously during deregulation and continue to change today. What does seem to be clear is that there has been a governmental safety response to economic deregulation.

For those countries contemplating a move toward economic deregulation, it seems clear that government infrastructure and oversight should respond to the new market (Panzar and Savage, Chapter 4). The type and magnitude of response may vary from country to country depending on present safety oversight and may include changes in on-the-road oversight and industry safety audits that should be scaled to the number of new entrants and the emerging economic environments. The process by which government oversight responds to economic deregulation is the important linkage for any country considering deregulation to understand.

AN INTERNATIONAL PERSPECTIVE

Debate on the safety implications of economic deregulation is international. The experience of nations differs in important ways. Some countries have already deregulated some or all of their modes of transportation; others are in the process of doing so. The pressures for regulatory reform are spreading worldwide.

In this section we present two international viewpoints. First, from an historical viewpoint, Albu and Glover assess the safety implications of the deregulation in 1970 of trucking in Great Britain. Second, Withers describes the safety-related precautions that Canada is implementing as it relaxes economic controls over the transportation sector.

These chapters show a common feature. Relaxation of economic controls in both Great Britain and Canada went hand in hand with a thorough review and revision of safety legislation. The reader is invited to contrast activities in these countries with the role, funding, and activities of the Federal Aviation Administration and the Federal Highway Administration in the early 1980s.

19

The Regulation of Road Freight Transport in Great Britain

MARTIN ALBU and MICHAEL GLOVER

The procedures for regulating road freight transport described in this chapter apply throughout Great Britain, that is, England, Scotland, and Wales. Northern Ireland has separate legal provisions, but in practice the regulatory system is broadly similar. Within Britain there are some 130,000 licensed carriers, of whom 70,000 operate "for hire," or in British terminology, "hire or reward" services. There are about 434,000 trucks, that is, heavy-goods vehicles exceeding 3.5 tonnes gross vehicle weight. They account for about 60% of all freight tonne-kilometers carried within Britain, compared with 9% by rail, 25% by inland and coastal shipping, and 11% by pipeline.

The principal basis for the regulation of road freight transport in Britain is the Transport Act of 1968. Before that Act came into force in 1970, access to the industry had been subject to a licensing system that severely restricted new entry to the for-hire market but was less restrictive of contract hire and private carriers. This had led to rapid growth in these categories, which distorted the market and rendered the control system obsolete. The 1968 Act removed all quantitative control on the operation of freight transport by road. Access to the market is now open to anyone, subject to a system of carrier licensing intended to safeguard the industry, its customers, and other road users by ensuring that operators are of suitable standing and that they observe the laws governing road transport operations.

As a further measure toward improving road safety, the 1968 Act introduced more stringent regulations on drivers' hours of work and rest periods. A further Act, in 1972, introduced new controls on vehicle standards and a new requirement of licensing for drivers of vehicles over 7.5

tonnes gross weight. Subsequent legislation, including European Economic Community (EEC) Regulations, have added to the system, which now comprises control of operators through goods operator, that is carrier, licensing; control of vehicles through construction and use regulations, type approval, annual testing and a national vehicle registration system; control of drivers through driver licensing, regulations on hours of work and rest periods, and the use of tachographs; and additional controls on the transport of dangerous goods.

This chapter describes these controls in more detail, outlines how they are enforced in practice, and then considers developments in the road haulage industry since 1970 and how they have affected road safety.

SYSTEM OF REGULATION

Operator Licensing

For the purpose of carrier licensing, Britain is divided into nine areas, each with a licensing authority appointed by the Department of Transportation (DOT) but which, in practice, is independent of the Department. Anyone who wishes to use a goods vehicle exceeding 3.5 tonnes (7,700 pounds) gross weight for carrying goods in connection with any trade or business is required to hold an operator's license. To obtain a license, carriers have to satisfy a licensing authority that (1) they are fit persons which normally means that they do not have a bad record of activities using goods vehicles and have not recently been convicted of certain offenses related to the use of vehicles; (2) that they have satisfactory maintenance arrangements and adequate financial resources to keep the vehicles fit and serviceable; and (3) that they have a suitable operating center.

To obtain a Standard License authorizing hire or reward (for-hire) operations, carriers must also satisfy the licensing authority that they are "of good repute," which allows a wider range of offenses to be taken into account; that they have adequate financial resources to ensure the proper operation of the road transport undertaking; and that they are, or have in full-time employment, someone who is professionally competent. The latter normally means that a Certificate of Professional Competence has been obtained by examination.

The requirements relating to "standard" licenses are based on Directives 74/561 and 77/796 of the EEC. Compliance with those requirements and holding a Certificate of Professional Competence covering international transport entitles the operator to operate anywhere within the EEC, subject to quantitative restrictions currently imposed by some member states.

An operator's license states the number and type of vehicle authorized to be kept and may include other conditions, for example, hours of use of the operating center. It implies an undertaking by the operators that they will ensure that their vehicles are properly maintained and not overloaded and that their drivers observe the rules on drivers' hours and records (see below). It is valid for up to five years, but may be curtailed or revoked at any time if the licensing authority considers that the conditions are not being observed. This power of the licensing authority is a most important sanction.

Construction and Use Regulations

The Construction and Use (C&U) Regulations developed from the statutory rules applied at the turn of the century to the use of motor cars and locomotives on the highway. Maximum weights and dimensions have always been carefully controlled, and the modern rules relate to the standards that have been applied in recent years to the construction of roads and bridges in Britain. The modern C&U Regulations also control braking efficiencies, tires, smoke and diesel engines, noise, equipment such as mirrors, and protective devices such as sideguards. The Regulations remain the prime statutory requirements for the use of vehicles. Failure to comply with them is the most common reason for prosecution of motorists. However, in recent years increasing attention has been given to the role of the manufacturer in ensuring that vehicles are fit for use on the roads, and this has resulted in the development of the type-approval system.

In 1981 a scheme of type approval for nonspecialized goods vehicles was introduced. To some extent the requirements were introduced as part of the government's plan to "civilize" the heavy lorry. They also have a trading justification in that Britain had become one of the few major manufacturing countries without some form of preregistration check on goods-vehicle construction standards. The goods-vehicle scheme requires individual systems approval for power-to-weight ratio, noise and silencers, exhaust emissions, radio interference, and brakes. It is linked to the determination of the design and operating weights of each vehicle, the latter being the weights that should not be exceeded when the vehicle is used on a road in Great Britain.

Most of the standards to be met are based on internationally agreed on standards such as directives issued by the EEC or regulations issued by the United Nations Economic Commission for Europe, and approval to the relevant international standard is acceptable as evidence of compliance with the national requirement. The remainder are derived from national requirements contained in the Construction and Use Regulations.

Plating and Testing

All goods vehicles that have been type approved, and all other goods vehicles and trailers after one year, must carry a DOT plate showing the gross weight and individual weights they are designed to carry. If these are higher than the maximum weights permitted under the Construction and Use Regulations, those weights are also shown on the plate.

All goods vehicles over 1,525 kilograms unladen weight, including trailers and semitrailers, are required to undergo an annual test at one of the DOT's 91 test stations. This test, which last some 45 minutes, covers all aspects of the vehicle's maintenance, including braking performance. In 1985–1986, some 924,000 tests were carried out, with a failure rate of 22.9%. Most of these failures were due to brake defects.

Vehicle Registration and Taxation

All motor vehicles in Britain have an index number that is registered at the national Driver and Vehicle Licensing Centre at Swansea, Wales. Details of all vehicles and their keepers are held on a computerized register. Some details are also held on the police national computer system. This makes it possible for the police to readily identify the keeper of any vehicle that comes to their attention.

All motor vehicles pay an annual tax called "vehicle excise duty." This ranges from £100 for cars and light vans to over £3,000 for the heaviest goods vehicles. The rate of duty for trucks is intended to be related to the highway damage costs imposed by different types of vehicle, so that, for example, a six-axle articulated truck pays less duty than a five-axle vehicle of the same weight.

Driver Licensing

All drivers of motor vehicles in Britain require a driving license, which is obtained after a rigorous practical test conducted by the DOT. The standard license covers the driving of cars and of goods vehicles up to 7.5 tonnes (16,530 pounds) gross vehicle weight. An additional heavy goods vehicle (HGV) license is required to drive a vehicle above that limit. To obtain this heavy goods vehicle license, a driver has to undergo a much more stringent driving test and also a medical examination. A license may be revoked if the driver's medical condition changes. Any offenses committed by drivers on the road that are recorded on their licenses will also be reported to the licensing authority, which can take immediate disciplinary action over the HGV licenses. Offenses against specific provisions relating to goods vehicles, for example, overloading or nonobservance of the rules of drivers' hours, may also be reported to the licensing authority and lead to disciplinary action against both the driver and the operator. There is a central computer-based record of all driving licenses and endorsements, to prevent duplication.

Drivers' Hours and Tachographs

The Transport Act of 1968 introduced a new regime of control of the hours worked, driving times, and rest periods of drivers of goods and passenger vehicles. It also required the keeping of written records in a logbook. Following Britain's accession to the EEC, most drivers came within the scope of EEC regulations, which laid down maximum driving times and minimum rest periods but did not deal with total duty time. Drivers on international journeys were subject only to the European regulations, but those on domestic journeys were subject also to the provisions of the British laws. In September 1986, major revisions of the EEC regulations came into effect. At the same time we took the opportunity of dispensing with most of the British provisions. So for the majority of drivers, only the EEC regulations now have effect.

Also, following Britain's accession to the EEC there was a gradual move from manual records to the use of tachographs for recording drivers' hours and rest periods. By 1980 it had become mandatory to fit and use tachographs in almost all goods vehicles and many passenger vehicles. Tachographs are now fitted as original equipment by most vehicle manufacturers, and the presence of a tachograph is checked at the annual roadworthiness inspections.

The main use of the tachograph is to ensure compliance with the EEC regulations governing drivers' hours and rest periods. But it also provides a record of speeds. This record, although not usable as uncorroborated evidence in a court of law, may be used by the police and traffic examiners as an indication that a vehicle has been speeding. Operators, too, find the tachograph record of use for purposes of fleet management. It indicates drivers who may be incurring unnecessary fuel and maintenance costs through driving too fast or irregularly. Although the drivers initially opposed the use of the tachograph, many of them now welcome the evidence of safe driving that it provides.

Dangerous Goods

The carriage of dangerous goods within Britain by road tanker or tank container is subject to regulations and an advisory code of practice. These require that the vehicles and containers be appropriate for the purpose, that drivers be adequately trained, that warning placards be displayed on the vehicle, and that instruction cards about the particular chemicals carried be in the driver's cab. Similar regulations govern the carriage of dangerous goods in packages. On international journeys, the European agreement on the carriage of dangerous goods by road applies. This requires the vehicle to be specially certified and the driver to have undergone an approved training course; it also specifies a different form of placarding of the vehicle.

ENFORCEMENT

Police

Responsibility for ensuring that operators and drivers comply with these various rules and regulations rests in the first place with the police. It is they who have the most direct contact with vehicles on the road. Although the police may, from time to time, arrange systematic vehicle checks at particular points on the road, more usually they encounter offenders in the course of other work. This may occur, for example, when police are called to the scene of an accident, or if they happen to notice a vehicle in a dangerous condition or not displaying operator's license and vehicle excise duty disks. The police can bring a prosecution against an offender; they also have the power to prohibit further movement of a vehicle that they believe to be in a seriously dangerous condition. All such cases should also be reported to the appropriate licensing authority for further action, but this does not always happen. There is no arrangement for collecting statistics about police activity in this area on a systematic basis, leading to underreporting in the statistics relating to such offenses.

Vehicle Inspectorate

The DOT's vehicle inspectorate carries out the annual HGV test. Its 91 test stations also certify vehicles to carry dangerous goods under the European agreement and to carry goods under the simplified customs arrangements that operate in most of non-EEC Europe and the Middle East. The inspectorate also carries out random tests on vehicles at the roadside or at operator's premises and has the power to prohibit the use of vehicles found to be in a defective condition. It also investigates accidents. Vehicle inspectors are also attached to traffic areas and assist the licensing authorities by inspecting and reporting on operators' maintenance facilities.

Traffic Examiners

The Department of Transport employs 235 traffic examiners, based in the traffic areas. They are concerned with ensuring compliance with all the laws relating to goods vehicles and passenger vehicles by systematic roadside checks. A particular concern is the weighing of goods vehicles, since overloading may lead to unsafe vehicles and to increased road damage and unfair competition. The Department has established a network of 62 sites situated near major roads where vehicles can be check weighed. This is usually by means of slow-speed dynamic axle weighing equipment, but other equipment is also used. Ten of these sites are situated at ports—higher weight limits apply on the continent than in Britain, so it is important to check incoming vehicles. Some include facilities for drivers to weigh their own vehicles if an inspector is not present. At the same time as they are weighed, vehicles may be checked for compliance with the laws on taxation and licensing, on dangerous goods if they are carried, and on drivers' hours.

Drivers' hours are checked through inspections, at the roadside, of the tachograph charts carried on the vehicle. In addition, police and traffic examiners analyze batches of tachograph charts taken from operators' premises. This process is greatly helped by the recent introduction of computer-aided analysis systems that enable large numbers of charts to be checked rapidly. In the year to September 30, 1986, some 250,000 charts were checked and 2,642 prosecutions made for offenses against the regulations. The tachograph charts also provide a record of speeds. Under existing law, this record may not be used as uncorroborated evidence. Nevertheless, it is used by the police and traffic examiners as an indication that a vehicle may have been speeding. Offending drivers may be warned and are at risk of losing their license if they appear to offend frequently.

Traffic examiners have powers to prohibit further movement of overloaded or dangerous vehicles. They may also initiate prosecutions against both operators and drivers for overloading and for drivers' hours offenses. But perhaps more important is their duty to report all offenses to the licensing authorities.

Licensing Authority

Operators who are reported frequently for inadequate maintenance of their vehicles or for other offenses concerning their operation, irrespective of whether the offenses have led to prosecution in the courts, may be summoned before the licensing authority. Initially they will probably be given a warning, but if the offenses continue they will be at risk of having their licenses curtailed or revoked. In the year to March 31, 1986, there were some 267 revocations or curtailments of operator licenses, and formal warnings were given in another 189 cases. More usually, drawing an operator's attention to problems in his or her organization can help to ensure that offenses do not recur. However, the threat of losing a license is an important deterrent.

Other Agencies

Two other agencies play a part in enforcement. Local Authority Trading Standards officers have the power to check weigh goods vehicles. They may do this using their own equipment or the Department's weigh-bridge sites referred to previously. The regulations on the carriage of dangerous goods are the formal responsibility of the Health and Safety Executive, through its factory inspectorate, which has wide-ranging powers in regard to all aspects of health and safety at work. This office may act on information provided by the police or traffic examiners from their routine inspections of vehicles on the road, but most of its activity takes place at loading or unloading points, since it is there that most problems occur.

DEVELOPMENT IN ROAD HAULAGE SINCE 1968

Table 19.1 shows the development of the road system and of truck operators and goods transported since 1968. The movement of goods by road

Table 19.1 Public Roads, Truck Operators, and Goods Carried in Great Britain

Year	Trunk Motorways[a] (km)	All Surfaced Roads (km × 10³)	Truck Operators (× 10³)	Goods Transported (tonne-km × 10⁹)
1968	890	317.7	—	79
1970	1,022	322.5	120.3[b]	85
1972	1,609	327.7	130.5[b]	87
1974	1,790	331.6	138.9	90
1976	2,062	333.4	138.0	92
1978	2,287	336.2	124.7	92
1980	2,445	339.6	130.0	92
1982	2,561	343.6	127.0	94
1984	2,702	347.1	128.0	100
1985	2,737	348.3	130.0	102

[a] Dual two- and three-lane highways with restricted access.
[b] Estimates.
SOURCE: U.K. Department of Transport (various years).

has increased by some 30%, from 79 billion tonne-kilometers in 1968 to 102 billion tonne-kilometers in 1985. Whereas deregulation may have played some part in this change, it seems likely that other factors were more significant. The development of a nationwide motorway network has made possible the development of new distribution systems. Decline in the heavy engineering industries has been accompanied by expansion in high-technology manufacturing, which places greater emphasis on the rapid movement of materials and components between suppliers. Retail distribution is increasingly based on large supermarkets and hypermarkets fed from out-of-town depots. While tonnage of goods carried by road declined from 1,707 million tonnes in 1970 to 1,435 million tonnes in 1985, the average length of haul increased from 46 kilometers in 1968 to 71 kilometers in 1985.

This increase in road freight tonne-kilometers was accompanied by a significant reduction in the number of heavy-goods vehicles. The reduction in the number of vehicles resulted largely from changes in vehicle dimensions and maximum permitted weights. In 1968, the length of an articulated vehicle was increased from 13 meters to 15 meters, thus allowing the 40-foot container to be carried on an articulated vehicle and allowing box van trailers to be constructed and used with similar capacity. Maximum width was set at 2.5 meters, allowing two standard European pallets each of width 1.2 meters to be placed side by side in the box. This increase in volume capacity had a strong influence in the development of the freight system. The number of heavy-goods vehicles decreased from 600,000 in 1968 to 436,000 in 1982. In 1983 the maximum gross vehicle weight of articulated vehicles was increased from 32.5 tonnes to 38 tonnes. Maximum axle weight was marginally increased from 10.17 tonnes to 10.5

tonnes. Maximum length of articulated vehicles was increased by 0.5 meters to allow for bigger tractor cabs. These changes do not appear to have had any significant effect on the total number of heavy vehicles in the period 1983 through 1986.

In addition to the increased maximum weight and dimension limits, there have been continuous improvements in engines and in the design of cabs and vehicles, which have made them easier and safer to drive. Minimum braking standards are set by an EEC directive, and this was recently amended to include a significant increase in braking performance for heavy-goods vehicles. We have introduced new requirements to have side-guards fitted, to prevent pedestrians and cyclists from falling under and being caught by the rear wheels of vehicles; and rear-guards, to protect occupants of following cars from underrunning in the event of an accident. We also now require vehicles to be fitted with spray-reducing devices.

THE ACCIDENT RECORD

Table 19.2 shows vehicle accident involvement rates over the past 15 years. It will be seen that for all vehicles there has been a marked improvement in accident rates. This is thought to reflect improvements in vehicles and in highways during this period, and in particular the increasing use of purpose-built motorways and of bypasses that keep through traffic out of built-up areas. The accident rate for heavy-goods vehicles was already significantly better than that for other vehicles in 1970, and the difference has increased since then. Part of this improvement may reflect the effects of the new operator licensing system, driving license requirements, and drivers' hours limits introduced in 1970, but it is not possible to quantify the amount of improvement due to each of these different factors. There has been no significant rise or fall in goods-vehicle accident rates since heavier vehicles were permitted in 1983.

Table 19.2 Vehicle Accident Involvement Rates per 100 Million Vehicle Kilometers (Great Britain) Distinguishing Fatal from Serious Accidents

Year	Heavy Goods			Cars			All Vehicles		
	All Severities	Fatal	Serious	All Severities	Fatal	Serious	All Severities	Fatal	Serious
1970	157			186			220		
1975	98	5.5	27.5	148	3.2	36.8	178	4.1	44.9
1980	76	4.2	21.8	133	2.7	32.3	158	3.4	40.6
1981	74	4.2	21.8	132	2.6	31.4	155	3.3	39.7
1982	76	4.7	21.3	132	2.6	31.4	155	3.3	38.7
1983	67	4.0	19.0	123	2.4	27.6	143	2.9	34.1
1984	68	4.1	18.9	126	2.3	27.7	146	3.0	35.0
1985	67	3.8	18.2	121	2.2	26.8	139	2.7	33.3

The DOT routinely collects and analyzes accident statistics and investigates serious accidents in detail. The results of this work have led to the introduction of several new safety measures in recent years, including the compulsory wearing of seatbelts in private cars and measures to curb drinking and driving. Emphasis is now placed on developing low-cost measures to improve existing road layouts and on applying safety principles to road design and maintenance in the longer term.

Although the frequency of accidents involving goods vehicles is relatively low, Table 19.2 shows that when heavy-goods vehicles are involved in accidents, the number of deaths is almost double compared with accidents in which such vehicles are absent. Attention has therefore been focused on improvements to vehicles to mitigate the effects of accidents. Thus it is estimated that the compulsory fitting of side-guards and rear underrun bumpers will save about 70 lives a year by the time that all goods vehicles have been so fitted. A further 60 lives could be saved if front underguards were also fitted, but so far this is not a legal requirement.

Long driving hours are commonly assumed to lead to excessive fatigue and hence to accidents, but there is as yet no firm evidence of a direct correlation. A study in 1977 of accidents involving heavy-goods vehicles on a motorway found that boredom was a more significant factor than long hours in determining fatigue, the critical distance for onset of boredom being between 100 and 200 miles of driving (Storie, 1984).

In 1974, when tighter speed limits were introduced as a temporary measure, evidence showed some improvement in road safety. However, the effect appeared to wear off quite quickly, and drivers resumed their previous habits. The government is considering introducing speed governors on passenger vehicles, but there is as yet no suggestion of similar measures for goods vehicles, which, unlike buses, are not allowed to use the lane closest to the central median on motorways.

CONCLUSIONS

The British system of regulation of the road freight industry is designed to maintain safety standards through a system of operator licensing, which in turn requires high standards of vehicle maintenance and observance of the regulations concerning construction and use, loading weights, and drivers' hours. Economic factors play no part in the regulatory system, except insofar as operators are required to have adequate financial resources to maintain their vehicles.

Since the regulatory system was introduced in 1968–1970, the safety record of the freight industry, which was already relatively good, has significantly improved. Although part of this improvement may be due to the operator licensing system and to the driver licensing system and drivers' hours rules introduced at the same time, most of it is probably attributable to improvements in the highway system and in vehicle construction and braking standards.

Road safety remains a dominant concern of the DOT. Improvements are likely to continue to come through better highway standards, better vehicles, and improved enforcement of the existing regulations, rather than through any major new changes in the legislation and regulatory systems.

20

Transportation Safety and Economic Regulatory Reform— The Canadian Perspective*

RAMSEY M. WITHERS

Since 1984 the Canadian government has set as one of its major objectives the removal of obstacles to economic growth. The Canadian transportation sector is seeing the effect of this objective primarily on two fronts.

First, as a result of the government's commitment to fiscal restraint, certain federal transportation programs and services are being reduced in some areas where it is felt public sector involvement is no longer required to support national or regional objectives. However, in most cases, resources have been reduced without significant reductions in levels of service through streamlining of operations and removal of duplication and program overlap.

Second, efforts are being made to reduce the amount of regulatory intervention or control over private transport enterprises. It is this major influence to which this chapter is devoted. Economic regulatory reform (or as it would be termed in the United States, "deregulation") is part of the Canadian government's overall plan to improve the competitiveness of Canadian firms. One of the first and most influential industrial sectors to undergo this regulatory reform is transportation where, in 1987, the Canadian Parliament approved two bills to amend the National Transportation Act of 1967 (NTA 1967) and the associated Motor Vehicle Transport Act of 1954 (MVTA 1954).

However, with the introduction of economic regulatory reform, concern has been expressed in some circles of Canadian society that with the resultant increase in competition, profits will be squeezed from the operating carriers, causing them to reduce their margin of safety so as to remain financially viable. In response to this, the government believes that no matter what the regulatory regime is, at any point in time transportation safety

* This chapter represents the legislative status as of the time of writing.

can only be guaranteed through the development and enforcement of effective safety standards. Furthermore, the government believes it has not been proven that economic regulation guarantees carrier profitability, or that profitability guarantees safety.

Consequently, the federal government, acting primarily through Transport Canada, sees the essential issue at stake as being: What is the best way for the federal government to continue its role in maintaining and improving transportation safety while reducing the burden of regulations on the transportation industry? Addressing this question first by assessing the economic regulatory reform-transportation safety relationship, and second by identifying the programs that have been introduced to enhance transportation safety, this chapter will outline how present high safety levels will be maintained, even in a less regulated and more competitive transportation services industry.

THE "NEW" NATIONAL TRANSPORTATION ACT

Whereas the NTA 1967 focused primarily on promoting *inter*model competition, the new legislation (hereinafter referred to as the NTA 1987 and MVTA 1987) goes one step further by also supporting *intra*modal competition. By encouraging increased competition within and among modes, it encourages changes in both the price and range of transportation services that will benefit producers, shippers, and ultimately the consumer. With less regulatory protection for carriers, this legislation is distinctly biased in favor of shippers and consumers. And, to ensure that this reduced regulation does not result in decreased safety, Transport Canada has undertaken a number of initiatives to ensure that carriers do not decrease safety levels.

Highlights of the regulatory reforms permitted as a result of the passage of this legislation by mode of transport are described hereafter. Rail transportation will also see regulatory reform. Railroads will be permitted to have confidential contracts with shippers, and there will be an end to collective rate making.

Air Transportation

Regulation of passenger fares and air cargo tariffs will be largely eliminated, encouraging greater price competition. The National Transportation Agency (Agency) which will be created as part of this reform package will be able to investigate complaints about unreasonable fare increases. For northern and remote areas, a special regime will ensure that essential services are not disrupted for these thin, widely dispersed markets. To protect against monopoly pricing, the Agency will have the power to examine fare levels and increases and disallow those judged unreasonable. Federal funding will be permitted to maintain essential air services in cases of necessity. A new license test based on safety and insurance requirements, entitled "fit, willing, and able," will give carriers meeting national safety standards the freedom to establish improved and expanded services.

Extraprovincial Trucking

As a result of a federal-provincial agreement signed in 1985, the MVTA 1987 established a uniform, nationwide entry test for extraprovincial trucking operators. The fit, willing, and able license test, based on safety and insurance requirements, became effective on January 1, 1988. A federal-provincial National Safety Code is being developed and will be fully in place by January 1, 1990, as a uniform basis for the fitness requirement.

For a five-year transition period, new service applications will also be subject to a public interest test, with the onus placed on objectors to prove that the public interest would not be served by any new operator. Rate regulation will be eliminated and other license conditions such as route and commodity restrictions will be removed at the end of the transition period.

A major review will be carried out in 1992 on all modes of the transportation system affected by the economic regulatory reform initiative. The Agency will also conduct annual reviews for four years on the technical operation of this reform legislation. Additionally the MVTA 1987 requires the Minister of Transport to carry out a review of the effects of the motor carrier reverse-onus tests and the transitional regime during the fourth year after the legislation comes into force.

It bears repeating that these initiatives deal only with reform of the current economic regulatory environment. Those regulations that establish the basis for safety requirements of carrier and operators are not being changed. In fact, separate initiatives are under way to improve safety and security. In the motor carrier area, the MVTA 1987 requires the federal government to provide Parliament with an annual report on the implementation and enforcement of safety regulations.

THE PROVISIONS OF SAFETY IN A
NEW COMPETITIVE ENVIRONMENT

Despite the concern expressed in some quarters that safety may diminish in the wake of increased competition, the government is confident that through effective program implementation, safety will not be compromised if effective safety standards are properly developed and enforced. Focusing on the air and road modes, as demonstrated below, the "safety net" of standards and regulations controlling these forms of transport have not been weakened. In fact, in some cases safety requirements have been strengthened.

Aviation Safety

In 1984, the Federal Aviation Administration (FAA) in the United States conducted a comprehensive study on the effects of deregulation. In its subsequent report, the FAA concluded that the vast majority of all carriers, including new entrants and established companies, were in compliance with applicable FAA requirements. However, where there was a compliance problem, one or more of the following characteristics were usually present: rapid expansion into areas of different operational environment; a relatively large amount of contract maintenance or training; inadequate internal auditing procedures; and management skills and philosophy incompatible with sound practices.

Additionally, the rapidly growing and changing air carrier industry has dramatically increased the demands on FAA inspector resources. The FAA concluded that there was a need to perform a comprehensive analysis of the overall air carrier safety regulatory structure, with more complete and timely information on air carrier operations and on inspection and surveillance management, and to improve communication between headquarters and field offices with respect to the interpretation and application of air safety policy and regulations.

On the basis of what has been observed in the United States, Transport Canada undertook a number of major aviation safety initiatives. Probably the most significant step was the modernization of the Aeronautics Act in 1985. The amended legislation strengthened the areas of airport security, government access to pilots' and engineers' records, owner and operator insurance requirements, standards for airworthiness, air traffic services and navigation aids, and medical reports for those holding licenses. More important, departmental officials were now provided with more power to enforce new, more efficient aviation safety regulations. Via the development of the new Canadian Aeronautics Code, these regulations were promulgated by Parliament in 1986. To the extent that safety could have been detrimentally affected, the revising or modernizing of this legislation was deemed imperative by the department in the wake of the recent changing aviation environment.

As a means of properly administering the new Aeronautics Act, Transport Canada is currently assessing how best to employ its certification, surveillance, monitoring, and enforcement resources. For instance, a Regional Master Surveillance Plan has been created to detect, investigate, and follow up on identified air carrier safety deficiencies. Utilizing inspectors, air traffic controllers, flight service station specialists, and so on, the program has successfully raised the compliance awareness of the industry. Additionally, a major review has been initiated of the present and forecast overload and assignment of personnel through the creation of an Aviation Human Resource Plan. This Plan is just about complete and will determine those areas where there is a need for further augmentation of aviation enforcement personnel.

Other recent safety initiatives that have been undertaken include the creation in 1984 of an independent Canadian Aviation Safety Board, to conduct investigations and make recommendations on aviation occurrences, and the establishment in 1985 of a Company Aviation Safety Management program (CASM) to introduce management and air carrier personnel to safety management systems. The CASM program spawned the creation in 1986 of the Confidential Aviation Safety Survey in which, at the request of interested air carriers, departmental officials will conduct a confidential survey of the carriers' safety operations toward rectifying deficiencies.

Trucking Safety

There is concern in certain quarters of Canada that economic regulatory reform will lead to a degradation of trucking safety, especially with respect to small trucking firms (i.e., one-truck, one-driver operation) competing with larger, more financially secure carriers. The fear is that safety will be compromised as companies reduce maintenance schedules, lengthen driver working times, and extend the life cycle of pieces of equipment in order to remain competitive or financially viable. As well, many are not convinced that governments (federal and provincial) have made available adequate personnel (e.g., police officers, safety inspectors, and auditors) to enforce safety regulations properly. To a large extent, the basis of this contention is the perception that the U. S. trucking industry has had a poorer safety record since the passage of the Motor Carrier Act of 1980.

Toward ensuring that potential problems would not, in fact, materialize, a major federal-provincial initiative was undertaken to develop a National Safety Code for Trucks and Buses. This Code is a consolidation and upgrading of the best safety standards in North America, some of which are already in place in Canada. Under the Code, increased uniformity and compatibility will be introduced into the existing regulations.

Although there is concern, primarily from certain segments of the trucking industry, that the National Safety Code will not be fully introduced concurrently with the MVTA 1987, four major programs are already in place:

1. A driver-license system consisting of different classes of licenses, qualification for which requires passing of medical, knowledge, and skill tests. No province allows a driver to hold more than one driver's license. Truck drivers' licenses are downgraded following identification of medical problems. There are demerit points and license suspension programs as well as enforcement of alcohol- and drug-impaired driving laws and rules of the road.
2. A program of vehicle inspections carried out on company premises and on the roadside. In addition to maintenance checks, load security compliance is enforced by all but one province and weight and dimension limits are enforced at all weighing scales.
3. All jurisdictions have a mandatory insurance requirement.
4. Uniform laws and standards are in effect across the country covering vehicles, driver training, and placarding for the movement of dangerous goods.

Already a greater uniformity between provinces has been achieved. The Canadian Council of Motor Transport Administrators (COMTA) single-license concept, forbidding drivers from holding licenses in more than one province, is now in place. It will enhance enforcement and control between provinces. Additionally, there is now a standardization between provinces of both the license classification systems and medical standards for drivers. COMTA model regulations for security of loads require minor modification of existing standards and participation by all jurisdictions. Additionally, Canada has joined the U. S. Commercial Vehicle Safety Alliance in order to achieve recognition and uniformity with respect to conducting roadside inspections. All of these changes supplement the carrier fitness, entry, and control standards contained in the MVTA 1987.

Remembering that the full transition to regulatory reform in the trucking industry is spread over a five-year period, the following measures are to be introduced by the end of the transition period. Indeed most of the measures are planned to be implemented by the end of 1989. Vehicle maintenance standards will be uniform in all provinces. Operator profiles will be introduced to which safety records will be appended for every company holding a license. Hours-of-service regulations will be in place in all provinces (legislation for this is required in some provinces) with new enforcement procedures for the audit of employers' records. Knowledge and performance tests for drivers will be uniform and standards will be raised. Driver-examiner training programs will have uniform standards and will be expanded to provide ongoing training, and self-certification standards and procedures for training drivers will be uniform in all jurisdictions. Driver profiles will be improved and the system will be made uniform in all provinces, and there will be criteria for short-term suspension of licenses of drink or drug impaired drivers.

Additionally, the new enforcement of an interprovincial record exchange will give each province access to the profile of all trucking companies and drivers. Clearly, the government recognizes that new safety standards, which are needed regardless of economic regulatory reform, will have little impact on enhancing safety unless provincial inspectors, police, and maintenance practices monitoring staff are adequately increased. However, identification of the programs to support these standards is a complex task requiring further intergovernmental agreement.

Initiatives such as the development of the National Safety Code for Trucks and Buses will go a long way toward enhancing safety. There are other reasons why Transport Canada believes that the perceived safety problems in the deregulated U. S. trucking industry will not be witnessed north of the border.

The U. S. Motor Carrier Act of 1980 almost instantaneously deregulated American interstate trucking, thereby resulting in a period of industry stability adjustment. The regulatory reforms contained in the MVTA 1987 involve a two-stage process that will effect a more gradual and stable adjustment to the increased competitive nature of the Canadian trucking industry. The initial stage involves the substitution of the reverse-onus test for the public convenience and necessity test. The subsequent stage involves total freedom of entry subject to a fitness test. Indeed, the transition to competition commenced prior to the MVTA 1987. In February 1985, all provinces, the territories, and the federal government agreed to an easing of entry standards in most segments of the Canadian trucking industry.

It also acknowledged that the protection provided to existing carriers by rate regulation prior to U. S. deregulation allowed marginal, inefficient operators to stay in business and efficient operators to make large profits. This resulted in the efficient carriers being able to reduce their rates and offer substantial discounts to large shippers following deregulation. By contrast, the absence of effective rate regulation in Canada and more relaxed entry regulation have already resulted in generally competitive rates with little room for excess profits. Industry sources indicate that in Canada industrywide, profit level is at just 3%.

With the aforementioned two-phase entry implementation approach, and the institution of the National Safety Code regulations prior to the fitness entry test being introduced, the Canadian government is confident its current and planned safety program initiatives will go a long way toward ensuring that the motor carrier industry continues to operate safely, even with the institution of economic regulatory reform.

CONCLUSION

Concern has been expressed in some quarters of society that safety, or the margin of safety, has diminished in the United States as a result of its economic deregulatory initiative, and that ultimately this situation will materialize after economic regulatory reform is legislated in Canada.

Our response to this is threefold. First, while there are indeed elements of the American transportation system that require more attention to safety, according to data currently available safety has not diminished significantly, if at all, in the United States. Second, the U.S. transportation market is significantly different in structure from that of Canada, and therefore not all of the alleged American deregulation safety-related problems can be said to be transferable to Canada. Third, primarily with respect to the slow and pragmatic manner in which economic regulatory reform is being introduced in Canada, the new policy not only is allowing the Canadian transportation industry sufficient time to adjust to the new environment, but by its very nature is uniquely tailored to Canadian geographic, economic, political, and social needs, not to those of the United States.

Despite the expressed concerns and the importance of transportation safety in general, the federal government believes the hypothesized causal link between increased competition and decreased safety is not convincing. Economic regulation does not guarantee carrier profitability, nor does carrier profitability guarantee safety. In point of fact, the department views relative safety among carriers as one of the main means of viable competition.

The Canadian government believes that safety, no matter what the prevailing regulatory environment might be, should be addressed through the development and enforcement of effective safety standards and education, not by increasing the degree of economic regulation.

21

Summary and Policy Implications

LEON N. MOSES and IAN SAVAGE

THE AIRLINE INDUSTRY

In the years when hearings were being held on the airline deregulation bill, supporters of the legislation argued strongly that competition would bring fares down. Many of them accepted the idea that there would be some reduction in quality of service, in part because an essential aspect of their argument for price competition was that airlines had engaged in excessive quality competition because it was the only form of competition open to them. Such competition, it was asserted, drove up costs of operation and fares without having lasting, favorable effects on industry profits. Supporters of the deregulation bill did not accept the idea that economic deregulation would lead to a reduction in safety.

The experts who gave opposing testimony on the bill emphasized quality of service, especially safety, and the impact that deregulation would have on the amount, as well as the quality, of service to small communities. The decline in the nonsafety aspects of service quality has been extensively documented. Deregulation has brought significant reductions in the comfort of travel. The number and duration of delays and the number of incidents of lost baggage have increased. Overbooking has increased, as has the density of passengers per flight. It would be hard to find experienced travelers who do not hold that the number and quality of the meals served has declined significantly.

The unfavorable effects on safety in the main corridors of travel were seen by critics of deregulation as developing from the financial pressure that increased price competition would put on firms to cut costs by skimping on safety investment, crew training, maintenance, replacement of aircraft, and so on. Small communities were seen as likely to suffer additional degradation of safety because they would be served by new, small companies that would fly small aircraft, staffed by less experienced pilots. Moreover, the operations of these small firms would frequently involve places that were far removed from Federal Aviation Administration (FAA) oversight. Questions were also raised about the new companies that would enter the industry with deregulation. It was argued that their managements would be inexperienced and they would have no record by which consumers could judge the safety of their operations.

Something not foreseen is that some travelers whose trips do not originate in and/or terminate in major travel markets need more time to complete their trips than they did in the days of economic regulation. This is the case because of the hub travel system that has emerged as a result of deregulation and the search by airlines for operational systems that reduce costs. Hub operations offer cost efficiencies because they use small aircraft to gather passengers from light travel markets and deliver them to a major airport, where they are then placed on large aircraft that are efficient for larger passenger loads and long-distance travel. In the days before deregulation, large aircraft were frequently used to pick up or deliver a small number of passengers in secondary travel markets. The hub system achieves cost economies for the airlines but increases travel time and travel uncertainty for passengers whose trips do not originate and terminate at the major hub airports.

The experience of almost a decade of economic deregulation of airlines has settled many of the issues that were debated in the hearings. Competition has occurred and continues, despite the recent merger movement. Real fares have on average fallen, and are a great deal lower than they would have been had the old regulatory regime remained in force. There is even evidence that real fares fell in the years from 1979 to 1985 in small (nonhub) markets, those being the markets that are commonly served by a single airline (Scocozza, 1987).

Fare structures exhibit a great deal more price discrimination than they did in the past, which is a source of considerable annoyance to some travelers. However, the adjusting of rates to take account of the differing elasticities of demand of different travelers is entirely rational in industries characterized by the kind of competitive situation that exists in air transportation. Moreover, even the coach fares paid by business travelers, the group normally subject to price discrimination, are lower in real terms than they would have been under the old regulatory regime.

It seems unlikely that decreased quality, as measured by the nonsafety quality variables, can figure as importantly as fares in decisions by individuals as to how much they will travel by air. If it did, passenger travel would not have increased as significantly as it has in the years following deregulation.

Would the increase in travel have taken place if deregulation had reduced travel safety, along with the other attributes of service quality? That is entirely possible, because while consumers value safety, they frequently make choices that involve more risk than other choices they could make in order to achieve cost economies. Such savings allow them to increase consumption of other things. For example, people travel by auto, and do so at speeds that exceed legal limits, when they could travel much more safely by air, bus, or train. However, one does not have to make the argument that travel might have increased even more if safety had not declined, because the rate of accidents, fatal accidents, and fatalities fell during the period of economic deregulation.

Comparison of the period 1979 through 1987 with 1970 through 1978 shows that for large jet carriers, accidents fell by 36%, fatal accidents by 40%, and fatalities by 32% (National Transportation Safety Board, various years). If the accident figures were expressed in relation to the number of flights, the percentage declines would be much greater. The improvements in safety have been shown to be statistically significant (see Chapter 10 in this volume).

Overall, people traveled more safely, though there are pockets of travel in which the trend of safety went the other way. Passengers whose trips originate or terminate in secondary travel markets do travel somewhat less safely than they would have in the regulated era. The accident rate for all commuter airlines is about three times as great as that of the large companies operating jet aircraft. The accident rate of even the top 20 commuter airlines is twice that of the large operators. The poorer safety record of the commuters is in substantial measure due to the fact that the crews they employ have considerably less experience than those of the jet operators, to the lower altitude at which they fly, to the nature of the airports into which they fly, the size of the aircraft they employ, and to economies of scale in maintenance of aircraft.

It should be borne in mind that commuter airlines account for less than 10% of all airline trips, and that most of that travel is handled by the largest 20 operators, whose safety record is somewhat closer to that of the jet operators than the remainder of the commuter airline industry. Another factor to be taken into account is that the substitution of commuter for jet service has reduced the number of intermediate takeoffs and landings that travelers in secondary markets experience between origin and final destination, and it is in the takeoffs and landings that most accidents occur. Research based on a survey of 60 secondary city pairs in which commuter airlines had replaced jet carriers indicated that the average number of intermediate stops had dropped by half. Finally on this point, the availability of low-cost travel may also have increased the safety of travel to and from secondary markets for those changing modes by reducing the amount of auto travel on rural highways. Such travel is held to be considerably less safe than travel by commuter airlines (Chapter 10).

The improvement in the overall accident record is not due to deregulation. The declines are largely the result of a 40-year record of technical improvements in aircraft and air traffic control in the United States and in such overseas areas as the European Economic Community. Still, the record stands. *The forecasts of some deregulation critics that price competition would cause an absolute decline in safety have proved incorrect* (see Chapter 8). That is the case because, so far at least, the reasoning that underlies the safety degradation arguments has not in the main been borne out.

Critics of deregulation argued that with freedom of entry there would be a flood of new jet airlines that would have inexperienced managements and less experienced and lower quality flying and support personnel. It was hypothesized that the new entrants would exhibit higher accident rates than the established firms. However, researchers have been unable to establish that new entrants have a statistically significant higher accident rate (see Chapter 9). One difficulty in such research involves criteria for deciding which firms (for example, the "new" Braniff Airways) should be treated as new entrants rather than as spinoffs or expansions of previously operating airlines.

The critics of deregulation held that price competition would reduce profit margins and that firms under financial pressure would skimp on various aspects of safety investment, such as maintenance and training of crews. It was also believed that they would be unable to replace old aircraft with modern, safer aircraft. Statistical investigations offer some support for the financial pressure argument, but it is very slim. A decrease in profitability, due to increased costs or possible losses in patronage, leads to a small, statistically significant (at the 5% level) increase in the accident rates of firms (Chapter 8).

To argue that deregulation causes firms to reduce the care with which they carry out their operations is to ignore the fact that in the days of regulation there were firms that lost money. The Civil Aeronautics Board (CAB) readily granted rate increases to compensate for cost increases and low profitability, but the adjustments were based on average performance for the industry. Except in the early years of the industry when subsidies were quite common, the CAB did not guarantee profitability for individual firms.

As indicated, research indicates a small effect of profitability on the safety record of firms. That it is small should not be surprising, because firms have incentives to avoid accidents. Airlines that experience serious accidents can lose business to their competitors and to other modes of travel for some period of time because of the reputation effect of accidents.

How serious the loss is depends on the history of the firm, the availability of competitors, and the accident records of competitors. Firms with bad accident records probably find it difficult to hold onto their most valuable employees and also to hire high quality workers. Material presented in this book also shows that after an accident the value of the involved airline's stock market shares declines. The average stock value decline after an accident was estimated at $4.5 million, whereas the demand loss—which typically lasts under three months—is equivalent to about 1.5% of a year's revenue. In dollar terms this is about $30 million for an airline the size of Continental Air Lines (Chapter 5). Thus, even those firms that are financially pressed by competition have incentives to avoid accidents. Nevertheless, there may be situations in which firms that are on the verge of financial failure take risks that endanger passengers, in order to stay in business.

Both proponents and critics of deregulation are interested in what the long-run safety situation will be. The difference between them is in the indicators that they believe should be used to make forecasts about the long run. Proponents use the actual record. Findings such as those on new

entrants and financial pressure just discussed support a forecast that deregulation will not reverse or slow the downward trend in accidents that the
nation has been experiencing for 40 years as a result of technical improvements. Effectively they argue that the future will be like the past.

This forecast is rejected by a large number of thoughtful people, not all
of whom have a vested interest in a return to economic regulation. The
difference of opinion hinges on what has been called the "safety margin"
or "safety buffer," terms used by Nance in Chapter 13. The basic idea
behind the safety buffer argument is that safety is *partly* a type of stock
variable in which airlines and the public invested more freely in the past
than they have since deregulation. Two types of elements enter into the
stock of safety: those that are largely in the hands of individual airlines,
and those that are primarily the responsibility of government. The former
include such things as the hiring of highly qualified personnel, extensive
programs of lifetime training of personnel, careful maintenance of aircraft,
and timely replacement of aircraft. As has been shown in numerous chapters in this volume, airline deregulation increased competition in price relative to competition in quality. One aspect of the reduced emphasis on
quality has been a significant increase in the number of years aircraft are
kept in use. Clearly, newness of a firm's fleet was a form of quality competition. The FAA's position on age of aircraft and safety has not changed.
It still holds that with proper investment in maintenance, aircraft can operate safely despite their age. However, recent accidents involving aged
aircraft have caused the agency to greatly increase its requirements of what
has to be done and how much money has to be invested to maintain safety
for aircraft as they age.

It is claimed that in the past airlines did more than they were required
to do by the Federal Aviation Administration, and that a stock of safety
was built up that provided substantial protection against increases in accident rates. The line of reasoning continues that at the present time, many
more firms tend to satisfy only minimum standards. Moreover, as the recent history of penalties imposed by the FAA shows, a number of large
firms, firms that account for significant percentages of total travel, fail to
meet even minimum standards, though part of this is due to stepped-up
and tougher enforcement by the FAA. These violations can occur because
there are many more firms for the FAA to oversee, and the resources the
FAA has devoted to inspections have until recently been less than they
were in the past (Kern, 1988). The latter, it is claimed by some, was the
result of the FAA's greater interest in certificating new carriers than in
overseeing maintenance and training standards of existing carriers. This
has changed somewhat since 1984 because certification is now a lesser
priority than surveillance. The position of the adherents of the safety stock
hypothesis is still that the stock of safety built up in the days of regulation
is being worn away.

The second type of element that enters into the safety stock argument
has to do with the amount of infrastructure—including airport capacity,

air traffic control (ATC), and collision avoidance systems—that government provides in support of air transportation. Enplanements have increased 52% since deregulation, but there are no new major airports and
the capacity of the ATC system is less than it was because of the reduced
number of controllers following the dismissal of illegally striking controllers in 1981. The system is straining the limits of capacity, especially at
major hub airports.

   Those who believe that deregulation is wearing away the stock of safety
also point to the hub-and-spoke system as a source of difficulty. With this
system there is a great deal of congestion at peak hours of travel as passengers are brought to hub airports from tributary airports by commuter
airlines and by connecting fights of the hubbing airline. The airspace in
the vicinity of these hubs is terribly crowded during those hours. The commuter pilots who perform the connecting function at peak hours are less
experienced than crews of the major airlines, because their wages are low
and turnover rates are high, and the supply of experienced new pilots from
the military has decreased. General aviation aircraft add significantly to
the amount of congestion and to the hazards of travel in the vicinity of
certain hub airports.

   Support for these ideas about airspace congestion and safety comes from
pilots. A recent survey conducted by the Air Line Pilots Association (ALPA),
which represents 85% of all pilots who fly large jet aircraft, reveals that
the primary concern pilots have is that growing airport congestion has
increased the number of near midair collisions (O'Brien, 1988). Forty percent of the pilots who responded to the ALPA questionnaire indicated that
their greatest safety concern was congestion and the inadequacies of the
air traffic control system. They mentioned these factors over 10 times as
frequently as they mentioned weather factors, including windshear (Fingerhut, 1986).

   *Adherents of the safety stock-congestion theory hold that the record of
accidents is insufficient to the task of predicting where the system is likely
to go in the future in terms of safety.* They argue that economists, for
example, do not base their judgments on the future direction of the economy solely by what happens to the unemployment rate. Instead, a set of
leading indicators is used to provide insights into direction of change (Lauber,
1988). They hold that the number of near midair collisions is one type of
leading indicator, and that the number of such incidents has increased (see
Chapter 14).

   The strength of the safety stock-congestion effect, assuming it exists, has
not as yet been evaluated. Even if the frequency of near midair collisions
(NMACs) is taken as an indicator of the erosion of safety stock, the theory
cannot be tested at the present time. The system for reporting NMACs is
imperfect and the data probably subject to considerable error. In part this
is the case because the rules governing immunity for commercial pilots
who report incidents in which they may be at fault tends to discourage
reporting. Pilots are guaranteed immunity, under a system run by NASA,

only for the first "at fault" incident in any five-year period. In addition, air traffic controllers have no immunity. The reporting of incidents should be made mandatory, substantial penalties imposed for failing to report, and the rule extended to general aviation pilots.

While the safety stock-congestion hypothesis cannot be accepted at present, neither should it be rejected because the data required for testing it are unavailable. It should not be rejected because it also has some logic on its side. Deregulation has been immensely successful in holding down real fares and in encouraging significant increases in the number of passengers and the number of flights. These increases, together with the adoption of the hub-and-spoke system, have greatly increased congestion at major hub airports. At the same time, the number of Full Performance Level air traffic controllers is below what it was seven years ago when most controllers were fired because of the illegal strike action. There are still fewer controllers today than there were at the time of the strike, whereas the number of flights per day is some 28% above what it was then. Surely it is not surprising that some scholars of the airline industry view the following combination of factors as a recipe for potential disaster:

A significantly greater number of aircraft in the air at major hub airports, especially during times when there are many connecting arrivals and departures

A smaller number of air traffic controllers, some of whom are still required to work a considerable number of overtime hours at major centers

The relatively low level of experience of commuter airline pilots because low wages and increased demand for pilots at major airlines cause commuter airlines to have high pilot turnover rates

A large number of private aircraft with relatively inexperienced pilots that attempt to land at major hubs or otherwise occupy air space in their vicinity

An alleged lower level of maintenance of an aging stock of aircraft

Too few field inspections of aircraft and airline operations due to an inadequate number of FAA inspectors.

Concerning the last point, the Air Florida (jet carrier) and the Bar Harbor Airlines, Henson Airlines, Simmons Airlines, and Air Illinois (commuter carriers) accidents might have been prevented had there been adequate surveillance of pilot training programs (Lauber, 1988).

If the adherents of the safety stock-congestion hypothesis are correct, and the number of accidents per commercial aviation flight does increase in the near future in a statistically significant way, Congress as well as the U. S. Department of Transportation (DOT) will have to bear a heavy measure of blame because they adopted a set of contradictory policies. That is, they adopted a program of economic deregulation because it promised to bring benefits to travelers. The program was a success: Fares were held down and the amount of travel increased substantially. On the other hand, government, which has much of the responsibility for the character, quality, and capacity of the system's staffing and infrastructure, failed to increase the capacity of the system, to adopt programs that would use the existing capacity more efficiently, and to take other needed actions to ensure safety. Indeed, when we take the number of air traffic controllers into account, the effective maximum capacity of the system is probably smaller than it was in the past.

The claim is also made that the Aviation Trust Fund, money collected from the taxes government imposes on airline travel, has over $5 billion of unexpended funds in it. DOT has been accused of failing to request the money to finance improvements and expansions in facilities and equipment because it has wished to cooperate with other branches of the Administration in an effort to make federal budget deficits appear smaller than they really are.

The issue of the Aviation Trust Fund is one about which there has been considerable debate, and about which most members of the airline industry feel very strongly. The Air Transport Association goes so far as to propose that the FAA's functions be turned over to a new quasi-independent group like the Federal Reserve System, and that such an agency be completely and freely funded by money collected in ticket taxes (Bolger, 1988).

The DOT has recently responded to claims that it has failed to spend the money in the Trust Fund by asserting that aviation funding has increased by $3 billion since 1982, but that a complicated provision in the authorizing legislation has forced much of the cost of the program to be financed from general tax revenues rather than the Trust Fund. DOT also asserts that Congress failed to appropriate over $1 billion that it requested for modernizing and increasing the capacity of the national airspace system. The basic difficulty between Congress and DOT is DOT's belief that 85% of FAA's current operating expenses should be paid out of the Trust Fund, with the remaining 15% paid by the military. However, in legislation passed by Congress, DOT was allowed to fund only 70% of FAA's current operating expenses from the Trust Fund. Moreover, Congress imposed a penalty, the complication referred to above—that the proportion of current expenses coming from the Trust Fund would be further reduced if DOT failed to expend authorized amounts from the Fund on research and development and improvements in the air traffic control system. This is, in fact, what has happened and it explains why there is unexpended money in the Trust Fund.

From the point of view of airline passengers and the public at large, it makes little difference who is right and who is wrong in the conflict between Congress and DOT. The simple fact is that *the federal government has not mounted a major campaign to significantly relieve congestion at major hub airports and in the airspace in their vicinity.* It has failed to bring the amount and quality of airport and ATC system capacity up to the levels required by the growth in numbers of commercial and general aviation flights. It has also failed to press for practices, such as congestion pricing, and rational landing fees, both of which are discussed at length below. By failing to do these things, the government has added to the probability of accidents, and made the safety stock-congestion argument a good deal more compelling than it would otherwise be.

THE AIRLINE INDUSTRY—POLICY RECOMMENDATIONS

Private Aircraft in Congested Hub Airports and Surrounding Areas

A significant factor in airport congestion, in near midair collisions, and in actual accidents is the number of private aircraft that land at hub airports or occupy airspace in their vicinity during peak travel hours. One of the reasons why such aircraft add significantly to the risk of accidents is that the pilots of many of the private aircraft are not sufficiently well trained. Another difficulty is that many small aircraft lack "mode C" transponders, devices that permit height as well as latitude and longitude to be recorded on radar. Without such information, controllers cannot determine whether a dangerously close situation is developing. Arguably, aircraft that are not equipped with such transponders should have been banned from highly congested hub airports several years ago. The FAA has recently issued such an order, but it is not clear at this time whether the rule is sufficiently extensive. We, as well as many others, hold that *mode C transponders should be required on any aircraft that operates within 50 miles of a major hub airport.*

Landing Fees, Congestion Fees, and the Efficient Utilization of Airport and Air Traffic Control Capacity

Airspace in the vicinity of congested hub airports and the physical facilities of such airports are scarce economic resources. These resources should be used efficiently so that the amount of public money that is needed to expand capacity is kept to a minimum. These resources are not currently being used efficiently, as evidenced by very significant peak hour overcrowding that exists. Airlines claim they set their schedules in response to the temporal pattern of consumer demand for travel, and that individually they can do little to change the pattern of demand. An individual airline that charged more for prime-time than for off-peak travel, and in doing so charged more than its competitors, would lose a great deal of business. It has been claimed that some headway on the congestion problem has been made as a result of recent voluntary scheduling agreements (see Chapter 11). However, such an approach should be avoided because it tends to lead to collusion on other matters.

The problem of congestion is exacerbated by the practice of airlines publishing arrival times that do not reflect the delays experienced. The point-to-point time of flight from New York to Chicago may be listed as the same whether the flight is at 8 a.m. or 11 a.m. As a result, the inexperienced traveler may choose to travel at a peak time, whereas another time might have been selected if the expected delay had been known at the time the flight was booked. Bailey and Kirstein in Chapter 11 in this book recommend that information on delays be made available to consumers as a way of encouraging airlines to publish more realistic schedules. Information on delays is now available to ticket purchasers. However, such information is not a panacea, as a simple look at our urban highways shows. Many are extremely congested, filled with commuters of long experience who know the average peak hour delays to be expected.

Two types of solutions to the peak load problem and the inefficiencies and safety hazards it entails are possible. The first is a nonmarket solution, one that has been suggested by some members of Congress. This solution would have the FAA determine the maximum number of landings and departures that could safely take place at each of the major hub airports in some time interval, say 30 minutes. Then firms that serve a given airport would be granted antitrust immunity and allowed to make agreements as to the allocation of peak-hour landing rights.

This solution to the congestion problem is unacceptable. In effect it resurrects the functions of the CAB by putting bureaucrats back into the position of determining how much competition and entry there should be, an arrangement it was the intent of the Airline Deregulation Act to end. Also, the solution would create monopoly conditions by in effect allowing existing firms to form legal cartels. New firms could be precluded from prime-time markets and existing firms could be encouraged to collude on prime-time landing rights, and perhaps on fares as well.

What is needed is a pricing system that allocates airport capacity and the airspace in the vicinity of congested hub airports in an efficient way. The economic model that underlies such a system of prices has been known for many years. It was first recommended as a cure for congested bridge travel in the early 1800s, and has been broadly and successfully applied around the world to such things as the pricing of telephone service and electric power. It has also been used with modern-day success in the pricing of facilities such as bridges and tunnels. In the case of airport landings, the essential idea behind such pricing is that the delays associated with increases in the number of scheduled landings per period would have a money value attached to them that would reflect the money cost of associated landing delays to airlines and to passengers. These costs would then be incorporated into a set of landing fees that would differ by time of day, day of week, season, and so on (see Chapter 12).

Under a congestion pricing system, a substantial part of the high fees that airlines would pay for landings at peak times would be passed on to passengers in the form of differentially high ticket prices. Such prices would cause some passengers to be willing to travel at less congested times. They would thereby increase the profitability to airlines of providing off-peak service. The peak would be spread out, less capacity would be required at those times, and existing capacity would be used more efficiently. The fees collected would also provide funds that would be under control of local authorities. These funds could be used to make investments in facilities and equipment that would increase airport capacity. Congestion fees would, of course, be imposed on private as well as commercial aircraft that land at peak times.

The issue of landing fees for private aircraft entails another aspect of rationality in the pricing of facilities that should be addressed. At the present time, landing fees are based on the weight of aircraft. The fees are much too low to begin with, and their being based on weight means that most private aircraft pay almost nothing to land, even at the most congested times. *Landing fees should reflect the amount of airport capacity that aircraft use in a landing.* Such usage is practically the same for all aircraft. The amount of time that a runway is tied up, and the amount of controller time required to assist a small aircraft to land, are certainly not less than the time required to land a large, commercial plane. In fact, since private aircraft frequently have pilots who are less experienced than those in commercial aviation and travel at slower speeds, they probably require more airport capacity than commercial craft. Some very small differential between private and commercial aircraft based on weight might be justified, because heavier aircraft do more damage to runways and other concrete surfaces than small aircraft, though it is unlikely that either type of aircraft accounts for as much damage to surfaces as weather and the passage of time.

A set of rational prices, one that takes congestion and utilization of airport capacity into account, would do more than increase the efficiency with which airport capacity is utilized and reduce the amount of investment in additional airport capacity needed to handle traffic safely. It would reduce the number of near midair collisions and runway incursions and the probability of accidents.

Up to now this policy recommendation has dealt with congestion at major hub airports. It has ignored the fact that there is also congestion in the air corridors in the vicinity of hubs that has nothing to do with hub takeoffs and landings. This second type of congestion, which is also serious and an important source of near midair collisions, is caused by the presence of private aircraft that are in transit in the vicinity of hub airports or that are taking off or landing at private airports located nearby. The application of a significant airspace user fee, one that might be administered through the private airports involved in the flights, would cause many private pilots to schedule their landings at less congested times and to detour around the congested areas. The safety and efficiency with which airspace is utilized by commercial and general aviation would thereby be increased.

Pricing techniques can increase the efficiency with which existing airport capacity is utilized. However, even with such pricing it is likely that the amount of capacity will have to be increased. Demand has risen rapidly in the last decade and continues to grow. Aside from Denver, no new major airport is in the advanced stages of planning anywhere in the nation. In part this is due to severe environmental constraints on the expansion of airport capacity.

Reporting of Near Midair Collisions

It seems indisputable that the number and trend of near midair collisions should enter into a judgment of how safe conditions are near a major airport, and how safe they may be in the future. That being the case, authorities need reliable information on the number of such incidents. *Reporting of near midair collisions should be made mandatory for general aviation as well as commercial pilots, and fines should be imposed for failure to report.*

The Ticket Tax Fund and Government Investment in Safety

Economists view safety as a desirable attribute of service. They recognize that its production requires the use of scarce economic resources, and that at any given time it may not be socially optimal to increase the level of safety (see Chapter 4). However, airline passengers have what amounts to a contract with government that it spend the money in the Aviation Trust Fund in ways that offer the greatest promise of increasing the safety and operating efficiency of airline operations.

In all fairness it must be mentioned that the FAA is now attempting to improve safety by hiring additional inspectors and increasing the number of field inspections. It has begun to identify carriers likely to violate the rules governing deferred maintenance, and has imposed substantial fines for violations of air safety regulations (Kern, 1988).

DOT is also involved in programs that offer substantial promise of improving safety. One of these is the National Airspace System (NAS) Plan which, when implemented, will upgrade the air traffic control system in a major way. It is unfortunate that DOT did not begin the program sooner. As has been pointed out by the General Accounting Office, introduction of the program could have been begun almost a decade ago as opposed to the 1981 publication of the first version of the NAS Plan. It is also unfortunate that the current implementation program is behind schedule, with the year 2001 as the completion date, assuming something that is unlikely—no further delays (McLure, 1988). To the extent that the schedule for NAS Plan implementation and the delays are due to insufficient funds rather than technical difficulties, *the money in the Aviation Trust Fund should be used to hasten completion of the NAS plan.*

The FAA is currently also carrying out tests of a collision avoidance system that is capable of recommending needed "climb or dive" evasive action if the intruding aircraft is equipped with a mode C transponder. This is another example of a technology that might have been introduced years ago.

Another, still more advanced, collision avoidance program would be capable of choosing the best evasive action, as between left-right and up-down. The FAA indicates that this system is at least several years away from certification. Again, to the extent that delays in advancing this system toward certification are due to inadequate funding rather than technological unknowns, increased expenditure of money from the Aviation Trust Fund might well be justified. DOT has expressed the opinion that technical problems, rather than insufficient funding, are the source of the delays in certifying this particular system.

The Number of Air Traffic Controllers

There are fewer Full Performance Level air traffic controllers today than there were in 1981, and the number of flights is 28% greater than immediately prior to deregulation. The inadequate number of fully qualified controllers leads to a great deal of overtime work for such controllers at the most congested airports. Overtime work, strain, and tiredness lead to errors in judgment and to accidents.

The number of air traffic controllers has to be increased. If that cannot be accomplished quickly by training new personnel, it is worthwhile to consider rehiring carefully selected, fired controllers. Among those who were fired are undoubtedly some who would be willing to return with loss of seniority, who are in the appropriate age group, and who could be retrained in much less time than is required to train inexperienced people.

THE MOTOR CARRIER INDUSTRY

The motor carrier industry was brought under the control of the Interstate Commerce Commission (ICC) in 1935. Thereafter, entry into, but not exit from, the industry was severely limited. Rate competition between individual carriers was severely limited. ICC-certificated general freight carriers could legally discuss and make agreements on rates, which were then presented to the ICC for approval. The main objective of the 1935 Motor Carrier Act was to achieve rate stability. It was a commonly held view in the 1930s that one way to cure the Great Depression was to control price cutting. The Motor Carrier Act and such legislation as the National Recovery Act mistakenly focused on price declines as a cause of the depression rather than recognizing them as a symptom.

The 1935 legislation achieved the objective of bringing about what was viewed as a more orderly industry. With rate competition between certificated carriers largely eliminated, the emphasis turned to competition in quality of service. Without a doubt, the certificated segment of the industry competed vigorously in service. The profitability and growth of individual firms depended on the skills of their management in containing costs and offering innovative and high-quality service. Despite the ease with which the ICC granted Tariff Bureau (the rate-setting truck operators cartel) requests for rate increases when costs increased, there were firms that failed to earn a reasonable return. In part this was the case because the rate increases the ICC granted were based on average performance.

Those firms whose costs were too high or who produced low-quality service had low returns. If their certificates granted them operating rights in valuable areas, their certificates were purchased by other carriers. The ICC readily approved mergers in the motor carrier industry. It was by such mergers and acquisitions that the companies with the greatest managerial skills grew to great size. Within such companies the emphasis on quality of service was even greater than in the industry as a whole. Cost containment, quality control, and marketing of service were the keys to success in the regulated era. The development of pricing strategies and competition in rates along with competition in service had to wait for passage of the Motor Carrier Regulatory Reform and Modernization Act of 1980.

The hearings that preceded passage of the 1980 Act were similar to those of the Airline Act in that some of the same kinds of forecasts were made about the effects of deregulation. It was claimed that freedom of individual carriers to quote rates, and the discounting that would take place with the loss of power by Tariff Bureaus would lead to chaos in rates; and reductions in profitabilty would reduce service quality. It was claimed that there had been a good deal of cross-subsidization of service under regulation. Competition would reduce profitability and eliminate such subsidization, with the result that there would be a loss of service to small communities and to firms with small shipments.

The rates now paid by small to medium-sized shippers that are located in small communities are probably higher on average than those paid by comparable shippers whose establishments are located in the main corridors of trade and transport. However, in the deregulated environment they have service and, in most instances, there is more competition for their business than there was in the past, so that service quality is better than in the past. Given the advantages that small communities have in land and labor costs, the transport situation of today is more favorable to their growth and development than it was under regulation when they had favorable tariffs but experienced great difficulty in securing service at those tariffs.

Those who opposed competition also offered opinions on safety that were similar to those that had been voiced in the airline hearings. It was claimed that severe price competition would greatly increase speed violations by truckers. Increased speeds would increase the rate and severity of highway accidents. It was also claimed that there would be more instances of drivers beginning trips with too little rest and staying behind the wheel for excessively long periods. It was believed that such behavior would also increase accident rates. Price competition and the cutting of profit margins would also force firms to reduce the amount spent on vehicle maintenance, and would force them to hold onto old vehicles much longer than they had in the past. These conditions were also supposed to lead to increases in accidents and reductions in quality of service.

No one can doubt that deregulation has benefited shippers by providing them with the lower rates that were promised by the legislation. In part, the lower rates are the result of what has happened to wages. Between 1970 and 1980, real wages increased by 0.5% per year. Between 1980 and 1985, real wages fell by 3% per year (U. S. Department of Labor, various years), at the same as the negative impacts on costs of restrictive work rules were somewhat mitigated.

Statistics that were presented in the introduction to this book indicated that real rates are now more favorable to shippers. This being the case, if there are societal disbenefits that are the result of economic deregulation, they must be in the areas of service quality and safety. Let us first take up the issue of quality of service, including accidents that involve loss and damage to freight but do not involve fatalities.

There is a near absence of complaints among shippers that the quality of service they receive has fallen since deregulation. One can read as widely as one wishes in the professional and business magazines that deal with logistics and physical distribution, traffic management, and materials handling, and fail to find evidence of the kind of consumer unrest that characterizes airline passenger travel. Price competition in the motor carrier industry has meant that shippers have a continuum of choices in rates and quality of service. If they wish they can have service quality as high as, or higher than, any they had in the regulated environment and pay a high price for it, or they can choose to pay less and have lower quality service.

In recent years many U. S. manufacturing firms have adopted a type of logistical strategy known as "just-in-time" production. It is characterized by an emphasis on the reduction of inventory carrying costs by, among other things, the use of very dependable trucking service that is closely integrated with a manufacturer's production schedule. Such close integration of transportation and production cannot be achieved without trucking service that meets tight delivery schedules and has very low loss and damage rates.

It should come as no surprise to anyone who has studied transportation, that where economic and technical conditions favor the existence of many carriers a type of competitive situation emerges in which a wide range of service qualities and charges are offered, and shippers are pleased with what the industry has to offer. After all, there are important, long-standing examples of such unregulated situations in U. S. transportation. They include the movement of almost all of the goods that are carried on the nation's inland waterways; the movement of agricultural goods by motor carriers that, at the insistence of shippers of agricultural goods, were exempt from the limits on competition imposed by the 1935 Motor Carrier Act; and freight transport in the Commercial Zones of cities. The latter, which were originally regarded as the areas from which rail terminals located inside cities derived their freight, have been areas of free competition from the days when goods moved to rail terminals by horse and wagon.

At various times there have been efforts to bring each of these transport sectors under economic regulation. However, such moves originated with, and were in the main supported by, carriers from other modes of transport or other branches of the same industry that were in competition with a less regulated sector. Such efforts did not receive significant support from the shippers who were served by the free-market sectors. Lack of shipper support for expanded regulation would not have been the case if price competition tended to denigrate service quality.

Data on accidents that involve loss and damage to freight lend support to the position that motor carrier firms can compete vigorously in price and still offer high quality service. If correction is made for changes in the value of goods, *the adjusted index of accidents per truck-mile fell from 100 in 1978 to 69 in 1985—a 30% reduction*. The nature of the adjustment requires some comment.

For many years, carriers were supposed to report all accidents involving $2,000 or more in property damage. Over time, the value of goods shipped and the cost of the repairs increased because of increases in prices. When prices go up, one should expect the number of reported accidents involving the fixed $2,000 limit to increase, even if the total number of accidents and mileage remain constant. Recently DOT carried out an investigation that involved use of a gross national product-based deflator. It recomputed the published data on accidents using a $2,000 real value instead of a $2,000 nominal value. The 30% reduction in accidents per mile reported above is the result of that adjustment (Schweitzer, 1988).

Such a reduction in the accident rate must be interpreted cautiously, because accident data that do not involve fatalities are probably flawed (see Chapter 18). They are based on reports that the interstate carriers themselves submit. Carriers have an incentive to make operations appear as safe as possible to shippers. They probably underreport accidents, especially those that involve relatively minor dollar amounts of damage. In addition, intrastate carriers are not required to report their nonfatal accidents to federal authorities.

In our opinion, these flaws should not be taken to mean that the decline in the adjusted-value property damage accident rate reported above is fictitious. If we look at the rate of fatal accidents, the reporting of which is not subject to the flaws that contribute to the property damage accident statistics, we find that the rate has fallen. It is difficult to believe that truck-related fatal accident rates could fall but property damage rates fail to do so. While it is true that the rate of survival in highway accidents has increased—due in part to increased seatbelt use—the aggregate levels of fatal and nonfatal accidents track each other closely over time.

The number of automobile fatalities in which trucks are involved has caused a great deal of concern. The public and some members of the media believe that automobile users of highways are considerably less safe than they were in the regulated environment when, because of restricted entry, there were many fewer truck-miles than there are today. It should be added that there are also many more cars on the road today than there were in 1980, and that lower fuel prices have led to an increase in average miles traveled per auto per year.

In point of fact, *the index of auto fatalities in truck-related accidents per mile of automobile usage fell by 21% from 1978 to 1985,* on the basis of data from the Federal Highway Administration and the National Safety Council. Per mile traveled, automobile users of the highways are safer with regard to accidents with trucks than they were in the more regulated environment. If fatalities are expressed in relation to truck- rather than auto-miles, the decline in the fatality rate is less pronounced, falling from 0.064 fatal accidents per million truck-miles in 1978 to 0.054 in the period 1983 through 1985, a 15% decline (Schweitzer, 1988). We believe that a more accurate picture of the risk of truck-related automobile fatalities is conveyed by the statistic in which truck-related automobile fatalities are in the numerator, and auto, rather than truck, mileage is in the denominator. It is auto mileage that determines amount of auto occupant exposure to situations that produce accidents. The statistic in which truck mileage is in the denominator is more appropriate for trucking firms that wish to insure themselves against claims that arise from fatal accidents, or for companies that insure them.

There is also at least one way in which regulatory reform has contributed substantively to safety. In the current environment, certificated carriers are free to choose the highways by which they travel in picking up and delivering freight. They are no longer required to follow a set of rules, which today look insane, involving what were known as gateway cities, and which influenced routings. In response to this aspect of truck regulatory reform, carriers shifted mileage to the interstate highway system, which has significantly lower accident rates than any other part of the highway network. The number of deaths from truck-auto head-on collisions would probably be greater today if trucks had not reduced their use of undivided highways in rural areas. In this regard, the motor carrier industry differs from airlines. In airlines, as we noted earlier, the decline in accident rates in the years since passage of the Act is more the result of long-run improvements in the quality of safety inputs and in safety technology than economic reform.

It is generally agreed that *economic deregulation has not led to an increase in the fatality rate. Neither has it increased the rate of industrial injuries and illnesses of trucking industry employees* (see Chapter 7). It has not done so despite the fact that some of the links between economic deregulation and safety measures, such as vehicle maintenance and compliance with federal regulations on driver qualifications and hours of driving, were found to exist by the researchers who prepared chapters for this book (Chow, Chapter 15; Corsi and Fanara, Chapter 16). However, while statistically significant, some of the linkage factors were found to have quite small effects. They explain only a small part of the total number of accidents.

Corsi and Fanara in Chapter 16 examine the safety record and the record of safety violations of new entrants into the motor carrier industry, and found that they have higher accident rates than established carriers. There was evidence of a learning curve concerning safety. The higher-than-average accident rates of new entrants fell rapidly as years in business increased. Thus, the accident record of a sample of new firms showed that in 1985, firms that had been established in that year had an accident record of 0.246 accidents per million vehicle miles, while the firms that had been established in 1980–1981 had a 1985 accident rate of 0.167.

It is difficult to foresee a detectable, long-run negative impact on safety from findings on the safety record of new entrants. Deregulation more than doubled the number of firms in the industry, but that effect is in the past. In each year in the future the number of new firms will be smaller than the number that entered in the first years of deregulation. The impact on the national safety record of the higher-than-average accident rate that new entrants tend to have in the first few years of their operations is unlikely to be detectable, because they will comprise a very small percentage of the total number of firms in the industry.

Another of the links between economic regulation and safety is the increased economic pressure that price competition puts on firms and which, some believe, can cause them to operate less safely in a variety of ways. Evidence presented in this book suggests that carriers close to bankruptcy spent relatively less on safety related items (Chapter 15). However, we cannot be certain from this evidence that these firms have an inferior accident experience. More recent evidence (Bruning, 1987) does show a negative relationship between profitability and accident rates. A 10% improvement in a firm's return on investment leads to a 3% decline in its accident rates. While this result is statistically significant, the work does suffer from a lack of theoretical structure to the estimated equation. This can lead to problems of bias in the calculated coefficients. Other researchers have reported on the results of a study in which drivers were asked to compare current safety conditions with those prior to 1980. Drivers did tend to agree that economic pressures led them to adopt less safe practices (Capelle and Beilock, 1988; see also Baker, 1985). It is difficult to assess the accuracy of the findings, because drivers were being asked to recall experiences over a six-year period. The results of such a survey have to be highly impressionistic and probably biased—recent events tend to be recalled more vividly than past events. Further, deregulation has had a negative impact on drivers' wages, and they should not be expected to be too pleased with it. Finally, on this point, an earlier 1970 survey reported that economic pressures had similar impacts on safety practices even then (Fellmuth, 1970).

It has also been suggested that the quality of drivers the industry has recently been attracting has been falling (Papai, 1988). In part the decline is due to the fact that in the new competitive environment, wage rates have failed to keep pace with their growth elsewhere in the economy. This situation may help explain the higher-than-average accident rates of new entrants into the industry. New, small firms tend to hire nonunion drivers. They pay lower-than-average wages and end up with drivers of lower-than-average quality. However, a second and equally important effect on driver quality comes from the fact that with deregulation there was a tremendous short-run increase in demand for drivers, and a short-run supply function that may have actually declined. However, the problem is one of the short rather than the long run. Over time, the low-quality drivers that have been hired will tend to be fired. In addition, wages will be bid up to the point where the relatively small number of new drivers that are needed each year will be of higher quality.

Regulatory reform and increased price competition in the motor carrier industry were seen by one researcher as a source of increased diversion of freight from rail to truck. Since trucking has a higher fatality rate than rail transport, such a shift was seen as a possible source of future declines in overall transport safety (see Chapter 17). Still, the record stands at this date. The fatality rate in the system as a whole has fallen during the years since passage of the Motor Carrier Act.

The essential conclusion regarding the motor carrier industry is that no objective evidence has been found to support a position that economic deregulation has caused a degradation of highway safety or of the quality of freight delivery services. Since it also seems clear that real transport rates have fallen, there is no basis for a return to economic regulation. *The trucking industry feels strongly that safety difficulties that are identified should be addressed by safety measures, not economic regulation.*

This conclusion is supported by the findings of leading groups that study accidents in which trucks are involved. Thus, in a recent publication (1987), the University of Michigan Transportation Research Institute (UMTRI) reported that accidents are strongly conditioned by the nature of the road on which travel takes place, and on driving conditions. The study points out that rural noninterstate roads account for 54% of all truck fatalities but only 37% of the travel. Fatalities on these roads are in the main the result of head-on collisions. Authors of the UMTRI report stated that the incidence of such collisions at dawn comprise a severe problem. As expected, on divided highways, rear-end collisions are more frequent than head-on collisions. They also tend to be concentrated in the evening hours and at dawn, when driver perceptions are poorest. These are not matters related to economic regulation.

The UMTRI report also raised questions about one of the supposed links between economic regulation and safety, namely, that increased price competition and financial pressure on firms might lead them to establish schedules that would force drivers to violate the rest and hours-of-driving regulations. Such violations would lead to increased driver fatigue and accidents.

The UMTRI report noted that more than 50% of all accidents that occur at dawn involve head-on collisions. However, it was found that in the dawn hours, drivers who had been driving for longer than 3½ hours comprised a smaller percentage than is the case for any other time period. The UMTRI report concluded that there is no simple link between fatigue and accidents.

THE MOTOR CARRIER INDUSTRY—POLICY RECOMMENDATIONS

Fatality Rates, High-Risk Zones, and Highway Improvements

State police organizations and local police are generally aware of the stretches of road in their areas that have a high incidence of serious accidents, both death and physical injury. *Money should be provided by a program that might be funded by states and the federal government to identify the nation's most hazardous zones and the time periods in which most of the accidents in those zones occur.* Amounts of travel, as well as numbers of accidents, would be taken into account in identifying the nation's riskiest stretches of highway. Program funds should be used to investigate these zones, and to make recommendations as to measures that would be most efficient and economical in reducing their accident rates.

In some cases the most effective measure in a cost-benefit sense might be increased police surveillance during certain travel periods. In other cases the recommendation might be to improve the physical characteristics of some stretch of highway. The cost of such improvements might have to be funded by states as well as the federal government. In still other cases the appropriate measure might be very high peak-hour tolls that would significantly reduce congestion in a hazardous zone, such as the Washington, D.C. Beltway, at certain times of the day.

A Central Computer Accident File Open to Shippers and Other Recognized Groups Such as the American Automobile Association

A program of providing consumers of airline service with information about delays in travel was recently introduced as a way of encouraging airlines to meet their announced schedules or to alter them in realistic ways. Information made available to consumers makes firms respond with quality improvements. The basic idea involved in that program should be carried over to the highway area.

A central computer file should be created in which data are stored on the fatal and property damage accident rates, and possibly the record of safety violations, of motor carriers. This file should be open to shippers, shipper groups, and other highway-user organizations. The availability of such information would significantly increase the marketplace incentives that trucking firms have to oversee and manage the safety aspects of their operations more carefully. The Commercial Vehicle Safety Alliance is already starting to computerize firm level accident and violation data to assist enforcement personnel.

Truck-related fatal accidents attract almost no media attention outside the local area in which they occur, because individual accidents involve very few deaths or injuries. As a result, there is little nationwide public awareness of truck accidents. The kind of file that is envisaged in the present proposal would increase such awareness by a group that is very important to carriers—shippers of freight. Before selecting a carrier, many shippers would use the central file to investigate its accident and safety violation record, perhaps because they might avoid carriers that could pose legal problems for them. Awareness of carriers' accident and safety records by shippers would increase the dollar incentives that trucking firms have to oversee and manage the safety of their operations more effectively. In evaluating this proposal, the costs of such a program should be taken into account.

Improved Data on Property Damage Accidents

The kind of program just described, and the amount of marketplace incentive for increased safety of trucking operations that it provides, depends on the extent and accuracy of the data included in the computer file. At the present time, only interstate carriers are required to file reports on accidents that involve property damage, and they probably underreport accidents. The reporting requirement for damage-only accidents was held constant at $2,000 in damages for many years. Ultimately with inflation, this tended to trivialize the activity and contribute to underreporting.

The minimum limit on dollar value of damages for accident reporting purposes has been raised to $4,400. It should periodically be increased as the price level increases. In addition, *the federal government should provide funds and other incentives for the states to gather comparable property damage accident statistics for intrastate carriers that would also be included in the federal data file.*

Speed Limit Monitoring

Everyone who studies accidents knows that the rate of serious injuries and deaths increases as the speed of vehicles increases. The American Trucking Associations opposed the recent increase in speeds on rural interstate highways because it believed that the result would be an increase in fatalities. Preliminary investigations have revealed that the accident rate has already increased measurably. *The effects of the increased speed limits should be carefully evaluated after they have been in effect for a suitable period,* say a year, in part to determine whether truck drivers violate the new higher speed limits, and if so what kinds of drivers tend to be the most frequent violators, and whether they have higher-than-average accident rates.

Commercial Zone Operations and Safety Regulations

Current safety regulations governing vehicle maintenance, driver licensing, and work hours do not apply to firms that operate strictly within the Commercial Zones of cities. Some knowledgeable people believe that these areas have become dumping grounds for bad drivers and poorly maintained equipment. This type of claim is much like those that were advanced against regulatory reform in the motor carrier industry. Whether or not the claim is valid, there are other reasons for changing the current situation. The open-entry, competitive environment in which the motor carrier industry functions has eliminated the need for the Commercial Zones. They were created and left free of entry restrictions and rate controls so that service to rail terminals would be efficient and low cost. There is no longer any reason for them to exist. *Commercial Zones should be eliminated.* The carriers that have been operating within them should be granted ICC certificates. If nothing else, such a change would eliminate any unfair cost advantage that Commercial Zone trucking firms now have in competing with ICC-certificated firms for business inside the zones. We are pleased to note that the DOT is currently repealing the Commercial Zone exemptions.

Monitoring of New Entrants and Financially Stressed Carriers

Research presented in this book finds that newest entrants into the motor carrier industry do tend to have a somewhat higher accident rate than firms with more years of experience. It is pointed out that the rate fell quickly as firms gained experience. Nevertheless, it might be worthwhile to maintain accurate accident records on new firms to determine whether it is worthwhile to subject them to special inspections concerning safety violations, vehicle maintenance, and so on. *A program of special monitoring of new firms should be adopted only if it can be proven to be in the social interest in a cost-benefit sense,* because the imposition of special burdens on new firms can act as a barrier to entry and as a way of restricting competition.

Other research in this book shows that firms on the verge of bankruptcy reduced their expenditure on safety-related areas of operations. It might be worthwhile following the accident rates of financially troubled firms to determine whether these rates are higher than average.

The Program of Road Inspections

With virtually unanimous agreement that accidents depend on the nature of the road, on the quality of the driver, and on driving conditions, rather than on the presence or absence of economic controls, it follows quite naturally that *we should look to safety measures for improvements in safety conditions.* We welcome the increase in safety oversight by states and the federal government. The Motor Carrier Safety Assistance Program, which received increased funding in 1986, is an example of such oversight. It provides funds for an almost 10-fold increase in road safety inspections. We also strongly support the objectives of the Commercial Motor Vehicle Safety Act of 1986, which in a few years will lead to a situation in which truck drivers will be able to hold only a single license. It will then be much more possible to identify and weed out drivers who have high accident rates, or who are found to be involved in drug and alcohol abuse. Progress is also being made in achieving uniformity in state vehicle inspection systems through the Commercial Vehicle Safety Alliance (Daust and Cobb, 1988).

Road inspections have been shown to reduce accidents, but it is not clear that a simple increase in police surveillance would not accomplish as much, at lower private and public cost. The nation is already embarked on a greatly increased program of road inspections. That program should be carefully evaluated in cost-benefit terms before further increases in it are made. In this regard an important general principle on the evaluation of safety programs has been expressed in this book: Evaluation of programs should be carried out by agencies other than the ones that are responsible for them (Chapter 6).

CONCLUDING REMARKS

Regulatory reform in the motor carrier industry has brought increased price competition. Shippers of freight and consumers of the products carried by motor carriers have benefited from truck rates and commodity prices that are lower than they would otherwise be. Both service quality and rates are now variables. Those shippers who require premium service and are willing to pay for it can obtain service equal to or better than anything that was available under regulation. However, there is now a continuum of quality and rates. Shippers can pick that combination of rates and services that best meets their needs, a situation that has led to major savings in logistical costs for firms.

Overall, there is no evidence that regulatory reform has had a negative impact on safety. However, there may be a few areas where increased truck-auto congestion has increased the fatality rate, and where increased surveillance, highway improvements, expansions, and appropriate pricing could be economically justified to cope with the problem.

Regulatory reform in the airline industry has also brought increased competition and price benefits to most airline travelers. Service quality has fallen. However, it seems indisputable that the declines in service have not been so great that travelers are worse off than they would be under the higher rates that would exist today were the system still under the economic control of the CAB.

There is no evidence that a reduction in economic regulation has caused an increase in airline fatality rates. Rather, the long-term downward movement of accident rates has continued during the years of economic deregulation. However, the success of deregulation in holding down rates has brought such an increase in airline travel that the system is currently straining at capacity, particularly at the major hub airports, where congestion during peak travel hours, from private as well as commercial aviation, is a serious problem. Such congestion poses a potential threat to the safety of airline passengers and may lead to increased fatalities in the future.

An important lesson learned from the United States' experience is that changes in the environment of economic regulation that achieve their economic goals also require that there be a careful and timely reevaluation of the role of government. Such reevaluation is required in the overseeing of safety, the use of pricing incentives to relieve congestion, and the provision of sufficient infrastructure.

Appendix: Background Information

This section is designed to give readers, especially those from outside the United States, some background information on the regulation and subsequent deregulation of the airline and motor carrier industries, and also on the various government agencies involved in the safety regulation of the industries.

An initial important point must be made. The United States, as the name implies, is a federal country. Therefore, in general, the individual states are responsible for the laws governing transportation movements within them (referred to as "intrastate"). It is therefore possible that regulations, both economic and safety related, may vary among states. The federal government becomes involved when traffic crosses state boundaries ("interstate"). The commonly referred to deregulation of the airlines and motor carriers in the late 1970s relates to removal of the federal economic restrictions.

AVIATION

Initial Regulation

Government involvement in aviation commenced in the mid-1930s as a result of interest in the delivery of mail by air. The Roosevelt Administration drafted legislation embodying the philosophy that airlines were a quasiutility requiring government control on entry, prices, and so on. The Civil Aeronautics Act of 1938 vested regulatory power in a new government agency that came to be known as the Civil Aeronautics Board (CAB). The CAB controlled routes, rates, and frequencies. The promulgation and regulation of safety standards was vested in another agency, the Federal Aviation Administration (FAA), a division of the U.S. Department of Transportation.

Deregulation

The Airline Deregulation Act of 1978 removed all economic controls from the industry, and also ultimately led to the disbanding of the CAB. This was the first major federal agency to actually be done away with in modern times, whose demise was not accompanied by the wholesale transfer of its functions to another agency. The requirements of the Act were phased in during the 1979–1981 time period. However, regulations regarding safety did not change; they continue to be enforced by the FAA.

Current Government Agencies

The FAA remains the sole government control agency for the airline industry. It sets the safety standards (Federal Air Regulations) and enforces them with a team of inspectors. The FAA also certifies new carriers in the safety area. An additional responsibility of the FAA is the provision of infrastructure. It employs air traffic controllers and operates the air traffic control system.

While individual airports are not federally owned, there are federal funds available for new or expanding airports. These monies are collected from a tax on airline tickets and are administered through the Aviation Trust Fund.

The other government agency involved in aviation safety is the National Transportation Safety Board (NTSB), a watchdog organization independent of the Department of Transportation. NTSB has a mandate to investigate and report on all aircraft accidents.

Mishap Data Bases

As the NTSB has a mandate to investigate all aircraft accidents, it maintains an accident data base. Additionally, the FAA maintains a data base on "incidents," which are mishaps with no injury or serious damage.

MOTOR CARRIERS
Initial Regulation

The Motor Carrier Act of 1935 created a regulated interstate motor carrier industry with two distinctive characteristics: (1) totally restricted entry and (2) collusive rate making relieved from the operation of antitrust law. The regulatory powers were vested in the Interstate Commerce Commission (ICC), a body that also had control over the railroads. Some traffic was not regulated under the system—"exempt" commodities (mainly agricultural), intrastate operations (although states could impose their own regulations), and traffic moving within defined "Commercial Zones" that surround major metropolitan areas.

Deregulation

The Motor Carrier Act of 1980, which took effect on July 1, 1980, was not as radical as the deregulation in the airline industry. Although there is no longer any control on entry, there is still some degree of collective rate making. The ICC continues to oversee agreements and continues to issue operating permits. However, most freight is carried at discounted rates negotiated directly between carrier and shipper.

The position regarding intrastate traffic is more mixed. In some states there is still quite strict regulation, whereas in other states, such as Arizona and Florida, there is complete deregulation.

Government Agencies

The current role of the ICC has been discussed above. Safety enforcement activity is a joint responsibility between the federal government (in the Office of Motor Carriers, a branch of the Federal Highway Administration, itself part of the Department of Transportation) and the individual states and state police. The federal government bears some of the cost of the state activity through the Motor Carrier Safety Assistance Program of 1982. Currently, safety regulations vary from state to state (such as the requirements for drivers' licenses), and there is an active movement to bring more standardization.

Mishap Data Bases

The NTSB is not the official keeper of truck accident statistics. NTSB confines itself to investigating only selected highway accidents. The Office of Motor Carriers maintains a record of reported truck accidents, which are accidents resulting in one or more of a fatality, an injury, or $4,400 property damage. An alternative data source is the National Highway Traffic Safety Administration (NHTSA) (the Department of Transportation's task force on highway safety) Fatal Accident Reporting System (FARS), that keeps a record of all accidents that cause fatalities. Data on nonfatal accidents is less comprehensive, because NHTSA's National Accident Sampling System (NASS) provides only a sample of all highway accidents.

References

Abruzzese, L. (1987). Industry backs plan to train truck drivers. *Journal of Commerce*, Oct. 29, p. 2B.

Advanced Technology, Inc. (1986). *An Evaluation of the Relationship between Aircarrier Financial Condition and Safety Posture*. Washington, D.C.: Federal Aviation Administration.

Air Transport Association of America (1987). *Air Transport 1987: The Annual Report of the U. S. Scheduled Airline Industry*. Washington, D.C.: Air Transport Association of America.

Akerlof, G. (1970). The market for lemons: Qualitative uncertainty and the market mechanisms. *Quarterly Journal of Economics* 84:488–500.

Allen Corporation of America (1985). *Job and Task Analysis of the Positions of: Aviation Safety Inspector—Airworthiness; Aviation Safety Inspector—Operations, Accident Prevention Specialist and Flight Inspection Procedures Staff Specialist*. Washington, D.C.: Office of Personnel Management.

Allen, F. (1984). Reputation and product quality. *Rand Journal of Economics* 15:311–27.

Altman, E. I. (1983). *Corporate Financial Distress: A Complete Guide to Predicting, Avoiding, and Dealing with Bankruptcy*. New York: John Wiley.

American Association of State Highway and Transportation Officials (1984). *A Policy of Geometric Design of Highways and Streets*. Washington D.C.: American Association of State Highway and Transportation Officials.

American Trucking Associations (various years). *American Trucking Trends*. Alexandria, Va.: American Trucking Associations.

American Trucking Associations (various years). *Motor Carrier Annual Report*. Alexandria, Va.: American Trucking Associations.

American Trucking Associations (1986) Statement before the Subcommittee on Surface Transportation of the Committee on Public Works and Transportation, U. S. House of Representatives. Washington, D.C.: U. S. Government Printing Office.

Association of American Railroads (1985). *Railroad Ten Year Trends*. Vol. 2. Washington, D.C.: Association of American Railroads.

Atkinson, A. B. and Stiglitz, J. E. (1980). *Lectures on Public Economics*. New York: McGraw-Hill.

Bailey, E. E. (1986). Deregulation: Causes and consequences. *Science* 234:1211–1216.

Bailey, E. E., Graham, D. R., and Kaplan, D. P. (1985). *Deregulating the Airlines*. Cambridge, Mass.: MIT Press.

Bailey, E. E. and Kirstein, D. M. (1987). Require airline truth in scheduling. *The New York Times*, May 27.

Bailey, E. E. and Williams, J. R. (1987). Sources of economic rent in the deregulated airline industry. (Working Paper 28-86-87.) Graduate School of Industrial Administration, Carnegie Mellon University, Pittsburgh, Pa.

Baker, F. (1985). *Safety Implications of Structural Changes Occurring in the Motor Carrier Industry*. Washington, D.C.: American Automobile Association Foundation for Traffic

Safety. Reprinted 1988 in Proceedings, Transportation Deregulation and Safety Conference, June 1987. Evanston, Ill.: Northwestern University Transportation Center.

Barnett, A., Abraham, M., and Schimmel, V. (1979). Airline safety: Some empirical findings. *Management Science* 25:1045–1056.

Barnett, A. and Higgins, M. (1987). Airline safety: The last decade. (Working paper.) Sloan School of Management, MIT, Cambridge, Mass.

Barnett, A. and Lofaso, A. J. (1983). After the crash: The passenger response to the DC-10 disaster. *Management Science* 29:1225–1236.

Becker, G. S. (1968). Crime and punishment: An economic approach. *Journal of Political Economy* 76:169–217.

Beilock, R. (1985). Are truckers forced to speed? *Logistics and Transportation Review* 21:277–291.

Beilock, R. and Freeman, J. (1985). *The Impact of Motor Carrier Deregulation on Freight Rates in Arizona and Florida*. Washington D.C.: U. S. Department of Transportation.

Blair, R. D., Kaserman, D. L., and McClave, J. T. (1986). Motor carrier deregulation: The Florida experiment. *Review of Economics and Statistics* 68:159–164.

Boburg, K. (1985). Track structure accounting and reported earnings of U. S. railroads. *Transportation Journal* 24:18–28.

Bolger, W. F. (1988). Deregulation: Past or prologue? (wither the infrastructure?). Proceedings, Transportation Deregulation and Safety Conference, June 1987. Evanston, Ill.: Northwestern University Transportation Center.

Borenstein, S. and Zimmerman, M. B. (1988). Market incentives for safe commercial airline operation. Proceedings, Transportation Deregulation and Safety Conference, June 1987. Evanston, Ill.: Northwestern University Transportation Center.

Boyer, K. D. (1987). The costs of price regulation: Lessons from railroad deregulation. *Rand Journal of Economics* 18:408–416.

Bruning, E. (1987). The relationship between profitability and safety performance. Paper presented at Allied Social Sciences Associations Annual Meeting, Chicago, Dec.

Bulow, J. and Shoven, J. (1978). The bankruptcy decision. *Bell Journal of Economics* 9:437–456.

California Highway Patrol (1987). *1986 Statewide Truck Involved and At Fault Accidents.* Sacramento: State of California.

Capelle, Jr., R. B. and Beilock, R. (1988). Deteriorating safety conditions and the results of reduced economic regulation of the motor carrier industry: The general freight/LTL trucking industry perspective. Proceedings, Transportation Deregulation and Safety Conference, June 1987. Evanston, Ill.: Northwestern University Transportation Center.

Carlin, A., and Park, R. E. (1970). Marginal cost pricing of airport runway capacity. *American Economic Review* 60:310–319.

Carlton, D. W., Landes, W. M., and Posner, R. A. (1980). Benefits and costs of airline mergers: A case study. *Bell Journal of Economics* 11:65–83.

Chalk, A. J. (1985). A new proposal for the reform of commercial air crash litigation. *Journal of Air Law and Commerce* 50:219–252.

Chalk, A. J. (1986). Market forces and aircraft safety: The case of the DC-10. *Economic Inquiry* 24:43–60.

Chalk, A. J. (1987). Market forces and commercial aircraft safety. *Journal of Industrial Economics* 36:61–81.

Chamberlin, E. (1965). *The Theory of Monopolistic Competition.* 8th ed. Cambridge, Mass.: Harvard University Press.

Chance, D. M. and Ferris, S. P. (1987). The effect of aviation disasters on the air transport industry. *Journal of Transport Economics and Policy* 21:151–165.

Chaumel, J. L. et al. (1986). Road accidents involving long-distance heavy trucks: The case of Eastern Quebec. (Working paper.) University of Quebec, Rimouski, Canada.

Chow, Garland, and Gritta, R. (1985). Motor carrier bankruptcy in an uncertain environment. *Transportation Law Journal* 14:39–57.

Chow, Garland, Gritta, R., Adrangi, B., and Ebelt, R. (1986). The financial status of U. S. general freight motor carriers before and after the Motor Carrier Act of 1980. Proceedings, Annual Meeting of the Decision Sciences Institute, Washington, D.C.

Chow, Garland, Vasina, M., Adrangi, B., and Gritta, R. D. (1987). The definition and measurement of financial fitness in the trucking industry. *Journal of the Transportation Research Forum* 27:319–324.

Chow, Gregory (1960). Tests of equality between sets of regressions. *Econometrica* 28:591–605.

Civil Aeronautics Board (various years). *Air Carrier Traffic Statistics.* Washington, D.C.: U. S. Government Printing Office.

Civil Aeronautics Board (Various years). *Handbook of Airline Statistics.* Washington, D.C.: U. S. Government Printing Office.

Corsi, T. M., Fanara, Jr., P., and Roberts, M. J. (1984). Linkages between motor carrier accidents and safety regulation. *Logistics and Transportation Review* 20:149–164.

Crandall, R. L. (1988). Luncheon address. Proceedings, Transportation Deregulation and Safety Conference, June 1987. Evanston, Ill.: Northwestern University Transportation Center.

Daust, J. E. and Cobb, D. L. (1988). The relationship between economic deregulation of the motor carrier industry and its effects on safety. Proceedings, Transportation Deregulation and Safety Conference, June 1987. Evanston, Ill.: Northwestern University Transportation Center.

Delaney, R. V. (1987). *The Disunited States: A Country in Search of an Efficient Transportation Policy.* CATO Institute Policy Analysis Paper No. 84. Washington, D.C.: CATO Institute.

Dixit, A. K. and Stiglitz, J. E. (1977). Monopolistic competition and optimum product diversity. *American Economic Review* 67:297–308.

Donohue, T. (1985). Impact of the MCA of 1980 (looking back after five years). Hearings before the Subcommittee on Surface Transportation of the Committee on Public Works and Transportation, U. S. House of Representatives. Washington, D.C.: U. S. Government Printing Office.

Dorfman, G. J. (1983). A model of unregulated airline markets. *Research in Transportation Economics* 1:131–148.

Douglas, G. W. and Miller, III, J. C. (1974). *Economic Regulation of Domestic Air Transport: Theory and Policy.* Washington, D.C.: Brookings Institution.

Drèze, J. and Stern, N. (1987). The theory of cost-benefit analysis. In A. Auerbach and M. Feldstein (Eds.), *Handbook of Public Economics.* Amsterdam: North Holland.

ENO Foundation (1985). Report of joint conference, Eno Foundation Board of Directors and Board of Consultants, October 17 and 18, 1985: Transportation in an era of deregulation. *Transportation Quarterly* 39:171–206.

Evans, M. K. (1987a). *Macroeconomic Implications of Trucking Deregulation.* Washington, D.C.: Coalition for Sound General Freight Trucking.

Evans, M. K. (1987b). Has deregulation helped or harmed the transportation sector? Paper presented at Allied Social Science Associations Annual Meeting, Chicago, Dec.

Fama, E. F., Fisher, L., Jensen, M. C., and Roll, R. (1969). The adjustments of stock prices to new information. *International Economic Review* 10:1–21.

Federal Aviation Administration (1980). *First Commuter Air Safety Symposium.* Washington, D.C.: U. S. Government Printing Office.

Federal Highway Administration (various years). *Accidents of Motor Carriers of Property.* Washington, D.C.: U. S. Government Printing Office.

Federal Highway Administration (various years). *Highway Statistics.* Washington, D.C.: U. S. Government Printing Office.

Federal Railroad Administration (various years). *Accident/Incident Bulletin.* Washington, D.C.: U. S. Government Printing Office.

Fellmuth, R. (1970). *The Interstate Commerce Omission—The Public Interest and the ICC.* New York: Grossman.

Fingerhut, V. (1986). The pilots view of air safety. *Air Line Pilot* 55:17–22.

Freeman, R. (1976). Individual mobility and union voice in the labor market. *American Economic Review Papers and Proceedings* 66:361–368.

Gander, J. P. (1986). Highway speed and uncertainty of enforcement: The travelling salesman (or trucker) case. *Logistics and Transportation Review* 22:43–55.

General Accounting Office (1984). *Safety Standards on Small Passenger Aircraft—With Nine Or Fewer Seats—Are Significantly Less Stringent Than On Larger Aircraft.* Report GAO/RCED-84-2. Washington, D.C.: U. S. Government Printing Office.

General Accounting Office (1987). *Trucking Regulation: Price Competition and Market Structure in the Trucking Industry.* Report GAO/RCED-87-16. Washington, D.C.: U. S. Government Printing Office.

Glaskowsky, N. (1986). *Effects of Deregulation on Motor Carriers.* Westport, Conn.: ENO Foundation for Transportation.

Golbe, D. L. (1981). The effects of imminent bankruptcy on stockholder risk preferences and behavior. *Bell Journal of Economics* 12:321–328.

Golbe, D. L. (1983). Product safety in a regulated industry: Evidence from the railroads. *Economic Inquiry* 21:39–52.

Golbe, D. L. (1986). Safety and profits in the airline industry. *Journal of Industrial Economics* 34:305–318.

Graham, D. R. and Bowes, M. (1979). *Do Finances Influence Airline Safety, Maintenance, and Services?* Alexandria, Va: The Public Research Institute of the Center for Naval Analysis.

Halvorsen, R. and Palmquist, P. (1980). The interpretation of dummy variables in semi logarithmic equations. *American Economic Review* 70:474–475.

Hansson, I. and Skogh, G. (1987). Moral hazard and safety regulation. *The Geneva Papers on Risk and Insurance* 12:132–144.

Harris, W. and Mackie, R. (1972). *A Study of the Relationships Among Fatigue, Hours of Service, and Safety of Operations of Truck and Bus Drivers.* Report Number BMCS-RD-71-2, Federal Highway Administration. Washington, D.C.: U. S. Government Printing Office.

Haseltine, P. (1985). Impact of the MCA of 1980 (looking back after five years). Hearings before the Subcommittee on Surface Transportation of the Committee on Public Works and Transportation, U. S. House of Representatives. Washington, D.C.: U. S. Government Printing Office.

Hausman, J., Hall, B. H., and Griliches, Z. (1984). Econometric models for count data with an application to the patents-R&D relationship. *Econometrica* 52:909–938.

Hirshleifer, J. (1988). *Price Theory and Applications.* 4th ed. Englewood Cliffs, N.J.: Prentice Hall.

Interstate Commerce Commission (various years). *Annual Report.* Washington, D.C.: U. S. Government Printing Office.

Interstate Commerce Commission Bureau of Operations (1978). Can they do that? Hot or exempt. Memorandum, May 19, 1978.

Jones, L. M. (1982). *Task Force for the Study of Management-Employee Relationships in the Federal Aviation Administration.* Washington, D.C.: U. S. Department of Transportation.

Joskow, P. J. and Rose, N. L. (1985). The effects of technological change, experience, and environmental regulation on the construction cost of coal-burning generating units. *Rand Journal of Economics* 16:1–27.

Jovanis, P. P. (1988). A perspective on motor carrier safety issues in the 1980s. Proceedings, Transportation Deregulation and Safety Conference, June 1987. Evanston, Ill.: Northwestern University Transportation Center.

Kaplan, D. P. (1986). The changing airline industry. In L. Weiss and M. Klass (Eds.), *Regulatory Reform: What Actually Happened.* Boston: Little Brown.

Keeler, T. E. (1983). *Railroads, Freight, and Public Policy.* Washington, D.C.: Brookings Institution.

Kennedy, P. E. (1981). Estimation with correctly interpreted dummy variables in semi-logarithmic equations. *American Economic Review* 71:801.

Kern, J. S. (1988). Effect of deregulation on the Federal Aviation Administration's inspection and surveillance efforts. Proceedings, Transportation Deregulation and Safety Conference, June 1987. Evanston, Ill.: Northwestern University Transportation Center.

Klein, B. and Leffler, K. B. (1981). The role of market forces in assuring contractual performance. *Journal of Political Economy* 89:615–641.

Koenker, R. W. and Perry, M. K. (1981). Product differentiation, monopolistic competition, and public policy. *Bell Journal of Economics* 12:217–231.

Koran, D. and Ogur, J. (1985). Marketable landing rights and economic efficiency. (Working paper 124.) Bureau of Economics, Federal Trade Commission, Washington, D.C.

Labich, K. (1987). The scandal of killer trucks. *Fortune,* Mar. 30, pp. 85–87.

Ladd, E. C. (1985). *The American Polity.* New York: Norton.

Lauber, J. K. (1988). Assessing the impact of deregulation: The benchmarks of airline safety. Proceedings, Transportation Deregulation and Safety Conference, June 1987. Evanston, Ill.: Northwestern University Transportation Center.

Lave, L. B. (1968). Safety in transportation: The role of government. *Law and Contemporary Problems* 33:512–535.

Lederer, J. F. and Enders, J. H. (1988). Aviation safety: The global conditions and prospects. Proceedings, Transportation Deregulation and Safety Conference, June 1987. Evanston, Ill.: Northwestern University Transportation Center.

Levine, M. E. (1969). Landing fees and the airport congestion problem. *Journal of Law and Economics* 12:79–108.

Levine, M. E. (1987). Airline competition in deregulated markets: Theory, firm strategy, and public policy. *Yale Journal on Regulation* 4:393–494.

Leyden, J. (1986). *Human Resource Information Systems.* Federal Aviation Administration Office of Human Resource Planning and Evaluation. Washington, D.C.: U. S. Government Printing Office.

Likens, J. D. (1976). The welfare costs of nonoptimal airport utilization. *Journal of Public Economics* 5:81–102.

Little, I. M. D. and McLeod, K. M. (1972). The new pricing policy of the British Airports Authority. *Journal of Transport Economics and Policy* 6:101–115.

Locklin, D. P. (1972). *The Economics of Transportation.* 7th ed. Homewood, Ill.: Richard Irwin.

McDole, T. L. and O'Day, J. (1975). *Effect of Commercial Vehicle Systematic Preventive Maintenance on Specific Causes of Accidents.* Highway Safety Research Institute report. Washington, D.C.: Federal Highway Administration.

McDonald, N. (1984). *Fatigue, Safety and the Truck Driver.* Philadelphia: Taylor and Francis.

McKenzie, R. B. and Shughart III, W. F. (1986). Has deregulation of air travel affected air safety? (Working paper 101.) Center for the Study of American Business, Washington University, St. Louis, Mo.

McKenzie, R. B. and Warner, J. T. (1987). The impact of airline deregulation on highway safety. (Working paper.) Center for the Study of American Business, Washington University, St. Louis, Mo.

McLure, H. R. (1988). The need for additional surveillance in the airline industry. Proceedings, Transportation Deregulation and Safety Conference, June 1987. Evanston, Ill.: Northwestern University Transportation Center.

Meyer, J. R. and Oster, Jr., C. V. (Eds.) (1981). *Airline Deregulation: The Early Experience.* Boston: Auburn House.

Meyer, J. R. and Oster, Jr., C. V. (1984). *Deregulation and the New Airline Entrepreneurs.* Cambridge, Mass.: MIT Press.

Meyer, J. R. and Oster, Jr., C. V. (1987). *Deregulation and the Future of Intercity Passenger Travel.* Cambridge, Mass.: MIT Press.

Mitchell, M. L. and Maloney, M. T. (1988). Crisis in the Cockpit? The role of market forces in promoting air travel safety. (Working paper.) Center for Policy Studies, Clemson University, Clemson, S.C.

Mohring, H. and Harwitz, M. (1962). *Highway Benefits: An Analytical Framework*. Evanston, Ill.: Northwestern University Press.

Mohring, H., Schroeter, J., and Wiboonchutikula, P. (1987). The values of waiting time, travel time, and a seat on a bus. *Rand Journal of Economics* 18:40–56.

Moore, T. G. (1986). Rail and Trucking Deregulation. In L. Weiss and M. Klass (Eds.), *Regulatory Reform: What Actually Happened*. Boston: Little Brown.

Morrison, S. A. (1983). Estimation of long-run prices and investment levels for airport runways. *Research in Transportation Economics* 1:103–130.

Morrison, S. A. and Winston, C. (1985). Intercity transportation route structures under deregulation: Some assessments motivated by the airline experience. *American Economic Review* 75:57–61.

Morrison, S. A. and Winston, C. (1986). *The Economic Effects of Airline Deregulation*. Washington, D.C.: Brookings Institution.

Motor Carrier Ratemaking Study Commission (1983). *Collective Ratemaking in the Trucking Industry*. Washington, D.C.: U. S. Government Printing Office.

Motor Vehicle Manufacturers Association (1986). *Motor Vehicle Facts and Figures*. Washington, D.C.: Motor Vehicle Manufacturers Association.

National Commission on Product Safety (1970). *Final Report of the National Commission on Product Safety*. Washington, D.C.: U. S. Government Printing Office.

National Highway Traffic Safety Administration (1982). *Large-Truck Accident Causation*. Technical Report DOT-HS-806-300. Washington, D.C.: U. S. Government Printing Office.

National Safety Council (various years). *Accident Facts*. Chicago: National Safety Council.

National Transportation Safety Board (various years). *Annual Report*. Washington, D.C.: U. S. Government Printing Office.

National Transportation Safety Board (1972). *Air Taxi Safety Study*. Report NTSB-AAS-72-6. Washington, D.C.: U. S. Government Printing Office.

National Transportation Safety Board (1980). *Special Study: Commuter Airline Safety, 1970–1979*. Report NTSB-AAS-80-1. Washington, D.C.: U. S. Government Printing Office.

National Transportation Safety Board (1981). *Air Traffic Control System*. Special Investigation Report NTSB/SIR-81/7. Washington, D.C.: U. S. Government Printing Office.

National Transportation Safety Board (1986a). *Safety Recommendation A-86-98 through -118*. Washington, D.C.: U. S. Government Printing Office.

National Transportation Safety Board. (1986b). *Aircraft Accident Report AAR-86/05, Delta Air Lines Accident of August 2, 1985 at Dallas/Fort Worth International Airport*. Washington, D.C.: U. S. Government Printing Office.

Nelson, P. (1970). Information and consumer behavior. *Journal of Political Economy* 78:311–329.

New York Times (1987). Delta offers refunds on safety issue. July 25.

O'Brien, J. E. (1988). Deregulation and safety: An airline pilot's view. Proceedings, Transportation Deregulation and Safety Conference, June 1987. Evanston, Ill.: Northwestern University Transportation Center.

Oi, W. Y. (1973). The economics of product safety. *Bell Journal of Economics* 4:320–329.

Oster, Jr., C. V. and Pickrell, D. H. (1986a). *A Study of the Regional Airline Industry: The Impact of Marketing Alliances*. Report for U. S. Department of Transportation, Washington, D.C.

Oster, Jr., C. V. & Pickrell, D. H. (1986b). Marketing alliances and competitive strategy in the airline industry. *Logistics and Transportation Review* 22:371–387.

Oster, Jr., C. V. and Zorn, C. K. (1982). *Commuter Airline Safety*. Report for U. S. Department of Transportation, Washington, D.C.

Oster, Jr., C. V. and Zorn, C. K. (1983). Airline deregulation, commuter safety, and regional air transportation. *Growth and Change* 14:3–11.

Oster, Jr., C. V. and Zorn, C. K. (1984). Deregulation and commuter airline safety. *Journal of Air Law and Commerce* 49:315–335.

Page, E. (1987). Statement before the Subcommittee on Surface Transportation of the Committee of Public Works and Transportation, U. S. House of Representatives. Washington, D.C.: U. S. Government Printing Office.

Pakes, A. and Griliches, Z. (1980). Patents and R&D at the firm level: A first look. *Economics Letters* 5:377–381.

Panzar, J. C. (1979). *Regulation, Service Quality, and Market Performance: A Model of Airline Rivalry.* New York: Garland.

Papai, R. F. (1988). Deregulation and safety: A truck load carrier's view. Proceedings, Transportation Deregulation and Safety Conference, June 1987. Evanston, Ill.: Northwestern University Transportation Center.

Pigou, A. (1920). *The Economics of Welfare.* London: Macmillan.

Pustay, M. (1983). Regulatory reform of motor freight carriage in the United States. *International Journal of Transport Economics* 10:259–280.

Roads and Transport Association of Canada (1986). *Manual for Geometric Design Standards for Canadian Roads 1986 Metric Edition.* Ottawa, Ontario, Canada: Roads and Transport Association of Canada.

Rose, N. L. (1985). The incidence of regulatory rents in the motor carrier industry. *Rand Journal of Economics* 16:299–318.

Schmalensee, R. (1977). The comparative static properties of regulated airline oligopolies. *Bell Journal of Economics* 8:565–576.

Schweitzer, R. P. (1988). The myth of economic deregulation and safety in the U. S. motor carrier industry. Proceedings, Transportation Deregulation and Safety Conference, June 1987. Evanston, Ill.: Northwestern University Transportation Center.

Scocozza, M. V. (1987). Testimony before the Senate Subcommittee on the Rural Economy and Family Farm of the Committee on Small Business, October 28, 1987. Washington, D.C.: U. S. Government Printing Office.

Scotchmer, S. and Wooders, M. H. (1987). Competitive equilibrium and the core in club economies with anonymous crowding. *Journal of Public Economics* 34:159–173.

Shapiro, C. (1982). Consumer information, product quality, and seller reputation. *Bell Journal of Economics* 13:20–35.

Shapiro, C. (1983a). Optimal pricing of experience goods. *Bell Journal of Economics* 14:497–508.

Shapiro, C. (1983b). Premiums for high quality products as returns to reputation. *Quarterly Journal of Economics* 98:659–680.

Shavell, S. (1979). On moral hazard and insurance. *Quarterly Journal of Economics* 93:541–562.

Shavell, S. (1984a). Liability for harm versus regulation of safety. *Journal of Legal Studies* 13:357–374.

Shavell, S. (1984b). A model of the optimal use of liability and safety regulation. *Rand Journal of Economics* 15:271–280.

Small, K. (1982). The scheduling of consumer activities: Work trips. *American Economic Review* 72:467–479.

Spence, A. M. (1975). Monopoly, quality and regulation. *Bell Journal of Economics* 6:417–429.

Spence, A. M. (1976). Product selection, fixed costs, and monopolistic competition. *Review of Economic Studies* 43:217–235.

Spence, A. M. (1977). Consumer misperceptions, product failure and producer liability. *Review of Economic Studies* 44:561–572.

Standard and Poors (1986). *Railroads and Trucking.* (Industry Surveys.) New York: Standard and Poors.

Stein, H. S. and Jones, I. S. (1987). *Crash Involvement of Large Trucks by Configuration: A Case-Control Study.* Washington, D.C.: Insurance Institute for Highway Safety.

Stiglitz, J. E. and Arnott, R. (1988). Safety, user fees and public infrastructure. Proceedings,

Transportation Deregulation and Safety Conference, June 1987. Evanston, Ill.: Northwestern University Transportation Center.

Storie, V. J. (1984). *Involvement of Goods Vehicles and Public Service Vehicles in Motorway Accidents*. Transport and Road Research Laboratory Report 1113. Crowthorne, United Kingdom: Transport and Road Research Laboratory.

Strotz, R. (1965). Urban transportation parables. In J. Margolis (Ed.), *The Public Economy of Urban Communities*. Baltimore, Md.: Johns Hopkins University Press.

Task Force on Railroad Productivity (1973). *Improving Railroad Productivity: Final Report to the National Commission on Productivity and the Council of Economic Advisors*. Washington, D.C.: U. S. Government Printing Office.

Transportation Policy Associates (various years). *Transportation in America: A Statistical Analysis of Transportation in the United States*. Washington, D.C.: Transportation Policy Associates.

Transportation Research Board (1986). *Twin Trailer Trucks: Effects on Highways and Highway Safety*. Special Report 211. Washington, D.C.: National Research Council.

Transportation Research Board (1987). *Designing Safer Roads*. Special Report 214. Washington, D.C.: National Research Council.

Tye, W. B. (1983). Balancing the ratemaking goals of the Staggers Rail Act. *Transportation Journal* 22:17–26.

Tye, W. B. (1987). *Encouraging Cooperation Among Competitors: The Case of Motor Carrier Deregulation and Collective Ratemaking*. New York: Quorum Books.

U. K. Department of Transport (various years). *Transport Statistics: Great Britain*. London: Her Majesty's Stationery Office.

U. S. Congress (1975). Committee on the Judicial Oversight of Civil Aeronautics Board Practices and Procedures, U.S. Senate. Washington, D.C.: U. S. Government Printing Office.

U. S. Department of Commerce (various years). *Statistical Abstract of the United States*. Washington, D.C.: U. S. Government Printing Office.

U. S. Department of Labor (various years). *Occupational Injuries and Illnesses in the United States by Industry*. Washington, D.C.: U. S. Government Printing Office.

U. S. Department of Labor (various years). *Employment and Earnings* (monthly). Washington, D.C.: U. S. Government Printing Office.

U. S. Department of Transportation (various years). *Air Carrier Operating Statistics*. Washington, D.C.: U. S. Government Printing Office.

U. S. Department of Transportation (various years). *National Transportation Statistics*. Washington, D.C.: U. S. Government Printing Office.

U. S. Department of Transportation (1987). *Heavy Truck Study*. Washington, D.C.: U. S. Government Printing Office.

University of Michigan Transportation Research Institute (1987). Large truck survey program. Proceedings, National Truck Safety Symposium, June 1987. Washington, D.C.: Motor Vehicle Manufacturers Association.

Vickrey, W. S. (1963). Pricing in urban and suburban transport. *American Economic Review Papers and Proceedings* 53:452–465.

Vickrey, W. S. (1971). Responsive pricing of public utility services. *Bell Journal of Economics and Management Science* 2:337–346.

Viscusi, W. K. (1979a). *Employment Hazards: An Investigation of Market Performance*. Cambridge, Mass.: Harvard University Press.

Viscusi, W. K. (1979b). The Impact of the Occupational Safety and Health Administration. *Bell Journal of Economics* 10:117–140.

Viscusi, W. K. (1983). *Risk by Choice: Regulating Health and Safety in the Workplace*. Cambridge, Mass.: Harvard University Press.

Viscusi, W. K. (1984). *Regulating Consumer Product Safety*. Washington, D.C.: American Enterprise Institute.

Viscusi, W. K. (1986). The impact of occupational safety and health regulation, 1973–1983. *Rand Journal of Economics* 17:567–580.

Wall Street Journal (1987). Gluttons for punishment? Fliers continue using airlines they hate. Nov. 19.

Walters, A. A. (1978). Airports: An economic survey. *Journal of Transport Economics and Policy* 12:125–160.

White, H. (1980). A heteroskedastic-consistent covariance matrix estimator and a direct test for heteroskedasticity. *Econometrica* 48:817–838.

Wilson, G. W. (1981). The relative importance of economic regulation of transportation vis-à-vis everything else. In K. D. Boyer and W. G. Shepherd (Eds.), *Economic Regulation: Essays in Honor of James R. Nelson*. East Lansing, Mich.: Institute of Public Utilities.

Winston, C. (1988). The incorporation of safety and government policy in the cost-benefit analysis of deregulation. Proceedings, Transportation Deregulation and Safety Conference, June 1987. Evanston, Ill.: Northwestern University Transportation Center.

Wyckoff, D. D. (1979). *Truck Drivers in America*. Boston: Lexington Books.

Index

DATE DUE

	DEC 9 3 1999	APR 1 2 2003	
		APR 3 0 2003	
APR 1 1 2001			
			Printed in USA

HIGHSMITH #45230